Problems in Applied, Industrial and Engineering Mathematics

Problems in Applied, Industrial and Engineering Mathematics

Edited by

H.K. KUIKEN and S.W. RIENSTRA

Eindhoven University of Technology,
Eindhoven, The Netherlands

Reprinted from *Journal of Engineering Mathematics*, Vol. 26, No. 1 (1992)

Springer-Science+Business Media, B.V.

ISBN 978-94-010-5076-0 ISBN 978-94-011-2440-9 (eBook)
DOI 10.1007/978-94-011-2440-9

Originally published by Kluwer Academic Publishers in 1992

TABLE OF CONTENTS

Twenty-five years of Engineering Mathematics

Now that the *Journal of Engineering Mathematics (JEM)* has just completed its first twenty-five years, it may be an appropriate time to reminisce about its history, describe its present state and look forward to the years ahead. More than a quarter of a century ago a group of Dutch applied mathematicians, associated with the Universities of Technology at Delft, Eindhoven and Twente and with the University of Groningen, got together under the inspired leadership of the late Reinier Timman, thinking that it would be a good idea to create a journal devoted primarily to mathematics arising from engineering contexts. In the mid-fifties, in Delft, it had been Timman himself who had promoted the concept of the mathematical engineer, whose foremost interest was in tackling problems of engineering significance. In those years there were only a few journals in the world, notably *QAM, QJMAM, SIAM J. Appl. Math.* and *ZAMP*, which sought to publish papers in the general area of applied mathematics. None of these were dedicated to serving the fuzzy area between applied mathematics and engineering. Therefore, the idea to create a new journal for the sole purpose of attracting papers of an engineering-mathematics nature seemed a good one, coming from mathematicians active at universities of technology and universities with applied-mathematics departments.

Despite its Dutch origins, *JEM* is now and in fact always has been a truly international journal. In the initial stages the Editorial Board was composed mainly of scientists who had their base in The Netherlands. Under the responsible guidance of the previous Managing Editor, the Board was rapidly internationalised, so that more than half of them are now from other countries. This trend is likely to continue. Most of our papers originate from the United Kingdom and the U.S.A., in that order, with The Netherlands taking a historically understandable third place. After these three we see countries such as Canada and Australia. Evidently, our journal shows a very strong bias towards the English-speaking scientific world. One can justifiably wonder why this is so. Is it true that our kind of mathematics, which relies more on intuition than on mathematical proof, is less popular outside the above-mentioned areas? Or is there a more mundane reason, namely that the possibilities of *JEM* have not yet been fully discovered in large sections of the world?

Over the years, the journal's editors have found that it is very difficult to define exactly what engineering mathematics is. First of all, some things which are indeed fully fledged engineering mathematics are unsuitable for *JEM*, since the mathematics of these papers is simply lacking in both depth and substance. Whenever such papers are sent to us, we advise authors to resubmit their work to a technical journal specialised in the engineering subject they address. On the other hand, papers which are clearly mathematical exercises for their own sake, devoid of all but the remotest reference to a field outside mathematics, are usually returned to the authors. Our advice is then for them to look for a journal devoted to applied mathematics per se. Ideally, papers published in *JEM* commence with a mainly verbal introduction in which the technical, i.e. non-mathematical, background is sketched. This is followed by a derivation or statement of the mathematical model. If at all possible, a paper should be concluded with a section in which the results of the paper are discussed, particularly in relation to the non-mathematical engineering background. The mathematics should be non-trivial.

Much as it is difficult to define clearly what engineering mathematics is, it is far from easy to delineate where engineering mathematics ends and applied mathematics begins. Many of the papers we publish, perhaps the majority of them, are sound applied mathematics. At the same time, much of what is called "applied mathematics" nowadays is not suitable for *JEM*. Even so, large sections of this particular branch of mathematics belong to the field we cover. This is why, henceforth, the term *applied mathematics* will appear on the cover of the journal.

We have also added the term *industrial mathematics*. This neologism was coined, possibly half a century ago, since it was recognized that a lot of mathematical modelling was and still is going on within large industrial laboratories, outside the direct sphere of influence of academia. Again, our journal does not intend to cover everything that comes under this heading. Industrial-mathematics papers which have a strong engineering-mathematics aspect (see Aims and scope) are welcome.

At twenty-five years old one can safely say that a scientific journal has come of age. If it has managed to survive for that period of time, it has proved its right to exist. Having said this, we believe that there is cause for celebration. We wish to do this in a way that seems to us the most suitable, which is to publish a special issue filled entirely with papers written by members of the Editorial Board. Who better than these people to show what is going on in engineering mathematics and its overlap region in both applied and industrial mathematics? It may be fitting to mention here that the Publisher also thinks there is cause for celebration, as it has allowed us to offer a greatly expanded first issue at no extra cost to the subscriber.

Finally, it may be good to try and look into the future. Ours is an expanding field. This is why we believe that *JEM* should expand commensurately. Although no definite date has been set as yet, we plan to increase the number of issues published annually from four to six in the not too distant future. Special issues will be produced as from 1993. Such issues will each be devoted to a single topic of great interest. Occasionally, review articles dealing with important areas in engineering mathematics will appear. We do hope that our regular audience will remain as loyal to our journal as ever and that their numbers will be ever-increasing.

H.K. KUIKEN
Editor-in-Chief

Journal of Engineering Mathematics **26**: 1–19, 1992.
H.K. Kuiken and S.W. Rienstra (eds), Problems in Applied, Industrial and Engineering Mathematics.

Three-dimensional convection in an inclined layer heated from below

F.H. BUSSE[1] and R.M. CLEVER[2]

[1]*Institute of Physics, University of Bayreuth, Universitätsstrasse 30, D-8580 Bayreuth, Germany*
[2]*Institute of Geophysics and Planetary Physics, University of California at Los Angeles, Los Angeles, CA 90024, USA*

Abstract. Longitudinal rolls aligned with the component of gravity parallel to the layer represent the preferred mode of convection at onset for a wide range of parameters of the inclined layer problem. As the Rayleigh number is increased beyond the critical value the longitudinal rolls tend to become unstable with respect to the wavy instability. The three-dimensional convection flows evolving from this instability are studied in this paper by numerical computations. The main effect of the wavy distortions of the longitudinal rolls is a decrease of the heat transport. The stability of the steady three-dimensional convection flow with respect to disturbances with the same periodicity interval in the plane of the layer is also investigated. Various instabilities are found in dependence on the Prandtl number P and the angle γ of inclination and their evolution is studied in a few cases.

1. Introduction

Convection in an inclined layer heated from below has been studied both experimentally and theoretically by numerous investigators during the past decades. In contrast to the case of a horizontal layer, the basic state of an inclined layer is characterized by a mean flow with a cubic profile. In the following we shall assume that the inclined layer is sufficiently extended that end effects do not affect the properties of convection flows very much. In particular we shall assume that the temperature distribution of the basic state is governed by conduction only and is approximately uniform throughout the plane of the layer.

The basic state of the inclined layer becomes unstable with respect to transverse or longitudinal rolls depending on the Prandtl number P of the fluid and on the angle of inclination. As discussed by Gershuni and Zhukhovitskii (1976) longitudinal rolls represent the preferred form of convection in high Prandtl number fluids for almost all inclinations as long as there exists a finite component of the temperature gradient opposite to the direction of gravity. Even in the case $P = 1$ longitudinal rolls remain preferred in an inclined layer heated from below up to an angle of 77° of inclination with respect to the horizontal. Only at much lower Prandtl numbers does the source of kinetic energy of the cubic profile flow contribute sufficiently to favor the onset of transverse rolls for all angles of inclination.

Longitudinal rolls are characterized by the property that their heat transport is independent of the angle γ of inclination if the normal component of gravity, $g \cos \gamma$, is used in the definition of the Rayleigh number. Other properties, however, such as the profile of the mean flow depend on the parameter γ. Accordingly, the stability properties of the longitudinal convection rolls also exhibit a strong dependence on the angle of inclination. In particular the wavy instability has been found to play a dominant role in restricting the region of the parameter space for which longitudinal rolls can be realized. For theoretical and experimental work on this topic we refer to the papers of Clever and Busse (1977, referred to in the following by CB77) and Ruth et al. (1980a). In the present paper we investigate finite amplitude properties of three-dimensional wavy rolls induced by the onset of the wavy instability.

Convection in an inclined layer heated from below is of considerable interest for engineering applications. In particular, in the design of solar energy collectors the efficiency of the convective heat transport plays an important role. The fact that this efficiency is reduced significantly after the onset of wavy convection rolls has already been noted in the experimental study of Ruth et al. (1980a). The numerical results of the present paper provide details on the parameter dependence of this effect. Another effect emerging from the theoretical study of this paper is the onset of subcritical finite amplitude wavy rolls in some cases. This property may be responsible in part for some of the remaining discrepancies between experimental observations and the prediction of linear theory.

The mathematical formulation of the problem and the numerical method of solution are described in Section 2. While the basic equations are the same as those treated in CB77, three-dimensional solutions and their stability properties are analyzed in the present paper in contrast to the study of two-dimensional solution and their stability in CB77. The properties of steady wavy rolls are shown in Section 3 in comparison with those of longitudinal rolls. The results of the stability analysis together with the calculations for two time-dependent solutions are presented in Section 4 and a concluding discussion is given in the final section.

2. Mathematical description of the problem

We consider a fluid layer between parallel no-slip boundaries inclined at an angle γ with respect to the horizontal. Constant temperatures T_1 and $T_2 (T_2 > T_1)$ are prescribed at the upper and lower boundaries. Using the thickness of the layer as length scale, d^2/κ as time scale where κ is the thermal diffusivity, and $(T_2 - T_1)/R$ as scale of the temperature we obtain the dimensionless form of the basic equations for a Boussinesq fluid as given by (2.1) of CB 77.

The basic state of the inclined layer as sketched in Fig. 1 is given by the dimensionless temperature distribution Θ_0 and the velocity field $\mathbf{U}_0$,

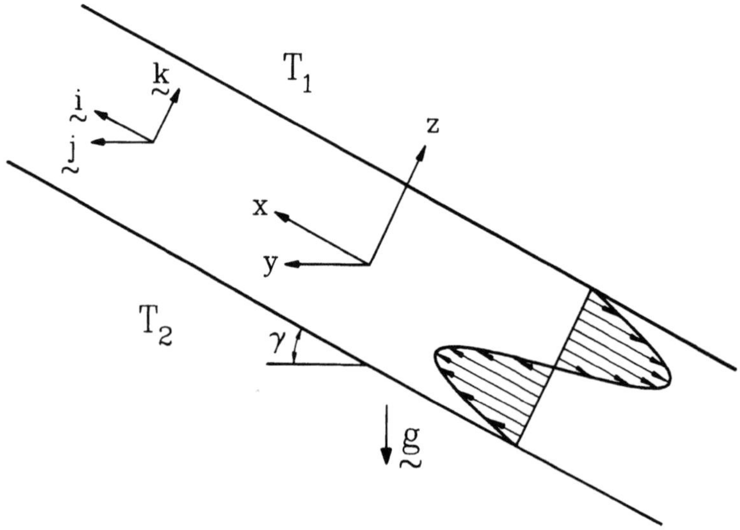

Fig. 1. Sketch of the geometrical configuration of the inclined layer heated from below.

$$\Theta_0 = -Rz + R(T_1 + T_2)/2(T_2 - T_1) , \qquad \mathbf{U}_0 = \mathbf{i}R\tfrac{1}{6}(z^3 - \tfrac{1}{4}z)\tan\gamma \equiv U_0\mathbf{i} , \tag{2.1a,b}$$

where the Rayleigh number R is defined by

$$R = \beta g \cos\gamma(T_2 - T_1)\,\mathrm{d}^3/\nu\kappa . \tag{2.2}$$

The coefficient β of thermal expansion, the acceleration g of gravity and the kinematic viscosity ν have been used in this expression. For the orientation of the coordinate system and the associated unit vectors **i**, **j**, **k** we refer to Fig. 1.

The secondary and tertiary solutions of the basic equations will be described as deviations from the primary solution (2.1),

$$\Theta = \Theta_0 + \vartheta , \qquad \mathbf{u} = U_0\mathbf{i} + U^{(x)}\mathbf{i} + \nabla \times (\nabla \times \mathbf{k}\varphi) + \nabla \times \mathbf{k}\psi \equiv U\mathbf{i} + \boldsymbol{\delta}\varphi + \boldsymbol{\varepsilon}\psi , \tag{2.3ab}$$

where we have assumed that the x, y-average of the velocity field is directed solely in the x-direction.

By operating with $\mathbf{k}\cdot\nabla \times [\nabla \times \cdots]$ and $\mathbf{k}\cdot\nabla\times$ onto the equations of motion and using the heat equation we obtain the following three equations for φ, ω and ϑ,

$$\nabla^4\Delta_2\varphi + \tan\gamma\,\partial^2_{xz}\vartheta - \Delta_2\vartheta = P^{-1}\{\boldsymbol{\delta}\cdot[(\boldsymbol{\delta}\varphi + \boldsymbol{\varepsilon}\psi)\cdot\nabla(\boldsymbol{\delta}\varphi + \boldsymbol{\varepsilon}\psi)] + (U\partial_x + \partial_t)\nabla^2\Delta_2\varphi - \partial^2_{zz}U\partial_x\Delta_2\varphi\} , \tag{2.4a}$$

$$\nabla^2\Delta_2\psi + \tan\gamma\,\partial_y\vartheta = P^{-1}\{\boldsymbol{\varepsilon}\cdot[(\boldsymbol{\delta}\varphi + \boldsymbol{\varepsilon}\psi)\cdot\nabla(\boldsymbol{\delta}\varphi + \boldsymbol{\varepsilon}\psi)] + (U\partial_x + \partial_t)\Delta_2\psi - \partial_z U\partial_y\Delta_2\varphi\} , \tag{2.4b}$$

$$\nabla^2\vartheta - R\Delta_2\varphi = (\boldsymbol{\delta}\varphi + \boldsymbol{\varepsilon}\psi)\cdot\nabla\vartheta + (U\partial_x + \partial_t)\vartheta . \tag{2.4c}$$

In addition an equation is needed for the component $U^{(x)}$ of the mean velocity field,

$$(\partial^2_{zz} - \partial_t)U^{(x)} + \partial_z\overline{[\Delta_2\varphi(\partial^2_{xz}\varphi + \partial_y\psi)]} + \bar{\vartheta}\tan\gamma = 0 , \tag{2.4d}$$

where the bar indicates the average over the x, y-plane. The symbol Δ_2 has been introduced for $\nabla^2 - \partial^2_{zz}$ and P is the Prandtl number, $P = \nu/\kappa$. The boundary conditions for φ, ψ, ϑ and $U^{(x)}$ are given by

$$\varphi = \partial_z\varphi = \psi = \vartheta = U^{(x)} = 0 \quad \text{at } z = \pm\frac{1}{2} . \tag{2.5}$$

Longitudinal rolls corresponding to x-independent steady solutions of the problem (2.4), (2.5) and their instabilities have been investigated in CB77. The variables φ, ϑ describing longitudinal rolls are the same as in a horizontal layer. The parameter γ only enters the solution for ψ and $U^{(x)}$; the convective heat transport as a function of the Rayleigh number as defined by (2.2) is thus independent of γ. The main instability of longitudinal rolls is the wavy instability which is the origin of some of the three-dimensional forms of convection observed in the experiments by Hart (1971), Ruth et al. (1980a) and others. The following analysis is devoted to the study of finite amplitude properties of this type of convection.

Since the imaginary part of the growth rate of the wavy instability vanishes we expect an evolution of the instability towards a steady three-dimensional solution of equations (2.4). Using the Galerkin method we expand the variables φ, ψ, ϑ and $U^{(x)}$ into complete systems of functions satisfying the boundary conditions (2.5),

$$\varphi = \sum_{\lambda,\mu,\nu} a_{\lambda\mu\nu} \begin{Bmatrix} \cos \lambda\alpha_x x \\ \sin \lambda\alpha_x x \end{Bmatrix} g_\nu(z)[p(\lambda)\cos \mu\alpha_y y + (1-p(\lambda))\sin \mu\alpha_y y] , \tag{2.6a}$$

$$\psi = \sum_{\lambda,\mu,\nu} c_{\lambda\mu\nu} \begin{Bmatrix} \cos \lambda\alpha_x x \\ \sin \lambda\alpha_x x \end{Bmatrix} \sin \nu\pi(z+\tfrac{1}{2})[p(\lambda)\sin \mu\alpha_y y + (1-p(\lambda))\cos \mu\alpha_y y] , \tag{2.6b}$$

$$\vartheta = \sum_{\lambda,\mu,\nu} b_{\lambda\mu\nu} \begin{Bmatrix} \cos \lambda\alpha_x x \\ \sin \lambda\alpha_x x \end{Bmatrix} \sin \nu\pi(z+\tfrac{1}{2})[p(\lambda)\cos \mu\alpha_y y + (1-p(\lambda))\sin \mu\alpha_y y] , \tag{2.6c}$$

$$U^{(x)} = \sum_{\nu} U_{1\nu} \sin 2\nu\pi(z+\tfrac{1}{2}) , \tag{2.6d}$$

where the summation runs through positive integers ν and non-negative integers λ and μ and where the function $p(\lambda)$,

$$p(\lambda) = 1 \quad \text{for even } \lambda , \qquad p(\lambda) = 0 \quad \text{for odd } \lambda \tag{2.7}$$

has been introduced in order to capture the symmetry of the convection flow generated by the interaction of the wavy instability with the longitudinal rolls. Another aspect of this symmetry manifests itself in the selection rule that the upper function in the wavy bracket must be chosen for even $\nu + \mu$ and the lower function for odd $\nu + \mu$. The functions $g_\nu(z)$ were first introduced by Chandrasekhar (1961, p. 635) and are also defined in CB77. After introducing the representations (2.6) into equations (2.4), multiplying the equations by the respective expansion functions and averaging them over the fluid layer, we obtain a system of nonlinear algebraic equations for the coefficients $a_{\lambda\mu\nu}$, $c_{\lambda\mu\nu}$, $b_{\lambda\mu\nu}$, $U_{1\nu}$. After truncating this infinite system of equations it can be solved by a Newton–Raphson method. As in the case of similar problems (Clever and Busse, 1987, 1989) the truncation condition in which all coefficients and corresponding equations satisfying

$$\lambda + \mu + \nu > N_T$$

are dropped offers an optimal combination of computational efficiency and economy. Almost all computations to be reported in the following have been carried out for $N_T = 6$, 8 and 10. In the plots the results obtained for $N_T = 10$ are shown provided they differ by less than a few percent from those obtained for $N_T = 8$. Only at the highest values of R $N_T = 12$ has also been used.

The Galerkin method offers the opportunity for a relatively simple analysis of the stability of the steady three-dimensional solutions. Arbitrary infinitesimal disturbances can be superimposed onto the steady solution and their growth rate can be determined. Here we shall restrict the analysis to those disturbances which fit the x, y-periodicity interval of the steady solution. This procedure excludes disturbances with finite Floquet exponents which are usually of lesser importance. The main reason for this restriction is that disturbances which fit the periodicity interval of the steady solution can be separated into four subsets differing in their symmetries with respect to the steady solution. This separation facilitates the stability analysis enormously since the eigenvalues of much smaller matrices must be evaluated. The infinitesimal disturbances that we are considering can thus be represented in the form

$$\tilde{\varphi} = \sum_{\lambda,\mu,\nu} \tilde{a}_{\lambda\mu\nu} \begin{Bmatrix} \cos \lambda\alpha_x x \\ \sin \lambda\alpha_x x \end{Bmatrix} g_\nu(z)[\tilde{p}(\lambda) \cos \mu\alpha_y y + (1 - \tilde{p}(\lambda)) \sin \mu\alpha_y y] \exp(\sigma t) , \tag{2.8a}$$

$$\tilde{\psi} = \sum_{\lambda,\mu,\nu} \tilde{c}_{\lambda\mu\nu} \begin{Bmatrix} \cos \lambda\alpha_x x \\ \sin \lambda\alpha_x x \end{Bmatrix} \sin \nu\pi(z + \tfrac{1}{2})[\tilde{p}(\lambda) \sin \mu\alpha_y y + (1 - \tilde{p}(\lambda)) \cos \mu\alpha_y y] \exp(\sigma t) , \tag{2.8b}$$

$$\tilde{\vartheta} = \sum_{\lambda,\mu,\nu} \tilde{b}_{\lambda\mu\nu} \begin{Bmatrix} \cos \lambda\alpha_x x \\ \sin \lambda\alpha_x x \end{Bmatrix} \sin \nu\pi(z + \tfrac{1}{2})[\tilde{p}(\lambda) \cos \mu\alpha_y y + (1 - \tilde{p}(\lambda)) \sin \mu\alpha_y y] \exp(\sigma t) , \tag{2.8c}$$

$$\tilde{U}^{(x,y)} = \sum_{\nu} \hat{p}(\nu)\tilde{U}_\nu^{(x,y)} \sin \nu\pi(z + \tfrac{1}{2}) \exp(\sigma t) , \tag{2.8d}$$

where the four classes *SE*, *SO*, *AE*, *AO* can be distinguished according to the following definitions:

$$S:\ \tilde{p}(\lambda) = p(\lambda) , \qquad A:\ \tilde{p}(\lambda) = 1 - p(\lambda) , \tag{2.9a}$$

$E(O)$: Upper (lower) functions in the wavy brackets correspond to even $\nu + \mu$ and lower (upper) functions correspond to odd $\nu + \mu$. (2.9b)

The disturbance mean flow (2.8d) for the classes S is directed solely in the x-direction, while for the classes A only a disturbance mean flow in the y-direction can exist. The mean flows are symmetric (antisymmetric) with respect to the plane $z = 0$ in the classes *SO* and *AE* (*SE* and *AO*). Accordingly $\hat{p}(\nu) = p(\nu)$ holds for *SO* and *AE* and $\hat{p}(\nu) = 1 - p(\nu)$ for *SE* and *AO*.

3. Steady three-dimensional convection flows

Three-dimensional solutions of the form (2.6) have been obtained for a variety of Prandtl numbers and angles of inclinations. The wavenumber α_y has been fixed at the critical value $\alpha_c = 3.117$ for the onset of longitudinal convection rolls because those rolls remain stable up to the Rayleigh number R_{II} for the onset of the wavy instability. Once the threshold value R_{II} is exceeded, wavy roll solutions exist for a broad band of wavenumbers α_x. Since the value α_x of the strongest growing disturbance increases significantly with R beyond the threshold value R_{II} according to CB77 and since typical observations (Hart, 1971) of experimentally realized wavy rolls exhibit relatively large values of α_x, we have chosen values of α_x which appear to be representative for the wavy roll regime.

For all cases that have been studied, the transition from longitudinal rolls to wavy rolls causes a drop in the convective heat transport. A typical example can be seen in Fig. 2 in which results for the case of Prandtl number $P = 12$ have been plotted. There thus exists a region for which the Nusselt number decreases with increasing Rayleigh number. For lower

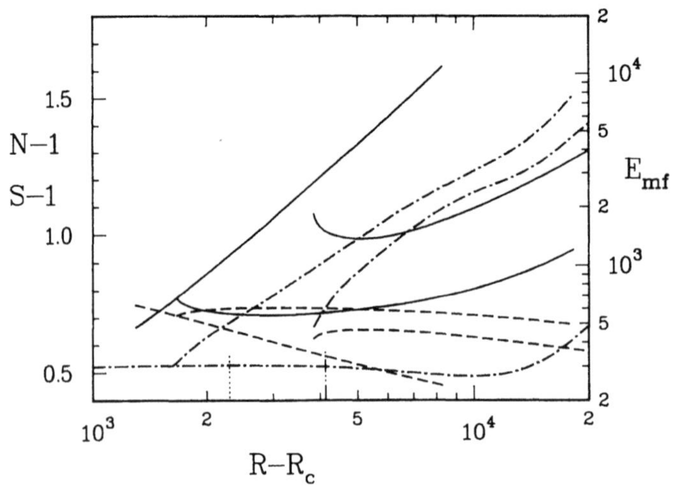

Fig. 2. Nusselt number N (solid lines), shear Nusselt number S (dashed lines) and kinetic energy of the mean flow E_{mf} (dash-dotted lines) for longitudinal rolls and wavy rolls with $\alpha_x = 1.5$ (bifurcating at $R - R_c \simeq 1.6 \cdot 10^3$) and with $\alpha_x = 2.0$ starting at about $R - R_c = 3.8 \cdot 10^3$. The wavy rolls become unstable at the Rayleigh numbers indicated by short lines (dotted for $\alpha_x = 1.5$, dash-double-dotted for $\alpha_x = 2.0$). The parameters $P = 12$, $\gamma = 60°$, $\alpha_y = 3.117$ have been assumed.

Prandtl numbers this effect is less pronounced, but the growth of the Nusselt number is always reduced in comparison with the case of longitudinal rolls.

In analogy to the Nusselt number N which is defined as the ratio of the heat transport with and without convection, the shear Nusselt number S can be defined as the ratio of the shear stresses at the walls in the presence and in the absence of convection,

$$S \equiv \left[\frac{\partial}{\partial z} U \Big/ \frac{\partial}{\partial z} U_0 \right]_{z=\pm\frac{1}{2}} . \tag{3.1}$$

Longitudinal convection rolls are very effective in reducing the shear stress primarily because an effective heat transport decreases the mean temperature gradient in the interior of layer which is responsible for driving the cubic profile flow. It is thus not surprising that an increase of S occurs together with a decrease of N. The kinetic energy of the mean flow,

$$E_{mf} \equiv \tfrac{1}{2}\langle U^2 \rangle \tag{3.2}$$

also increases with the onset of wavy rolls. Here the angular brackets indicate the average over the fluid layer.

Another interesting effect that can be seen in Fig. 2 is the subcritical onset of wavy rolls for larger values of α_x, say $\alpha_x \gtrsim 2$. Because on the inverted part of the solution branch the numerical scheme does not converge, the three-dimensional solution in the case $\alpha_x = 2.0$ is not connected in the figure with a bifurcation point on the longitudinal roll solution. The numerical scheme mimics the physical system in this respect since the inverted part of the bifurcating branch can also not be realized because of its instability.

In Fig. 3 the kinetic energies of the fluctuating components of the velocity field are shown which are defined by

$$E_{\text{pol}} \equiv \tfrac{1}{2}\langle |\nabla \times (\nabla \times \mathbf{k}\varphi)|^2 \rangle , \tag{3.3a}$$

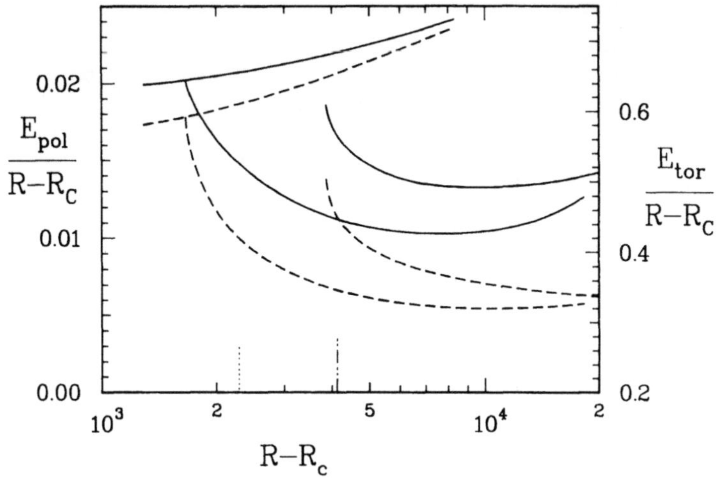

Fig. 3. Same as in Fig. 2 except that E_{pol} (solid lines) and E_{tor} (dashed lines) have been plotted.

$$E_{tor} \equiv \tfrac{1}{2}\langle |\nabla \times \mathbf{k}\psi|^2 \rangle \,. \tag{3.3b}$$

As is evident from the figure, these quantities increase roughly in proportion to $R - R_c$ for longitudinal rolls, but vary much less rapidly with R after the onset of wavy rolls.

In Figs 4 through 6 results for water ($P = 7$) are shown for two different angles of inclination. The change in the heat transport caused by the onset of wavy rolls is less dramatic than in the case $P = 12$ and a tendency towards a recovery of the transport efficiency can be noticed at higher Rayleigh numbers. This latter tendency is especially pronounced at the high inclination of $\gamma = 80°$ where it has also been found that the energy of the poloidal component of flow can grow to values much higher than the corresponding values of longitudinal rolls. This result may indicate a tendency towards the realisation of transverse rolls which are a strong competitor for longitudinal rolls at high inclinations according to linear theory.

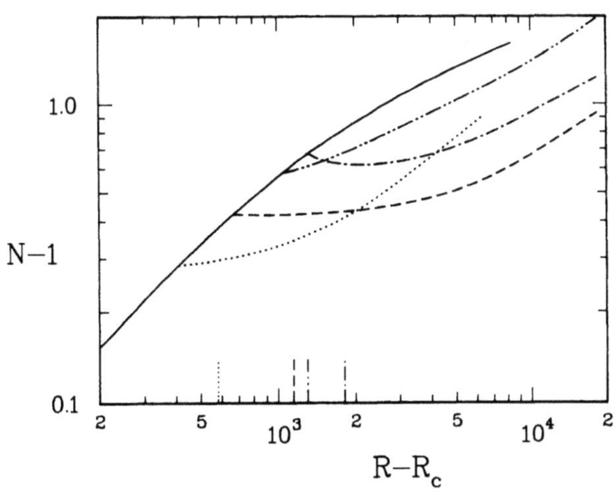

Fig. 4. Nusselt number N in the case $P = 7$, $\alpha_y = 3.117$ for longitudinal rolls (solid line) and wavy rolls with $\alpha_x = 1.5$ (dashed line) and $\alpha_x = 2.0$ (dash-dotted) for $\gamma = 60°$ and with $\alpha_x = 1.5$ (dotted) and $\alpha_x = 2.0$ (dash double-dotted) for $\gamma = 80°$. The Rayleigh numbers at which the respective wavy rolls become unstable are indicated at the abscissa.

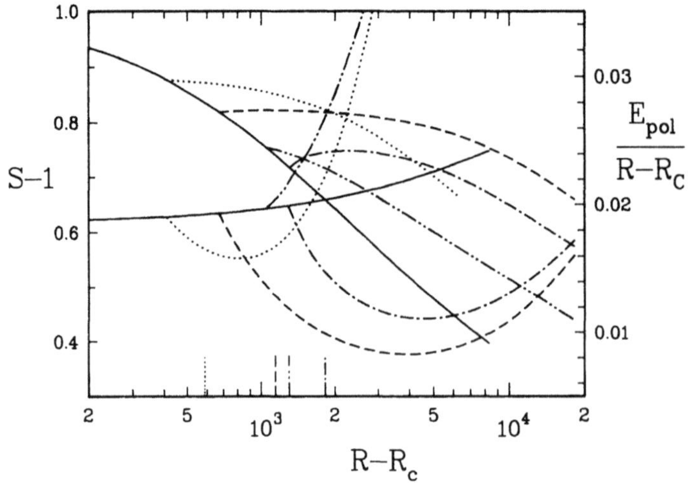

Fig. 5. Shear Nusselt numbers (descending solid line and attached branches) and E_{pol} for the same cases as in Fig. 4.

While the Prandtl number of water at room temperature is 7, it decreases rapidly with increasing temperature and reaches a value of about 2.5 at 60°C. This property has motivated the computations for the case $P = 2.5$ displayed in Figs 7 and 8. It is evident from these figures that the results resemble qualitatively those obtained in the higher Prandtl number cases. Since the difference $R_{II} - R_c$ between the thresholds for wavy rolls and longitudinal rolls increases almost proportional to P^2 according to Fig. 10 of CB77, the curves shown in Figs 7 and 8 are shifted to much lower values of R in comparison to those of the previous figures.

Because of the general similarity of the evolution of wavy rolls for different values of γ and P, we show only a few representative plots of the flow structure. The left-hand column of Fig. 9 shows the relatively smooth dependence of the flow field in the case of lower inclination, lower Prandtl number, and lower Rayleigh number, while all of these parameters assume higher values in the case of the right-hand column, where much smaller flow structures become visible. It is remarkable that the isotherms are almost identical to the lines

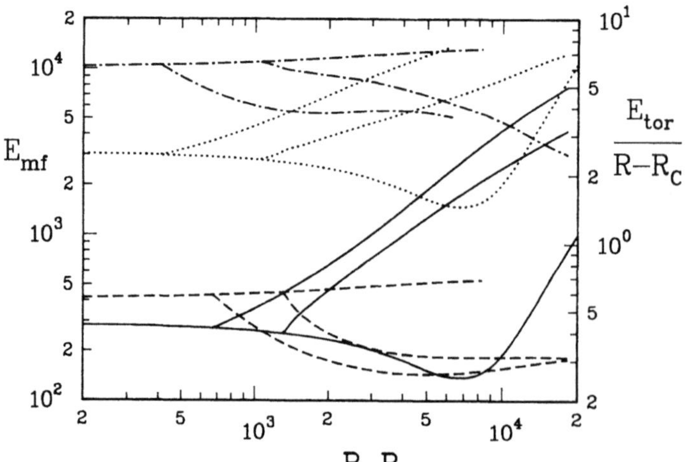

Fig. 6. E_{mf} (solid lines for $\gamma = 60°$, dotted lines for $\gamma = 80°$) and E_{tor} (dashed lines for $\gamma = 60°$, dash-dotted lines for $\gamma = 80°$) for the same parameter values as in Fig. 4.

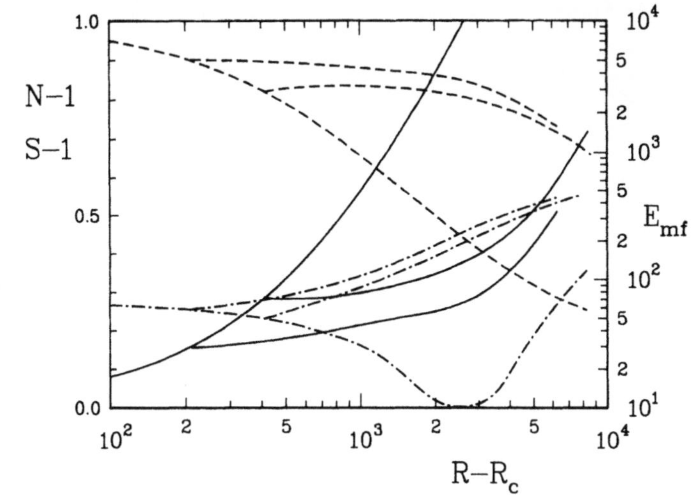

Fig. 7. Nusselt number N (solid lines), shear Nusselt number S (dashed lines), and E_{mf} (dash-dotted) for longitudinal rolls and for bifurcating wavy rolls with $\alpha_x = 1.5$ (bifurcating at $R - R_c \simeq 200$) and with $\alpha_x = 2.0$ in the case $P = 2.5$, $\gamma = 40°$, $\alpha_y = 3.117$.

of constant velocity in the z-direction in the case of lower Prandtl numbers. While lines of positive and negative z-velocity exhibit the expected symmetry in the center plane, $z = 0$, a considerable asymmetry develops as one moves to planes closer to the boundaries. The plots in the middle of the figure exhibit this property quite clearly. The features of wavy rolls in the case $P = 2.5$ are also visible in the case of $P = 0.71$ shown in Fig. 10. In addition, the toroidal component of the velocity has been plotted in this figure which shows the effect of the advection of mean flow by the z-component of the velocity field.

Figures 11 through 15 display results obtained for an inclined air layer ($P = 0.71$). In Fig. 11 the Nusselt number is plotted for a number of different values of γ and α_x. It is interesting to see, that the recovery of the Nusselt number for wavy rolls to values comparable to those of longitudinal rolls is almost complete in the case $\gamma = 25°$. But at higher angles of inclination the Nusselt number deficiency for wavy rolls persists. The change of the shear Nusselt number S induced by wavy rolls and plotted in Fig. 12 is even more dramatic than at the

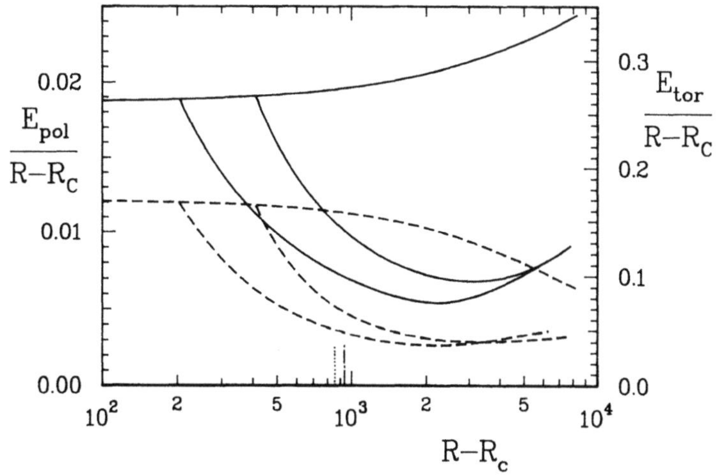

Fig. 8. E_{pol} (solid lines) and E_{tor} (dashed lines) for the same parameter values as in Fig. 7. The Rayleigh numbers at which the wavy rolls become unstable are indicated at the abscissa (dotted for $\alpha_x = 1.5$, dash-dotted for $\alpha_x = 2.0$).

Fig. 9. Lines of constant normal velocity, $-\Delta_2\varphi = \text{const.}$, in the planes $z = 0$ (top row) and $z = -0.3$ (middle row) and isotherms in the plane $z = 0$ (bottom row) for the cases $P = 2.5$, $\gamma = 40°$, $R = 3500$ (left column) and $P = 7$, $\gamma = 60°$, $R = 10^4$ (right column). $\alpha_x = 2.0$, $\alpha_y = 3.117$ and $N_T = 10$ have been used in all cases. Solid lines describe positive values, dashed lines correspond to negative values and solid line adjacent to the dashed lines indicates zero. The y-direction is upward, the x-direction towards the right.

Fig. 10. Lines of constant normal velocity, $-\Delta_2\varphi = \text{const.}$, in planes $z = 0$ (upper left) and $z = -0.3$ (lower left), isotherms in the plane $z = 0$ (upper right) and streamlines of the toroidal component of the velocity field, $\psi = \text{const.}$, in the plane $z = 0$ (lower right) for $P = 0.71$, $R = 2490$, $\gamma = 40°$, $\alpha_x = 2.4$, $\alpha_y = 3.117$. Solid (dashed) lines indicate positive (negative) values and the solid line adjacent to the dashed lines describes zero. The y-direction is upward, the x-direction towards the right.

higher Prandtl numbers mainly because the decrease of S with increasing R owing to the actions of longitudinal rolls becomes more pronounced at lower Prandtl numbers. The changes in the energies of the fluctuating and of the mean velocity fields shown in Figs 13 through 15 are very similar to those found in the case of higher Prandtl numbers except for a shift owing to the strong dependence of R_{II} on P which we have mentioned above.

In the last two Figs 16 and 17, profiles of the mean flow $U(z)$ are shown. The amplitude of mean flow is reduced in comparison with the cubic profile flow (2.1b), but not as much as in

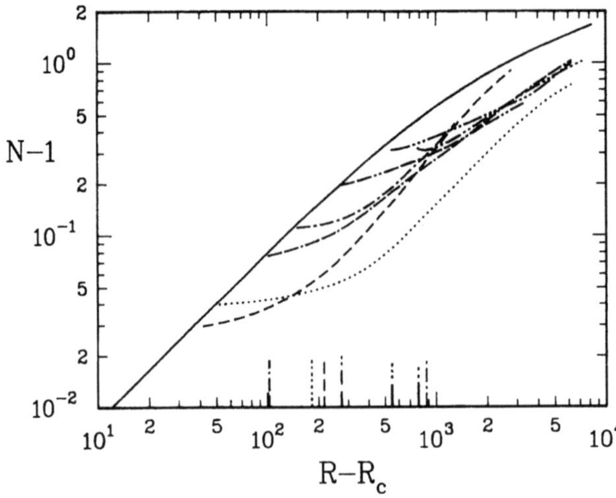

Fig. 11. Nusselt number N for longitudinal rolls (solid line) and wavy rolls with $\alpha_x = 1.5$ (dashed line) and $\alpha_x = 2.1$ (dash-dotted) for $\gamma = 25°$, with $\alpha_x = 1.5$ (dotted line) and $\alpha_x = 2.4$ (dash double-dotted) for $\gamma = 40°$, and with $\alpha_x = 1.2$ (double-dash-dotted), $\alpha_x = 1.8$ (double-dash double-dotted) and $\alpha_x = 2.1$ (dash triple-dotted) for $\gamma = 60°$. The parameters $P = 0.71$ and $\alpha_y = 3.117$ have been assumed. The Rayleigh numbers at which the respective wavy rolls become unstable have been indicated at the abscissa.

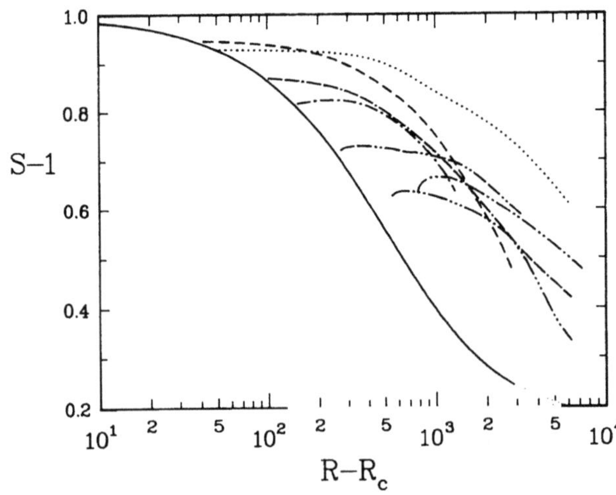

Fig. 12. Same as Fig. 11 for the shear Nusselt number.

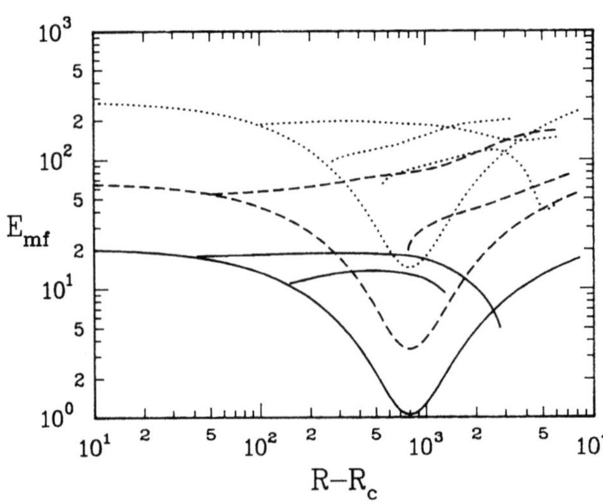

Fig. 13. Kinetic energy of the mean flow E_{mf} for longitudinal rolls and bifurcating solutions of wavy rolls in the cases $\gamma = 25°$ (solid lines), $\gamma = 40°$ (dashed lines), and $\gamma = 60°$ (dotted lines). The parameter values are the same as in Fig. 11.

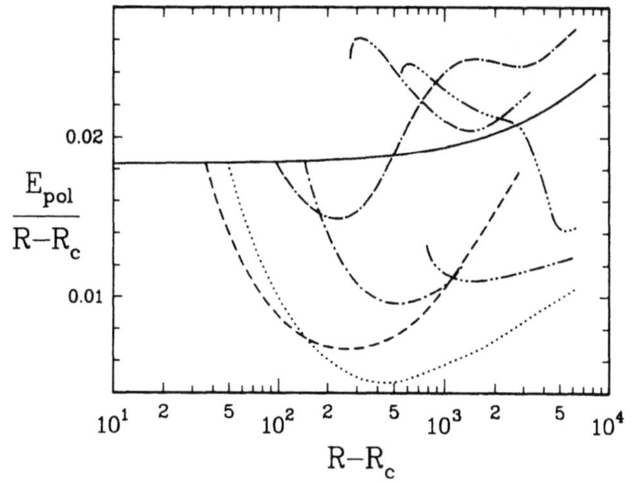

Fig. 14. Same as Fig. 11 for $E_{pol}/(R-R_c)$ instead of $N-1$.

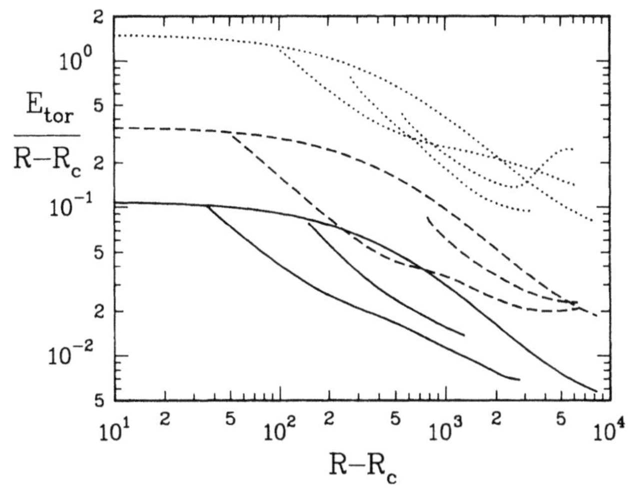

Fig. 15. Same as Fig. 13 for $E_{tor}/(R-R_c)$ instead of E_{mf}.

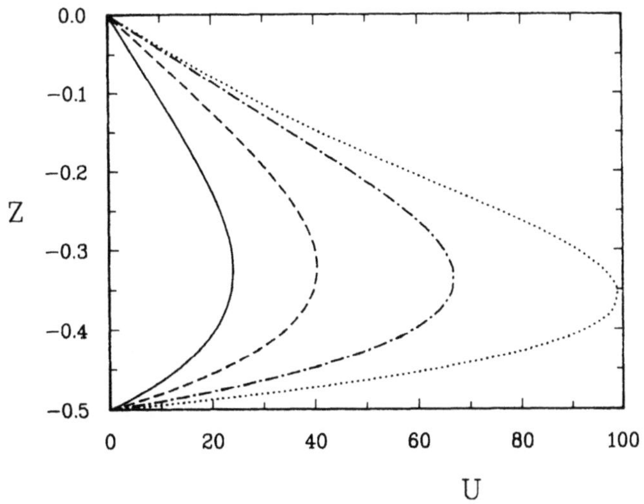

Fig. 16. Mean flow profile $U(z)$ for wavy convection rolls in an inclined layer with $P=7$, $\gamma=60°$, $\alpha_x=2.1$, $\alpha_y=3.117$. The curves correspond (left to right) to $R=3100$, 5000, 10^4, $2\cdot10^4$.

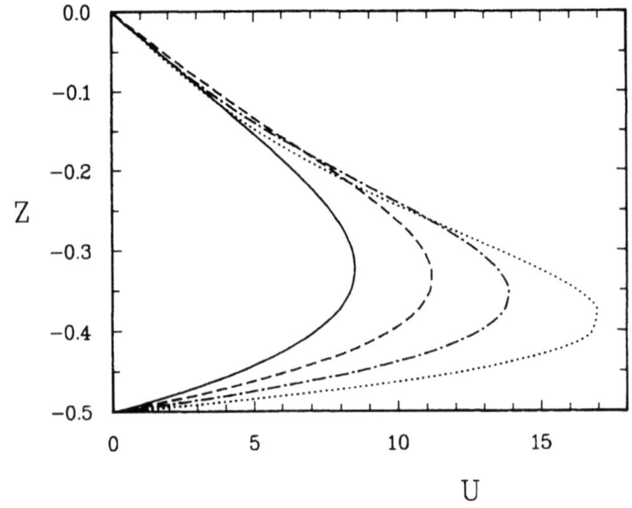

Fig. 17. Same as Fig. 16, but for $P = 0.71$, $\gamma = 40°$. The curves correspond to $R = 2000$ (solid), 3000 (dashed), 5000 (dash-dotted), 10^4 (dotted).

the case of longitudinal rolls (see Figs 3 and 4 of CB77 for comparison). The maximum of mean flow profile is shifted towards the boundary with increasing Rayleigh number since the x, y-average of the temperature in the interior of the layer becomes nearly isothermal and the buoyancy force driving the mean flow becomes located in the thermal boundary layers. As must be expected from the Prandtl number dependence of equations (2.4), the convection induced effects on the mean flow are much more dramatic in the case of air than in the case of water.

4. Instabilities of wavy convection rolls

The procedure for the search of growing disturbances has been outlined at the end of Section 2. Growth rates σ have been obtained for all four classes of disturbances SE, SO, AE, AO. The results of the computations show that disturbances of the class SE, usually yield the highest real part σ_r of σ for $P \gtrsim 1$. In the case of air, $P = 0.71$, disturbances of the classes SE, AE and AO are close competitors and for $P = 0.3$ AO disturbances dominate.

The instability in the case $P \gtrsim 1$ always exhibits a finite imaginary part σ_i of the growth rate. The period corresponding to the frequency σ_i appears to be roughly related to the circulation time in the rolls. Because the growing disturbances do not change the symmetry of the steady wavy roll solution, it is relatively easy to follow its evolution by a forward integration in time. For this purpose the coefficients a_{lmn}, b_{lmn} c_{lmn} in expressions (2.b) are assumed to be functions of time and a Crank–Nicholson scheme is used for the numerical integration. After transients have died away it is found that a periodic oscillation occurs in which the flow appears to vacillate between the state of longitudinal rolls and a state of nearly transverse oriented vortices as is evident from the example shown in Fig. 18. The mean quantities plotted as a function of time in Fig. 18a are periodic with half the period of oscillation as can be seen from the comparison with the flow field shown in Fig. 18b. The interval of growing transport of heat and momentum corresponds to the state of nearly aligned longitudinal rolls. In the sudden transition to modulated transverse vortices kinetic energy of the toroidal component is converted into kinetic energy of the poloidal component

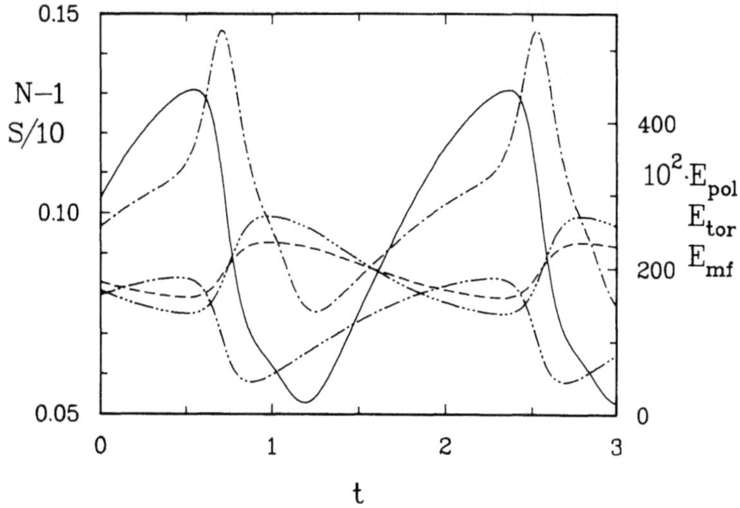

Fig. 18a. The Nusselt number N (solid line), the shear Nusselt number S (dashed), the kinetic energies of the poloidal (dash-dotted), toroidal (dash double-dotted) and mean (dash triple-dotted) components of the velocity field as a function of time for vacillating convection with $R = 1900$, $P = 0.71$, $\gamma = 60°$, $\alpha_x = 1.2$, $\alpha_y = 3.117$.

of the velocity field. But since the longitudinal orientation of the rolls is lost, the transports are decreased in this process. After the decay of the modulated transverse vortices the longitudinal rolls start to grow again.

Since the cases where the vacillation instability occurs are also characterized by relatively high angles of inclination where the critical Rayleigh numbers for the onset of longitudinal rolls and for transverse vortices get relatively close, the vacillation between these two states is perhaps not surprising. It must be taken into account in this connection that two-dimensional transverse vortices are not very stable according to the analysis of Nagata and Busse (1983) and that they are replaced by three-dimensional vortices exhibiting significant structure in the transverse direction similar to that of the second picture from the top of Fig. 18b in the middle column.

While wavy rolls in an air layer with 40° inclination and $\alpha_x \geqslant 1.8$ are unstable with respect to the vacillation instability, the instability with the AO symmetry and with vanishing σ_i predominates at lower angles of inclination and low values of α_x, $\alpha_x \leqslant 1.5$. For intermediate cases such as $\gamma = 25°$, $\alpha_x = 2.1$ and $\gamma = 40°$, $\alpha_x = 1.5$ the preferred instability is of the AE-type. As is evident from Fig. 11, the region in which wavy rolls are stable in the case of 25° inclination is so small that we feel forced to conclude that the wavy rolls observed in the experiments of Ruth et al. (1980a) are not the original wavy rolls analyzed in this paper, but they are instead those which are already modified by the AE-type instability which appears to be most relevant for the parameters of the experiment. Obviously the effects of this modification are not very dramatic since the observed convection exhibits a close resemblance to wavy rolls. Consistent with this interpretation is the fact that Ruth et al. report slight lateral motions whenever the angle of inclination reaches a value $\gamma \gtrsim 20°$. This property must be expected in the case of an AE-type instability according to the discussion at the end of Section 2.

In order to test this hypothesis, we have done a forward integration in time in the case $\gamma = 40°$, $\alpha_x = 1.5$, $\alpha_y = 3.117$, $R = 2000$. Since the number of coefficients that must be determined as function of time is twice the number for steady wavy rolls due to symmetry considerations, the computations become rather time consuming. The remarkable result that

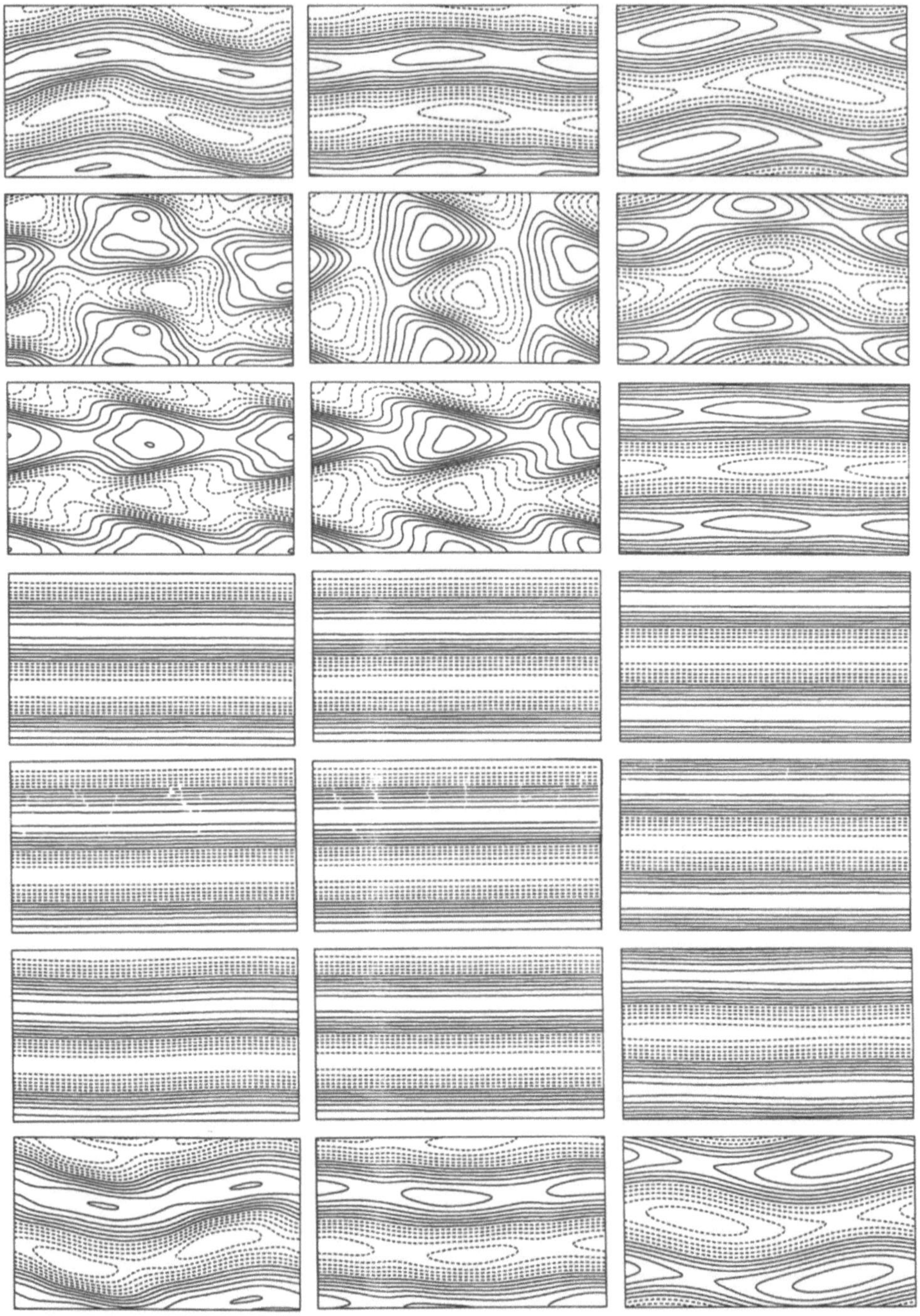

Fig. 18b. Lines of constant normal velocity at $z = -0.3$ (left column) and at $z = 0$ (middle column) and lines of constant ψ at $z = 0$ (right column) are shown at equal time steps ($\Delta t = 0.3034$) starting at the time 0.585 of Fig. 18a such that half a cycle is completed at the time 2.405 corresponding to the bottom row. Solid (dashed) lines indicate positive (negative) values except for the solid line adjacent to the dashed lines which indicates zero. The y-direction is upward, the x-direction towards the right. Parameter values are the same as in Fig. 18a.

Fig. 19. Lines of constant velocity u_z (left column) and constant toroidal stream function ψ (right) in the plane $z = 0$ for steady wavy rolls (top row) and for drifting asymmetric wavy rolls (middle and bottom rows) at two points in time, $\Delta t = 1.0$ apart. The parameter values $R = 2000$, $P = 0.71$, $\gamma = 40°$, $\alpha_x = 1.5$, $\alpha_y = 3.117$, $N_T = 8$ have been used.

has been found is displayed in Fig. 19. The growing disturbances do indeed change the original wavy roll pattern. But after a short transient, a stationary pattern of asymmetric wavy rolls is again attained except for a constant drift in the transverse direction. This direction depends on the sign of the critical disturbances of AE-type. Thus, left or right drifting patterns can be obtained, depending on the choice of initial conditions.

5. Concluding discussion

Although the problem of inclined layer convection has received considerable attention from experimenters, there exist few opportunities for a quantitative comparison of the theoretical results of this paper with laboratory observations. The general result that less heat is transported by wavy rolls than by longitudinal rolls is borne out by the experimental data of Ruth et al. (1980a). For a more detailed discussion of the heat flux data see Ruth et al. (1980b). But a quantitative comparison between theory and experiment is difficult for a number of reasons. First, the finite extent of experimental layers causes end effects and leads to a longitudinal component of the temperature gradient in the basic state which exerts a considerable influence on the dynamics of the system (Hart, 1971; Bergholz, 1978). Second, the finite lateral extent of the layer and its influence on the mean flow are not represented in the theory. The finite conductivity of the bounding surfaces of the enclosed fluid and deviations from the Boussinesq approximation also can give rise to asymmetries which may cause discrepancies between theory and experiment. For instance, the stationarity of wavy rolls which depends on the antisymmetry of the mean flow with respect to the center plane of the layer will give way to a time dependence once this symmetry is broken.

A detailed discussion of the problem of convection in inclined enclosures can be found in the book by Schinkel (1980) who also presents new theoretical and experimental results for the case of inclined rectangular boxes of finite aspect ratio filled with air. The theoretical computations are restricted, however, to the case of two-dimensional transverse rolls. Three-dimensional solutions for wavy patterns are described in the present paper for the first time. While finite aspect ratios and deviations from the Boussinesq approximation will undoubtedly affect quantitative aspects of these solutions, they clearly represent the wavy structure seen in experiments as demonstrated by the close similarity between the photographs of Hart (1971) and the plots of Figs 9 and 10. The analysis of the time-dependent three-dimensional solutions which bifurcate from the wavy rolls according to stability analysis are of special interest. The few preliminary cases that we have reported in this paper have already exhibited a number of unexpected features and a more systematic study appears to be warranted. Perhaps those surprising features such as the vacillations and the transverse drift will stimulate new experimental investigations.

The findings of this paper may have implications for the design of solar heat collectors. The presence of convection in an inclined layer strongly reduces the mean circulation which transports a considerable amount of heat in inclined layers of finite extent. Secondly, the heat transport by wavy rolls is much reduced in comparison with longitudinal rolls. It may thus be advantageous to realize the Rayleigh number regime where both the Nusselt number and mean flow strength are relatively low.

Acknowledgement

The research reported in this paper has been supported by the Atmospheric Sciences Section of the U.S. National Science Foundation.

References

Bergholz, R.F., Instabilities of steady natural convection in a vertical fluid layer. *J. Fluid Mech.* 84 (1978) 743–768.

Chandrasekhar, S., *Hydrodynamic and Hydromagnetic Stability*. Oxford: Clarendon Press (1961).
Clever, R.M. and Busse, F.H., Instabilities of longitudinal rolls in an inclined layer. *J. Fluid Mech.* 81 (1977) 107–127.
Clever, R.M. and Busse, F.H., Nonlinear oscillatory convection. *J. Fluid Mech.* 176 (1987) 403–417.
Clever, R.M. and Busse, F.H., Nonlinear oscillatory convection in the presence of a vertical magnetic field. *J. Fluid Mech.* 201 (1989) 507–523.
Gershuni, G.Z. and Zhukhovitskii, E.M., *Convective Stability of Incompressible Fluids*. Translated from the Russian by D. Louvish. Keter Publications, Jerusalem (1976).
Hart, J.E., The stability of flow in a differentially heated inclined box. *J. Fluid Mech.* 47 (1971) 547–576.
Nagata, M. and Busse, F.H., Three-dimensional tertiary motions in a plane shear layer. *J. Fluid Mech.* 135 (1983) 1–26.
Ruth, D.W., Hollands, K.G.T. and Raithby, G.D., On free convection experiments in inclined air layers heated from below. *J. Fluid Mech.* 96 (1980a) 461–469.
Ruth, D.W., Raithby, G.D. and Hollands, K.G.T., On the secondary instability in inclined air layers. *J. Fluid Mech.* 96 (1980b) 481–492.
Schinkel, W.M.M., *Natural convection in inclined air-filled enclosures*. Dutch Efficiency Bureau, Pijnacker (1980) 342 pp.

Journal of Engineering Mathematics **26**: 21–37, 1992.
H.K. Kuiken and S.W. Rienstra (eds), Problems in Applied, Industrial and Engineering Mathematics.

Marangoni convection in V-shaped containers

H.W. HOOGSTRATEN

Department of Mathematics, University of Groningen, P.O. Box 800, 9700 AV Groningen, The Netherlands

and

H.C.J. HOEFSLOOT* and L.P.B.M. JANSSEN

Department of Chemical Engineering, University of Groningen, Groningen, The Netherlands

Abstract. This paper presents a numerical study of the time evolution of Marangoni convection in two V-shaped containers involved in the microgravity experiments reported in Hoefsloot et al. [7]. First the case of the triangular container with a plane gas/liquid interface is considered, next the container having the shape of a circular sector with a curved interface is dealt with. The numerical results show the same behaviour as observed experimentally: convection caused by macroscale effects in the former, and microconvection in the latter case.

1. Introduction

Marangoni convection, or surface tension driven convective flow, may occur when a solute evaporates from a liquid at a gas/liquid interface and the liquid's surface tension depends on the solute concentration. Perturbations of the mass transfer across the interface create local differences in surface tension, and the liquid at the interface will flow from locations with low surface tension towards locations with high surface tension. If this mechanism is sufficiently strong to overcome the counteracting viscous resistance, a convective flow is created in the liquid. In many cases a characteristic roll-cell pattern is seen to develop. Very often a second mechanism that can give rise to convective flow is present: buoyancy effects as a result of density variations (Rayleigh instability). In practice it is sometimes difficult to determine whether an observed convective flow is the result of Marangoni or Rayleigh instability, or of a combination of the two effects. We mention that both types of instability can also occur in the case of heat transfer across a gas/liquid interface, viz, when surface tension and/or density are temperature-dependent.

In chemical process industry Marangoni convection is considered to be an important phenomenon in mass transfer equipment, like packed-bed columns, where thin liquid films are in contact with a gas phase. It has been shown (see e.g. Nield [13]) that Marangoni instability determines the convective flows in thin liquid films, whereas Rayleigh instability is dominant in thick layers. Since the presence of roll-cell activity in a chemical reactor enhances mass transfer, thus leading to an improved performance, the study of the Marangoni effect is of practical importance.

Experimental work on Marangoni convection in thin liquid layers is somewhat problematic, because flow measurements in such layers are very difficult and in thick layers the simultaneous presence of Rayleigh convection is an obstacle. Therefore experiments are usually carried out in a microgravity environment, so that buoyancy effects are negligible and

*Present address: Department of Chemical Engineering, University of Amsterdam, Amsterdam, The Netherlands.

the restriction to thin liquid layers is removed. A number of years ago the Chemical Engineering Department of the University of Groningen (The Netherlands) started a program of microgravity experiments to study Marangoni convection. In all experiments the driving force for the convective flows was the evaporation of acetone into air from an acetone in water solution. For this liquid the surface tension is a decreasing function of the solute concentration. This function is usually taken to be a linear one for the range of concentration values under consideration. Marangoni convection near a plane gas/liquid interface was studied during the D1-Spacelab flight of 1985 (Lichtenbelt et al. [11], Dijkstra and Lichtenbelt [2]). Experiments concerning surface tension driven flow around a ventilated air bubble in the acetone in water solution, carried out during two sounding-rocket flights (1987, 1988) in Kiruna (Sweden) and during two parabolic-flight campaigns (1986, 1988) in Houston (USA), were meant to investigate the influence of interface curvature (Dijkstra and Lichtenbelt [2], Lichtenbelt [10], Hoefsloot and Janssen [3, 4], Hoefsloot et al. [9]). The latest experiments were performed in 1990 with narrow V-shaped containers on board of the Caravelle 234 airplane at the Centre d'Essais en Vols in Bretigny (France), see Hoefsloot et al. [7].

The experiments showed that it makes sense to distinguish between two types of Marangoni convection: convection created by overall surface tension gradients occurring initially in the system (macroscale convection or, shortly, macroconvection) and convection which is the result of hydrodynamic instability (microscale convection or microconvection). We emphasize that this difference concerns only the mechanism governing the *onset* of the convective flow, and not necessarily the final flow development towards a more or less regular roll-cell pattern. Macroscale effects are mostly due to geometrical factors which lead to inhomogeneous mass transfer across the interface, whereas microconvection occurs when evaporation is uniform. A linear stability analysis describing theoretically the initial stage of microscale convection has been presented by Hoefsloot et al. for a cylindrically curved interface [5] and for the spherical interface separating a ventilated air bubble from the surrounding liquid [6, 8]. For the latter system numerical results describing the longer-time evolution of Marangoni convection are reported in Hoefsloot et al. [9].

In this paper we present a numerical study of the time evolution of Marangoni convection in two of the V-shaped containers involved in the experiments described in [7]. The flows in this type of container are believed to be representative for those occurring in dead zones (stagnant liquid zones) in packings in mass transfer equipment. Two different containers will be considered: a triangular one with a plane gas/liquid interface, in which convective flow due to macroscale effects is to be expected, and another having the shape of a circular sector with a curved interface, which will show microscale convection. In Section 2 the experimental setup and results concerning these two containers are briefly outlined. The mathematical formulation of the two-dimensional coupled fluid-flow/mass-diffusion problem is given in Section 3, together with a discussion of the numerical solution technique. In Section 4 the numerical results are presented, and the paper ends with Section 5 in which some concluding remarks are made.

2. Results of microgravity experiments

The experimental setup as well as the experimental results are fully described in Hoefsloot et al. [7]. For completeness a brief summary will be given here. The experiments were performed during a number of parabolic flights, in each of which a microgravity level of less than $0.05g$ was attained for a period of about 17 seconds. The main parts of the setup were narrow V-shaped glass containers, filled with a 5 wt% acetone solution in water in which, due to the evaporation of the acetone, Marangoni convection will start. The liquid contained tracer particles which were monitored by both a video and a photo camera. Figure 1 presents two photographs that are representative for the whole experiment. The triangular container with the flat gas/liquid interface shows two large roll cells (Fig. 1a). In the container with the curved interface (Fig. 1b), a number of smaller roll cells are visible near the interface,

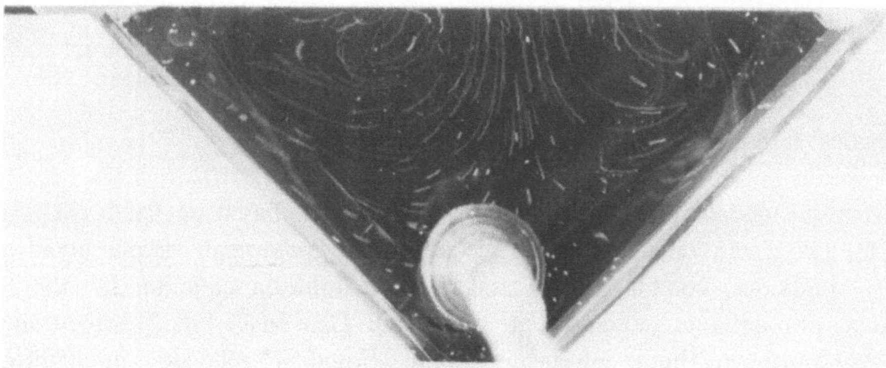

(a)

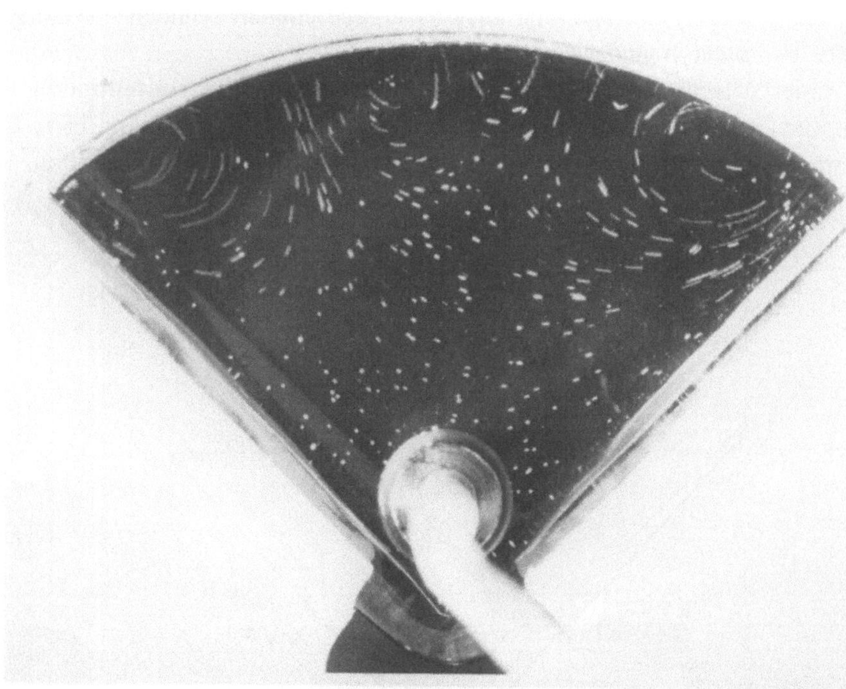

(b)

Fig. 1. Roll cells observed in a triangular container with plane interface (a) and a container with curved interface (b). The largest dimension of both containers is 28.3 mm. Camera shutter time: 1 s. Photograph (a) was taken 7.4 s after injection of the liquid into the container, (b) after 9.3 s.

whereas the liquid bulk remains more or less quiescent. The maximum flow velocities were about 4 mm/s in the former case, and about 1.5 mm/s in the latter.

The difference between these two flows has already been explained by Hoefsloot et al. [7]. The geometry of the triangular container leads to a depletion of acetone in the corner regions, creating a surface tension gradient which forces the liquid at the interface to flow towards the corners (macroscale effect). For the container with the curved interface the mass transfer by evaporation across the interface is homogeneous, and the onset of the convective flow is entirely due to hydrodynamic instability (microscale convection). This type of instability often tends to create a large number of small roll cells at the initial stage, with various coalescences or breakups of cells at later times, contrary to the rather stable two-cell pattern of the macroscale case. Apart from this, the flow caused by the macroscale effect is characterized by much larger flow velocities than in the case of microconvection.

3. Mathematical formulation and numerical solution method

The mathematical modelling of the convective flows will be based on the Navier–Stokes equations for two-dimensional flow of an incompressible Newtonian viscous liquid under zero-gravity conditions, combined with a convection/diffusion equation for the solute concentration. Two different geometries are considered. Case I (see Fig. 2) is the triangular container with a plane gas/liquid interface of length $2H$ and two rigid sides of length $H\sqrt{2}$. This case corresponds to the experimental container depicted in Fig. 1a. Case II (see Fig. 3) corresponds to the experiment shown in Fig. 1b. The liquid region has the shape of a sector of a circle with radius H and opening angle $\pi/2$. The circular boundary is the gas/liquid interface, the two plane boundaries are rigid walls.

Length, time, velocity, stream function, vorticity and solute concentration are non-dimensionalized by scaling with respectively, H, H^2/ν, ν/H, ν, ν/H^2 and c_{in}. Here ν denotes the kinematic liquid viscosity and c_{in} is the (uniform) initial solute concentration.

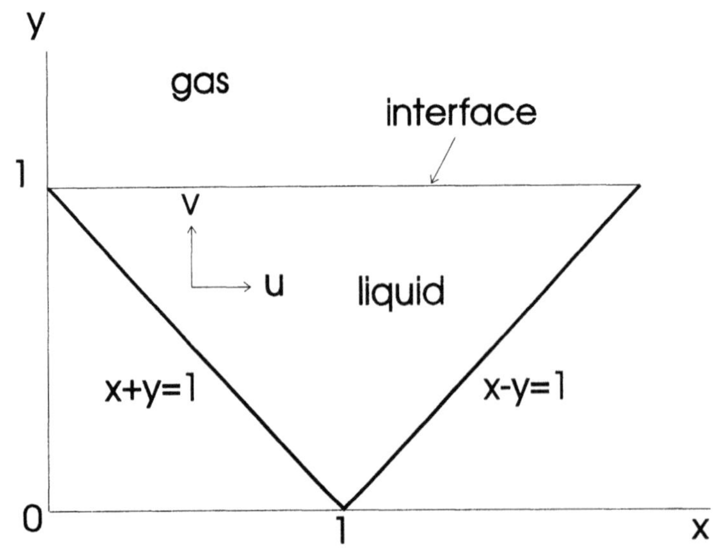

Fig. 2. The triangular container with plane gas/liquid interface in the dimensionless x, y-plane (case I).

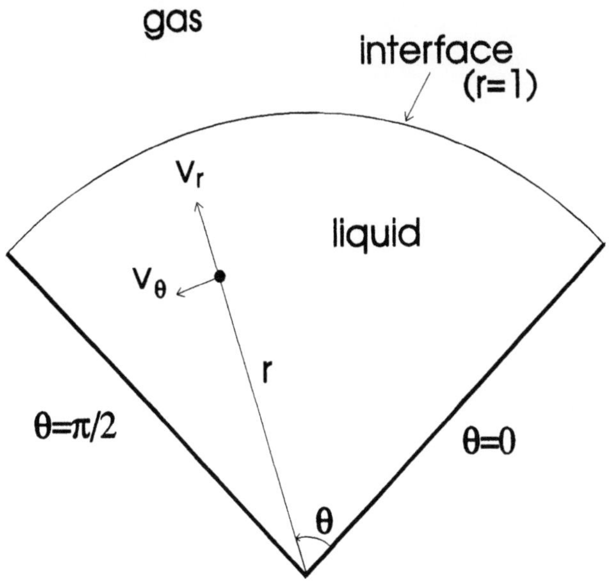

Fig. 3. The circular-sector container with curved gas/liquid interface in the dimensionless r, θ-plane (case II).

3.1. *Case I (plane interface)*

In the non-dimensional formulation the liquid region in the x, y-plane is given by $x + y > 1$, $x - y < 1$, $y < 1$ (see Fig. 2). After introduction of the stream function Ψ and the vorticity ω by the relations

$$u = -\frac{\partial \Psi}{\partial y}, \qquad v = \frac{\partial \Psi}{\partial x}, \qquad \omega = \frac{\partial u}{\partial y} - \frac{\partial v}{\partial x},$$

where (u, v) is the velocity vector, the governing equations for Ψ, ω and solute concentration c can be written as

$$\frac{\partial \omega}{\partial t} - \frac{\partial \Psi}{\partial y}\frac{\partial \omega}{\partial x} + \frac{\partial \Psi}{\partial x}\frac{\partial \omega}{\partial y} = \nabla^2 \omega, \tag{1}$$

$$\omega = -\nabla^2 \Psi, \tag{2}$$

$$\frac{\partial c}{\partial t} - \frac{\partial \Psi}{\partial y}\frac{\partial c}{\partial x} + \frac{\partial \Psi}{\partial x}\frac{\partial c}{\partial y} = \frac{1}{\mathrm{Sc}} \nabla^2 c, \tag{3}$$

where t is the time variable, ∇^2 is the two-dimensional Laplacian $\partial^2/\partial x^2 + \partial^2/\partial y^2$ and Sc denotes the Schmidt number $\nu/\mathbb{D}$ with $\mathbb{D}$ the coefficient of mass diffusion in the liquid.

At the solid walls we have the boundary conditions

$$\Psi = \frac{\partial \Psi}{\partial x} + \frac{\partial \Psi}{\partial y} = \frac{\partial c}{\partial x} + \frac{\partial c}{\partial y} = 0, \quad \text{for } x + y = 1,$$

$$\Psi = \frac{\partial \Psi}{\partial x} - \frac{\partial \Psi}{\partial y} = \frac{\partial c}{\partial x} - \frac{\partial c}{\partial y} = 0, \quad \text{for } x - y = 1,$$

implying zero flow velocity (no-slip condition) and mass impermeability. Assuming that the gas/liquid interface does not deform, we may put

$$\Psi = 0 \quad \text{at} \quad y = 1 .$$

Further we impose the mass transfer condition

$$\frac{\partial c}{\partial y} + \mathrm{Bi}\, c = 0 \quad \text{at} \quad y = 1 ,$$

where Bi denotes the Biot number defined as $kH/\mathbb{D}$ with k the mass transfer coefficient of the gas phase. The tangential stress balance at the interface gives (see Dijkstra and Van de Vooren [1])

$$\omega\left(= -\frac{\partial u}{\partial y}\right) = \frac{\mathrm{Ma}}{\mathrm{Sc}}\frac{\partial c}{\partial x} \quad \text{at} \quad y = 1 ,$$

where Ma is the Marangoni number

$$\mathrm{Ma} = -\frac{(\mathrm{d}\gamma/\mathrm{d}\tilde{c})c_{\mathrm{in}}H}{\mu\mathbb{D}} .$$

The derivative $\mathrm{d}\gamma/\mathrm{d}\tilde{c}$ of the (dimensional) surface tension γ with respect to the dimensional solute concentration $\tilde{c}$ is assumed to be equal to a negative constant.

Finally, we have the initial conditions

$$\Psi = \omega = 0 , \qquad c = 1 \quad \text{for } t = 0 .$$

The above initial/boundary-value problem has been solved numerically by the well-known Alternating Direction Implicit (ADI) finite difference method (see e.g. [12] and [14]). The discretization for this method leads, after linearization, to sets of linear algebraic equations with a tridiagonal coefficient matrix which are relatively easy to solve. The time-discretized form of equations (1)–(3) reads

$$\frac{\omega^{m+1/2} - \omega^m}{\Delta t} = \frac{1}{2}\left(\frac{\partial^2\omega^{m+1/2}}{\partial y^2} - \frac{\partial\Psi^m}{\partial x}\frac{\partial\omega^{m+1/2}}{\partial y}\right) + \frac{1}{2}\left(\frac{\partial^2\omega^m}{\partial x^2} + \frac{\partial\Psi^m}{\partial y}\frac{\partial\omega^m}{\partial x}\right), \tag{4}$$

$$\frac{\omega^{m+1} - \omega^{m+1/2}}{\Delta t} = \frac{1}{2}\left(\frac{\partial^2\omega^{m+1/2}}{\partial y^2} - \frac{\partial\Psi^m}{\partial x}\frac{\partial\omega^{m+1/2}}{\partial y}\right) + \frac{1}{2}\left(\frac{\partial^2\omega^{m+1}}{\partial x^2} + \frac{\partial\Psi^m}{\partial y}\frac{\partial\omega^{m+1}}{\partial x}\right), \tag{5}$$

$$\frac{\Psi^{n+1/2} - \Psi^n}{\tau} = \frac{1}{2}\left(\frac{\partial^2\Psi^{n+1/2}}{\partial y^2} + \frac{\partial^2\Psi^n}{\partial x^2} + \omega^{m+1}\right), \tag{6}$$

$$\frac{\Psi^{n+1} - \Psi^{n+1/2}}{\tau} = \frac{1}{2}\left(\frac{\partial^2\Psi^{n+1/2}}{\partial y^2} + \frac{\partial^2\Psi^{n+1}}{\partial x^2} + \omega^{m+1}\right), \tag{7}$$

$$\frac{c^{m+1/2} - c^m}{\Delta t} = \frac{1}{2}\left(\frac{1}{\mathrm{Sc}}\frac{\partial^2 c^{m+1/2}}{\partial y^2} - \frac{\partial\Psi^{m+1}}{\partial x}\frac{\partial c^{m+1/2}}{\partial y}\right) + \frac{1}{2}\left(\frac{1}{\mathrm{Sc}}\frac{\partial^2 c^m}{\partial x^2} + \frac{\partial\Psi^{m+1}}{\partial y}\frac{\partial c^m}{\partial x}\right), \tag{8}$$

$$\frac{c^{m+1}-c^{m+1/2}}{\Delta t}=\frac{1}{2}\left(\frac{1}{\mathrm{Sc}}\frac{\partial^2 c^{m+1/2}}{\partial y^2}-\frac{\partial \Psi^{m+1}}{\partial x}\frac{\partial c^{m+1/2}}{\partial y}\right)+\frac{1}{2}\left(\frac{1}{\mathrm{Sc}}\frac{\partial^2 c^{m+1}}{\partial x^2}+\frac{\partial \Psi^{m+1}}{\partial y}\frac{\partial c^{m+1}}{\partial x}\right), \tag{9}$$

where the superscript m denotes the time level $t=m\,\Delta t$. In equations (6) and (7) a pseudo-time discretization has been introduced with time step τ and counter n. All derivatives with respect to the spatial variables x and y occurring both in the equations and in the boundary conditions have been replaced by second-order difference quotients.

A nonuniform rectangular grid has been used. The grid lines parallel to the x-axis are given by

$$y=y_i=-0.99\,(i/N)^2+1.99(i/N)\,,\quad i=0,1,\ldots,N\,.$$

The grid lines $x=$ constant run through the points of intersection of the lines $y=y_i$ with the boundaries $x\pm y=1$. In this way we have the highest density of grid points in the region near the interface where the largest velocity and concentration gradients will occur.

The numerical computation is carried out as follows. Assuming that Ψ, ω and c are known at time level m, first the value $\omega^{m+1/2}$ at the intermediate level $m+1/2$ is computed from (4) and using this value of $\omega^{m+1/2}$ in (5), the value of ω^{m+1} is computed. Next, Ψ at level $m+1$ is computed iteratively from (6) and (7) with n as iteration counter and using the stop criterion $|\Psi^n-\Psi^{n+1}|<10^{-6}$. During this iteration, the pseudo-time step τ is not kept fixed, but it can be enlarged when n increases, thus saving computer time. Finally, the computed value Ψ^{m+1} is used in (8) and (9) to compute subsequently the concentration values $c^{m+1/2}$ and c^{m+1}. At this stage Ψ, ω and c are known at time level $m+1$, so m can now be changed into $m+1$ and the procedure is repeated. This method is probably not the most cost-efficient one, but it is reliable and works well also for large values of Ma and Sc.

3.2. *Case II (curved interface)*

In this case we use polar coordinates r and θ in terms of which the liquid region is given by $0<r<1$, $0<\theta<\pi/2$ (see Fig. 3). The velocity components v_r and v_θ are now related to Ψ and ω in the following way:

$$v_\theta=\frac{\partial\Psi}{\partial r}\,,\qquad v_r=-\frac{1}{r}\frac{\partial\Psi}{\partial\theta}\,,\qquad \omega=-\frac{1}{r}\frac{\partial v_r}{\partial\theta}+\frac{1}{r}\frac{\partial(v_\theta r)}{\partial r}\,,$$

and the equations to be solved for Ψ, ω and c read

$$\frac{\partial\omega}{\partial t}=\frac{\partial^2\omega}{\partial r^2}+\frac{1}{r^2}\frac{\partial^2\omega}{\partial\theta^2}+\frac{1}{r}\frac{\partial\omega}{\partial r}-\frac{1}{r}\left[\frac{\partial\Psi}{\partial r}\frac{\partial\omega}{\partial\theta}-\frac{\partial\Psi}{\partial\theta}\frac{\partial\omega}{\partial r}\right], \tag{10}$$

$$\omega=\frac{\partial^2\Psi}{\partial r^2}+\frac{1}{r^2}\frac{\partial^2\Psi}{\partial\theta^2}+\frac{1}{r}\frac{\partial\Psi}{\partial r}\,, \tag{11}$$

$$\frac{\partial c}{\partial t}=\frac{1}{\mathrm{Sc}}\left[\frac{\partial^2 c}{\partial r^2}+\frac{1}{r^2}\frac{\partial^2 c}{\partial\theta^2}+\frac{1}{r}\frac{\partial c}{\partial r}\right]+\frac{1}{r}\left(\frac{\partial\Psi}{\partial\theta}\frac{\partial c}{\partial r}-\frac{\partial\Psi}{\partial r}\frac{\partial c}{\partial\theta}\right). \tag{12}$$

The boundary conditions at the solid walls are

$$\Psi = \frac{\partial c}{\partial \theta} = 0\,, \qquad \omega = \frac{\partial^2 \Psi}{\partial^2 r}\,, \quad \text{at } \theta = 0 \text{ and } \theta = \pi/2\,.$$

Analogous to the preceding case, we impose three boundary conditions at the gas/liquid interface $r = 1$:

$$\Psi = 0\,, \qquad \frac{\partial c}{\partial r} = -\mathrm{Bi}\, c \quad \text{at} \quad r = 1\,,$$

and

$$\omega - 2\frac{\partial \Psi}{\partial r}\left(= \frac{\partial v_\theta}{\partial r} - \frac{v_\theta}{r}\right) = -\frac{\mathrm{Ma}}{\mathrm{Sc}}\frac{1}{r}\frac{\partial c}{\partial \theta} \quad \text{at} \quad r = 1\,.$$

The initial conditions at $t = 0$ are:

$$\omega = 0\,, \qquad \Psi = 0 \quad \text{and} \quad c = 1\,.$$

Since the above initial/boundary value problem possesses the motionless solution $\Psi = \omega = 0$, $c = c_0(r, t)$, a very small initial perturbation of the interfacial value of c will be introduced to trigger off numerically the onset of micro-scale convection.

As in case I, the numerical solution of the above problem has been obtained by means of an ADI method. The time-discretized version of equations (10)–(12) is

$$\frac{\omega^{m+1/2} - \omega^m}{\Delta t} = \frac{1}{2}\left(\frac{1}{r^2}\frac{\partial^2 \omega^{m+1/2}}{\partial \theta^2} - \frac{1}{r}\frac{\partial \Psi^m}{\partial r}\frac{\partial \omega^{m+1/2}}{\partial \theta}\right) + \frac{1}{2}\left(\frac{\partial^2 \omega^m}{\partial r^2} + \frac{1}{r}\left(1 + \frac{\partial \Psi^m}{\partial \theta}\right)\frac{\partial \omega^m}{\partial r}\right), \tag{13}$$

$$\begin{aligned}\frac{\omega^{m+1} - \omega^{m+1/2}}{\Delta t} = {} & \frac{1}{2}\left(\frac{1}{r^2}\frac{\partial^2 \omega^{m+1/2}}{\partial \theta^2} - \frac{1}{r}\frac{\partial \Psi^m}{\partial r}\frac{\partial \omega^{m+1/2}}{\partial \theta}\right) \\ & + \frac{1}{2}\left(\frac{\partial^2 \omega^{m+1}}{\partial r^2} + \frac{1}{r}\left(1 + \frac{\partial \Psi^m}{\partial \theta}\right)\frac{\partial \omega^{m+1}}{\partial r}\right),\end{aligned} \tag{14}$$

$$\frac{\Psi^{n+1/2} - \Psi^n}{\tau} = \frac{1}{2}\left(\frac{1}{r^2}\frac{\partial^2 \Psi^{n+1/2}}{\partial \theta^2} + \frac{\partial^2 \Psi^n}{\partial r^2} + \frac{1}{r}\frac{\partial \Psi^n}{\partial r} - \omega^{m+1}\right), \tag{15}$$

$$\frac{\Psi^{n+1} - \Psi^{n+1/2}}{\tau} = \frac{1}{2}\left(\frac{1}{r^2}\frac{\partial^2 \Psi^{n+1/2}}{\partial \theta^2} + \frac{\partial^2 \Psi^{n+1}}{\partial r^2} + \frac{1}{r}\frac{\partial \Psi^{n+1}}{\partial r} - \omega^{m+1}\right), \tag{16}$$

$$\begin{aligned}\frac{c^{m+1/2} - c^m}{\Delta t} = {} & \frac{1}{2}\left(\frac{1}{\mathrm{Sc}}\frac{1}{r^2}\frac{\partial^2 c^{m+1/2}}{\partial \theta^2} - \frac{\partial \Psi^{m+1}}{\partial r}\frac{1}{r}\frac{\partial c^{m+1/2}}{\partial \theta}\right) \\ & + \frac{1}{2}\left(\frac{1}{\mathrm{Sc}}\left[\frac{\partial^2 c^m}{\partial r^2} + \frac{1}{r}\frac{\partial c^m}{\partial r}\right] + \frac{1}{r}\frac{\partial \Psi^{m+1}}{\partial \theta}\frac{\partial c^m}{\partial r}\right),\end{aligned} \tag{17}$$

$$\begin{aligned}\frac{c^{m+1} - c^{m+1/2}}{\Delta t} = {} & \frac{1}{2}\left(\frac{1}{\mathrm{Sc}}\frac{1}{r^2}\frac{\partial^2 c^{m+1/2}}{\partial \theta^2} - \frac{\partial \Psi^{m+1}}{\partial r}\frac{1}{r}\frac{\partial c^{m+1/2}}{\partial \theta}\right) \\ & + \frac{1}{2}\left(\frac{1}{\mathrm{Sc}}\left[\frac{\partial^2 c^{m+1}}{\partial r^2} + \frac{1}{r}\frac{\partial c^{m+1}}{\partial r}\right] + \frac{1}{r}\frac{\partial \Psi^{m+1}}{\partial \theta}\frac{\partial c^{m+1}}{\partial r}\right),\end{aligned} \tag{18}$$

where the superscript m denotes the time level $t = m\,\Delta t$. As in the discretized equations for case I, τ denotes the pseudo-time step and n the pseudo-time level. All derivatives with respect to the spatial variables r and θ occurring in the basic equations and in the boundary conditions have been replaced by second-order differences. An equidistant grid has been used in the θ-direction, whereas in the r-direction a non-uniform grid has been used with a higher density of grid points near the interface $r = a/H$ where the concentration and velocity gradients are large. The non-equidistant grid points have been defined as:

$$r_i = -0.9(i/N)^2 + 1.9(i/N)\,, \quad i = 0, 1, \ldots, N\,.$$

The numerical computation is carried out in the same way as in case I.

3.3. *Accuracy and convergence of the numerical method*

The ADI method which has been applied to solve the problems numerically has a first-order convergence rate in the time variable and second-order convergence with respect to the spatial variables. To check the correctness of the numerical algorithms a number of tests has been carried out. Since the convergence behaviour for case II is somewhat more intricate than for case I, we begin with some tests concerning case II.

Table 1 shows values of the maximum of the stream function (Ψ_{max}) and the velocity at the interface at $\theta = 22.5°$ (denoted by $V_{22.5}$) at the point of time $t = 1.25 \times 10^{-3}$ for the parameter values Ma $= 10^5$, Sc $= 100$ and Bi $= 55–100(\theta/\pi)$. Note that Bi has been chosen here as a function of θ along the interface, so that an artificial macroscale effect has been introduced and therefore no initial disturbance is required to start the convection. The computation has been done for three $\theta \times r$ grids (41×41, 81×81 and 161×161 points) and three Δt values (5×10^{-5}, 2.5×10^{-5} and 1.25×10^{-5}). The values of Ψ_{max} and $V_{22.5}$ show the expected convergence behaviour.

In the next test for case II the Biot number has been taken constant (Bi $= 20$) and there is an initial perturbed interfacial concentration given by

$$c = \begin{cases} 0.999 + 0.001(j/m)\,, & j = 0, \ldots, M\,, \\ 1.001 - 0.001(j/M)\,, & j = M, \ldots, 2M\,, \end{cases}$$

where $2M + 1$ is the number of grid points in θ-direction and j runs from $j = 0$ (corresponding to $\theta = 0$) to $j = 2M$ ($\theta = \pi/2$). To make this disturbance compatible with respect to the three spatial grids which have been used in this test, the first row below the interface in the 81×81 grid is given the linearly interpolated value between the undisturbed value ($c = 1$) of the second row and the interfacial value of c. For the 161×161 grid this interpolation is done for three rows, and for the 321×321 grid for seven rows. The computations were performed with Ma $= 10^5$, Sc $= 100$ and $\Delta t = 1.25 \times 10^{-5}$. The results for Ψ_{max} and $V_{22.5}$ at time $t = 2.5 \times 10^{-4}$, given in Table 2, demonstrate the desired convergence behaviour.

Table 1. Test results for case II: θ-dependent Biot number, no initial concentration perturbation.

grid	41×41		81×81		161×161	
Δt	Ψ_{max}	$V_{22.5}$	Ψ_{max}	$V_{22.5}$	Ψ_{max}	$V_{22.5}$
5×10^{-5}	0.0987	−3.768	0.0849	−3.397	0.0817	−3.314
2.5×10^{-5}	0.1013	−3.832	0.0874	−3.464	0.0842	−3.384
1.25×10^{-5}	0.1027	−3.864	0.0886	−3.497	0.0854	−3.416

Table 2. Test results for case II: constant Biot number, perturbed initial concentration (see text).

grid	81 × 81		161 × 161		321 × 321	
t	Ψ_{max}	$V_{22.5}$	Ψ_{max}	$V_{22.5}$	Ψ_{max}	$V_{22.5}$
2.5×10^{-4}	2.12×10^{-4}	−0.00631	2.52×10^{-4}	−0.00751	2.61×10^{-4}	−0.00776

A further computation, making use of an 81×161 grid and keeping the remaining parameter settings unchanged, gave a value for $V_{22.5}$ that differed only 2×10^{-7} from the 161×161 value of Table 2. Changing the time step to $\Delta t = 2.5 \times 10^{-5}$ and keeping all other settings the same gave $\Psi_{max} = 2.59 \times 10^{-4}$ on the 161×161 grid.

The above tests show that the convergence behaviour is as expected and accuracy is good for small time. To see how the accuracy is for larger values of t, computations have been carried out for case II on two $\theta \times r$ grids (161×321 and 321×641 points) with $\Delta t = 2.5 \times 10^{-5}$, $\mathrm{Ma} = 10^5$ and $\mathrm{Sc} = 100$ and $\mathrm{Bi} = 20$. The initial perturbed concentration along the interface is the same as above, with linearly interpolated c-values for the first three rows below the interface for the coarser grid, and for the first seven rows in the case of the finest grid. The results for Ψ_{max} are given in Table 3, from which it can be observed that accuracy deteriorates gradually in time, but it remains acceptable for $t \leqslant 7.5 \times 10^{-3}$.

For case I, being less difficult than case II, only one test is presented here with the parameter values $\mathrm{Ma} = 10^5$, $\mathrm{Sc} = 100$ and $\mathrm{Bi} = 20$. Three spatial ($x \times y$) grids were used with time step $\Delta t = 2.5 \times 10^{-5}$. Table 4 gives results for Ψ_{max}. From this table it can be seen that accuracy is reasonable as long as t is smaller than 1.25×10^{-2}.

3.4. *Choice of parameters*

In the numerical computations of which the results will be presented in the next section, the following parameter values have been used: $\mathrm{Ma} = 10^5$, $\mathrm{Sc} = 100$ and $\mathrm{Bi} = 20$. Theoretically, the Marangoni number would be about 10^8, but under experimental circumstances the value will mostly be several orders of magnitude smaller. The choice of $\mathrm{Sc} = 100$ instead of the actual value of acetone in water (which is 787), is a compromise. With the real value of Sc unacceptably small time steps would have been required to suppress numerical instabilities in

Table 3. Test results for case II: constant Biot number, perturbed initial concentration (see text).

grid	161 × 321	321 × 641
t	Ψ_{max}	Ψ_{max}
6.25×10^{-4}	0.00039	0.00041
1.25×10^{-3}	0.00053	0.00057
2.5×10^{-3}	0.00070	0.00076
5.0×10^{-3}	0.00087	0.00094
7.5×10^{-3}	0.00096	0.00104
1.0×10^{-2}	0.00103	0.00177
1.25×10^{-2}	0.00249	0.00397

Table 4. Test results for case I: constant Biot number, no initial concentration perturbation.

grid	161 × 81	321 × 161	641 × 321
t	Ψ_{max}	Ψ_{max}	Ψ_{max}
6.25×10^{-4}	0.0011	0.0007	0.0006
1.25×10^{-3}	0.0043	0.0030	0.0025
2.5×10^{-3}	0.0140	0.0109	0.0095
5.0×10^{-3}	0.0345	0.0265	0.0237
7.5×10^{-3}	0.0483	0.0418	0.0415
1.0×10^{-2}	0.2020	0.2180	0.2059
1.25×10^{-2}	0.5381	0.5323	0.4609

the solution (like wiggles). Lowering Sc increases the weight of the diffusive terms with respect to the convective terms, thus improving numerical stability. The chosen value of Bi is in the usual range of experimental values.

The spatial grid used for case I was the $x \times y$ grid of 641×321 points given in Table 4, and for case II the $\theta \times r$ grid of 321×641 points given in Table 3 was used. In all computations Δt was equal to 2.5×10^{-5}.

4. Numerical results

The numerical results are presented in the form of contour plots of the stream function Ψ for various points in time. Figure 4 shows some results for case I at times $t = t_1 = 2.5 \times 10^{-3}$, $2t_1$, $3t_1$, $4t_1$ and $5t_1$. The corresponding values of Ψ_{max} can be found in Table 4. It is seen that the characteristic roll-cell pattern for this macroscale case emerges: a quite stable pattern of two large roll cells, in accordance with the experimental observations. At $t = 3t_1$, an additional small cell is seen in each corner region, but it vanishes again soon afterwards. Although the plot for $t = 5t_1$ is not to be completely trusted from a numerical point of view (see Table 4), it has nevertheless been included since it confirms the global impression of a stable two-cell pattern for case I.

The first set of results for case II is represented by Fig. 5. The slightly perturbed initial concentration distribution, which is symmetric with respect to the line $\theta = \pi/4$, is the same as the one used in the second and third numerical test for case II (see previous section). This initial concentration has been chosen because the initial flow pattern consists of two large roll cells and resembles the flow pattern created by the injection of the liquid into the container during the experiment. The points in time for which the plots have been made correspond to those of Table 3: $t = t_1 = 2.5 \times 10^{-3}$, $2t_1$, $3t_1$ and $4t_1$. The initial two-cell pattern is clearly not persistent: the first stage of breakup into smaller cells is seen to occur in Figs 5c,d. This tendency towards a pattern of many small cells near the interface, along with a nearly motionless liquid bulk, is typical of microconvection. Comparing the values of Ψ_{max} for this case (see Table 3) with those for case I (see Table 4), it is clear that the macroscale effects of case I lead to higher liquid flow velocities than the microscale effects of case II.

Figures 6 and 7 show two simulations with initial disturbances leading to a flow pattern with a relatively large number of roll cells. In the case of Fig. 6 the initial concentration has

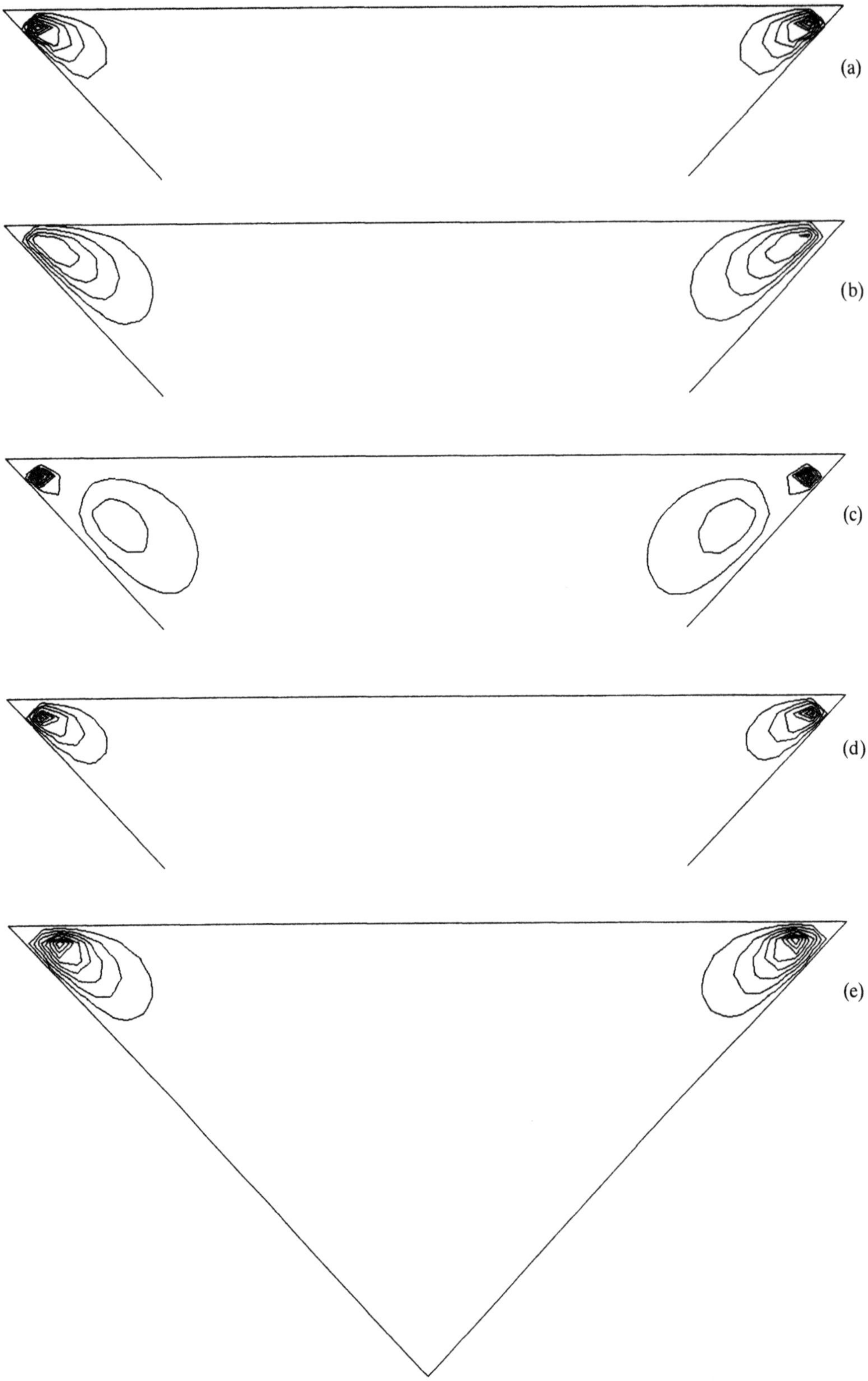

Fig. 4. Contour plots of the stream function for case I at dimensionless times t_1 (a), $2t_1$ (b), $3t_1$ (c), $4t_1$ (d) and $5t_1$ (e), where $t_1 = 2.5 \times 10^{-3}$.

been perturbed by assigning the value $c = 0.999$ to the five grid points $j = 0$, 80, 160, 240 and 320 on the interface, whereas in Fig. 7 this has been done for the eleven grid points $j = 0, 32, \ldots, 288, 320$. Although the time evolution of the patterns is slow in both cases (which, of course, is a typical property of microconvection), a tendency towards an increase of the number of cells is visible. These microconvection simulations would require extremely large computing times to monitor their full evolution in time. For each set of results presented above the computing time amounted to several hours on a Cyber 962 mainframe.

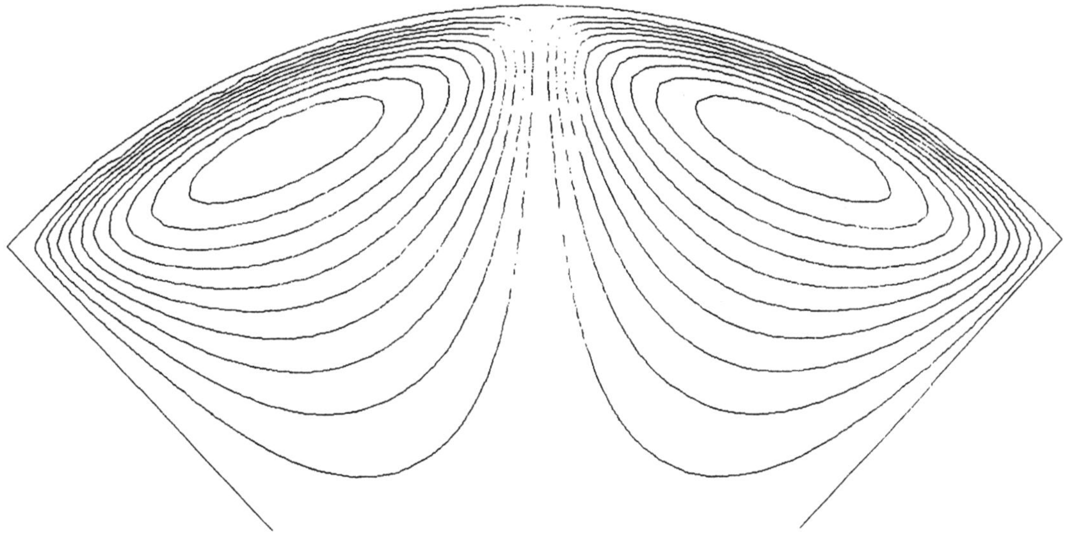

(a)

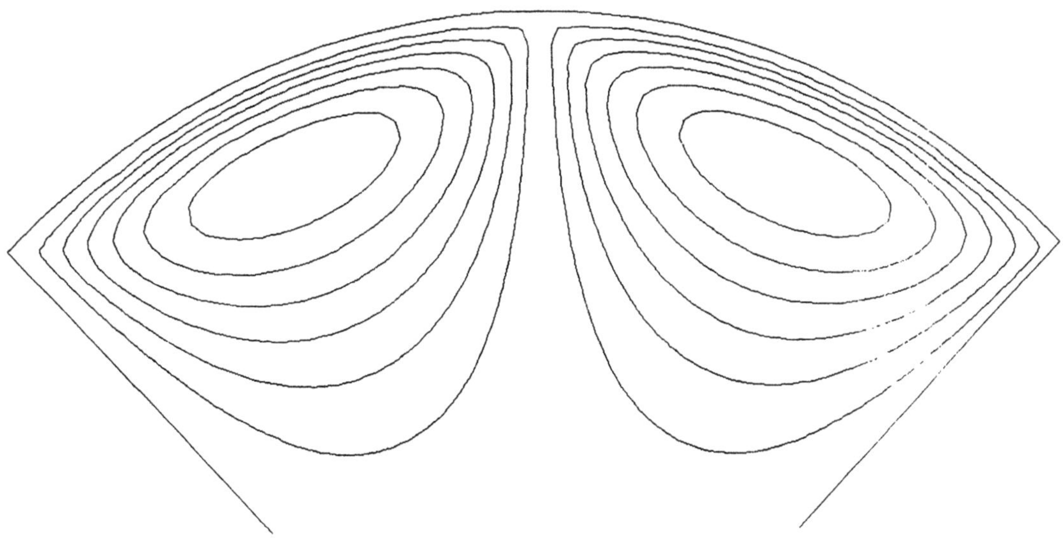

(b)

Fig. 5. Contour plots of the stream function for case II at dimensionless times t_1 (a), $2t_1$ (b), $3t_1$ (c) and $4t_1$ (d), where $t_1 = 2.5 \times 10^{-3}$, computed from the symmetrically perturbed initial concentration distribution described in the text.

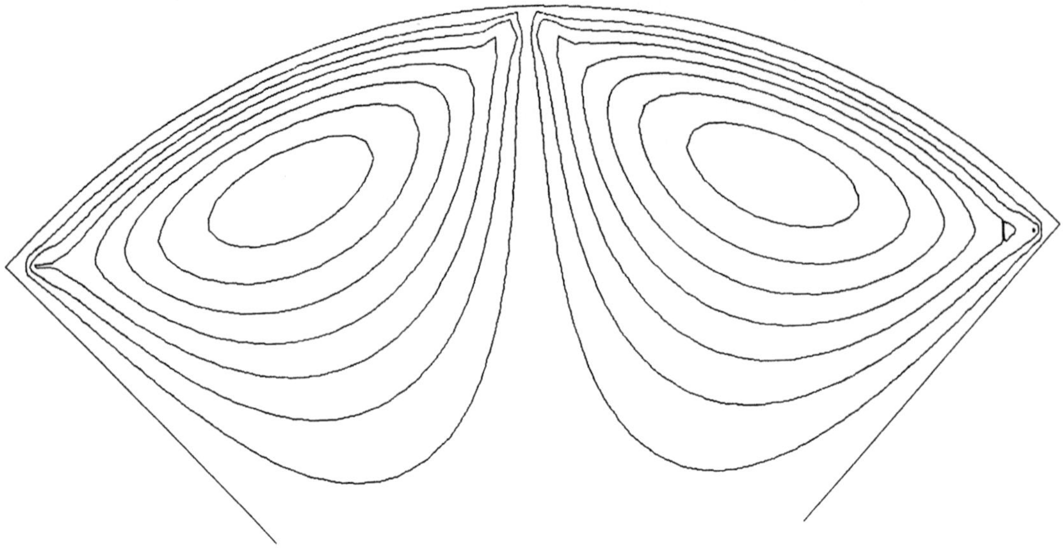

(c)

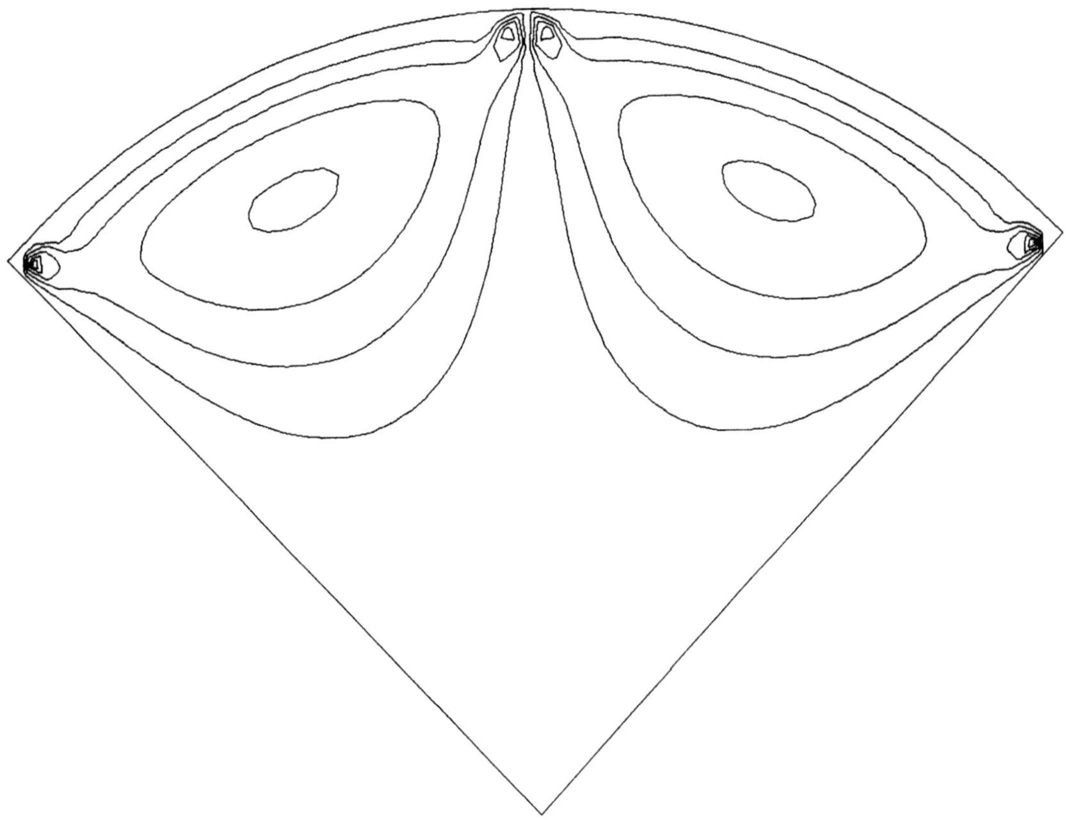

(d)

Fig. 5 (cont.).

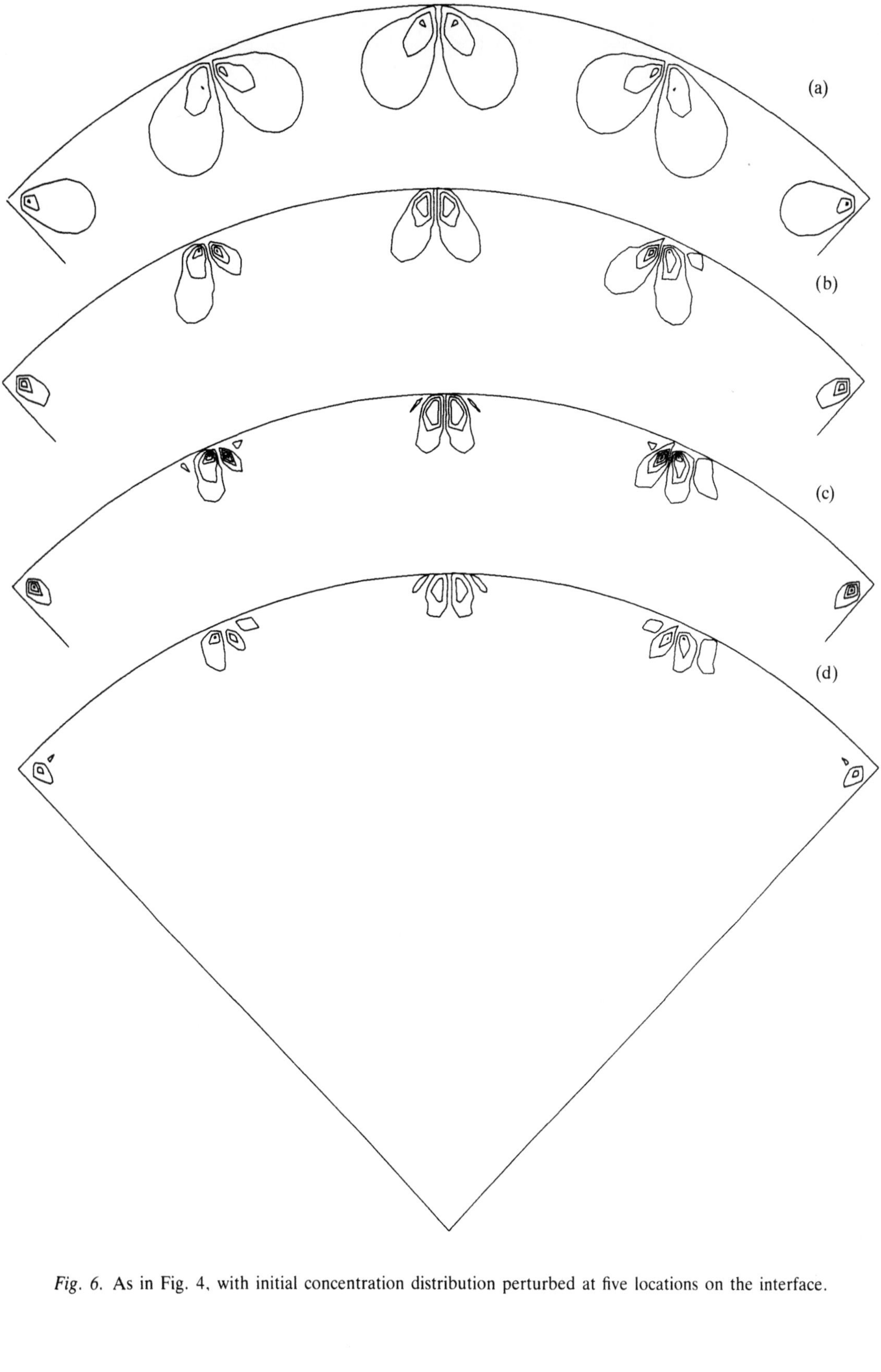

Fig. 6. As in Fig. 4, with initial concentration distribution perturbed at five locations on the interface.

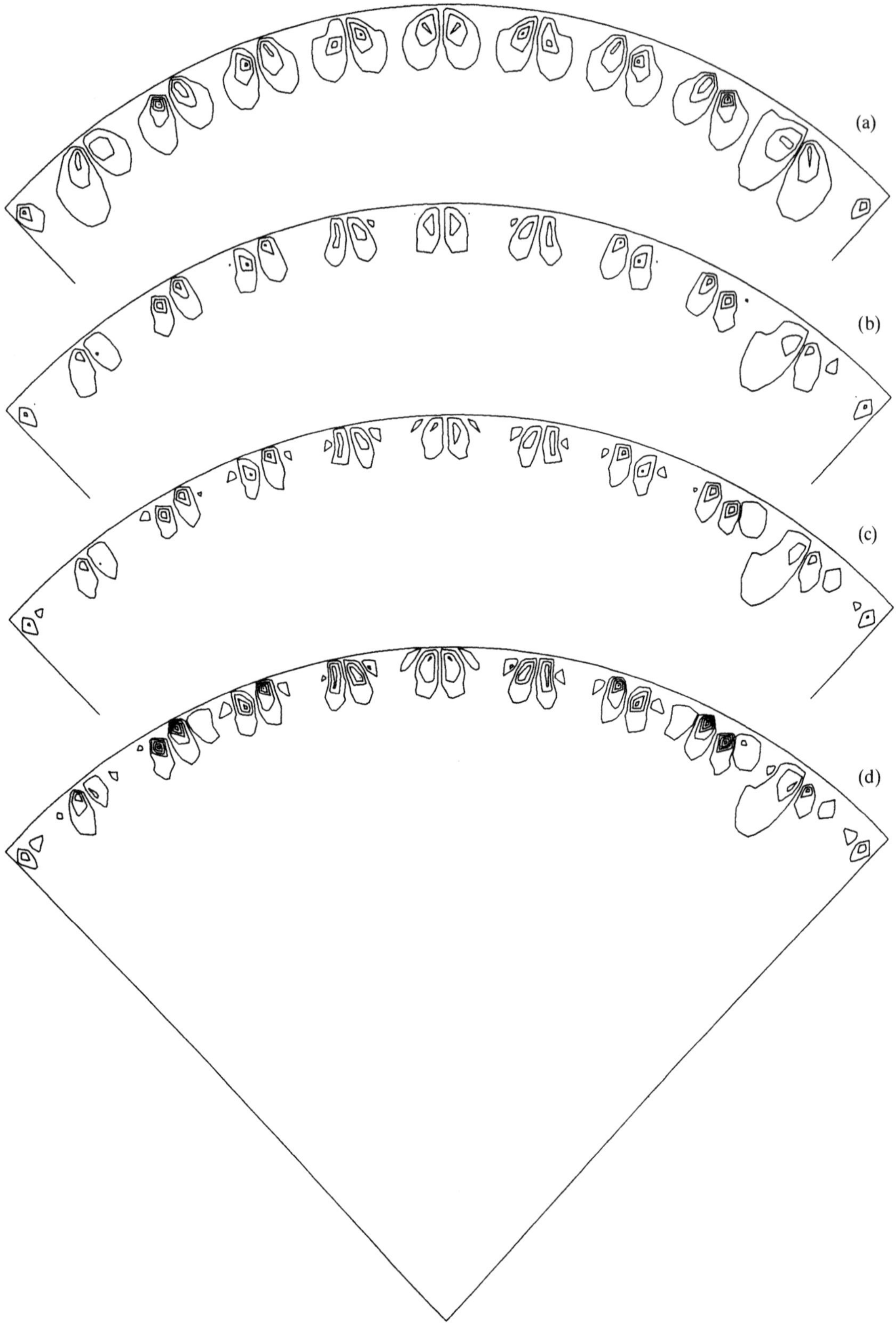

Fig. 7. As in Fig. 4, with initial concentration distribution perturbed at eleven locations on the interface.

5. Concluding remarks

The numerical results presented in this chapter confirm the experimentally observed phenomena for Marangoni convection in V-shaped containers as described by Hoefsloot et al. [7]. The triangular container (case I) shows a flow behaviour which is characteristic for macroconvection: a rather stable roll-cell pattern consisting of a small number of large cells, extending into the bulk region. The time development of the pattern is more rapid than in the case of microconvection and also the flow velocities are larger. The computations involving the circular-sector container (case II) give rise to typical microconvection behaviour: small flow velocities, slow development of flow pattern and nearly motionless liquid bulk. The initial tendency is towards breakup of cells, but after longer times it is to be expected that coalescence of cells will lead to a more or less persistent pattern of somewhat larger cells. However, an accurate and numerically stable simulation of this long-term behaviour would require the availability of a considerable amount of computing time on a very large (super)computer.

References

1. H.A. Dijkstra and A.I. van de Vooren, Initial flow development due to Marangoni convection in a mass transfer system. *Int. J. Heat Mass Transfer* 28 (1985) 2315–2322.
2. H.A. Dijkstra and J.H. Lichtenbelt, Mass transfer driven Marangoni convection under microgravity. *Appl. Microgravity Technology* 1 (1988) 180–187.
3. H.C.J. Hoefsloot and L.P.B.M. Janssen, MASER-2 report, in: *Experiment reports from flight opportunities on MASER-1 and -2 supported by ESA*, J.P.B. Vreeburg (ed.), National Aerospace Laboratory NLR, Amsterdam, The Netherlands (1989).
4. H.C.J. Hoefsloot and L.P.B.M. Janssen, Marangoni convection mass transfer, in: *Microgravity experiments during parabolic flights of KC-135 aircraft, sixth ESA campaign August 1988*, A. Gonfalone, V. Pletser and D. Frimout (eds), ESTEC, Noordwijk, The Netherlands (1989).
5. H.C.J. Hoefsloot, H.W. Hoogstraten, A. Hoven and L.P.B.M. Janssen, Marangoni instability in a liquid layer bounded by two coaxial cylinder surfaces. *Appl. Sci. Res.* 47 (1990) 1–21.
6. H.C.J. Hoefsloot, H.W. Hoogstraten and L.P.B.M. Janssen, Marangoni instability in a liquid layer confined between two concentric spherical surfaces under zero-gravity conditions. *Appl. Sci. Res.* 47 (1990) 357–377.
7. H.C.J. Hoefsloot, L.P.B.M. Janssen, R.T. Sibbald and H.W. Hoogstraten, Experimental results from Marangoni convection in V-shaped containers under microgravity conditions. *Microgravity Sci. Technology* 4 (1991) 55–59.
8. H.C.J. Hoefsloot, H.W. Hoogstraten, L.M.B.M. Janssen and J.W. Knobbe, Growth factors for Marangoni instability in a spherical liquid layer under zero-gravity conditions. *Appl. Sci. Res.* (to appear).
9. H.C.J. Hoefsloot, H.W. Hoogstraten and L.P.B.M. Janssen, Marangoni convection round a ventilated air bubble under microgravity conditions. submitted to *Chem. Engineering Sci.*
10. J.H. Lichtenbelt, Improvement of flight hardware and isothermal Marangoni convection under microgravity conditions. *Adv. Space Res.* 5 (1986) 97–100.
11. J.H. Lichtenbelt, A.A.H. Drinkenburg and H.A. Dijkstra, Marangoni convection and mass transfer from the liquid to the gas phase. *Naturwissenschaften* 73 (1986) 356–359.
12. A.D. Myshkis, V.G. Babskii, N.D. Kopachevskii, L.A. Slobozhanin and A.D. Tyuptsov, *Low-gravity Fluid Mechanics*. Springer Verlag, Berlin/New York (1987).
13. D.A. Nield, Surface tension and buoyancy effects in cellular convection. *J. Fluid Mech.* 19 (1964) 341–352.
14. R. Peyret and T.D. Taylor, *Computational Methods for Fluid Flow*. Springer Series in Computational Physics, Springer Verlag, Berlin/New York (1983).

Journal of Engineering Mathematics **26**: 39–50, 1992.
H.K. Kuiken and S.W. Rienstra (eds), Problems in Applied, Industrial and Engineering Mathematics.

Structure of the temperature profile within a high-pressure gas-discharge lamp operating near maximum radiation efficiency

H.K. KUIKEN
Philips Research Laboratories, P.O. Box 80.000, 5600 JA Eindhoven, The Netherlands

Abstract. A singular perturbation technique is used to describe the temperature profile in a wall-stabilized high-pressure gas-discharge arc in the limit of extremely high radiation efficiency. The analysis shows that the profile is characterized by three different regimes, one of which is a transition layer.

Introduction

In this paper we shall investigate a certain aspect of the temperature field produced by a high-pressure gas-discharge arc. The arc is assumed to be enclosed within an elongated cylindrical tube, i.e. it is a so-called wall-stabilized arc. Such a configuration can be technologically important as a high-yield light source. A simplified sketch of such a system is given in Fig. 1.

Two electrodes stick into the tube at both ends. When the lamp is operating, these electrodes are extremely hot, so that they are easily induced to emit electrons. The electric field causes these electrons to accelerate, thereby imparting large amounts of energy to them. The tube is filled with a gas such as mercury or sodium. Because of the high pressure in the tube (from 5 up to 100 atmospheres), the mean free path of the gas molecules is very short, thus collisions with the hot electrons are extremely frequent. During a collision between an electron and a gas molecule the electron loses part of its energy which is transferred to the gas molecule. The electrons soon regain their previous energy levels through the interaction with the electric field.

When a gas molecule absorbs energy as the result of a collision with an electron, the energy transfer may be, roughly speaking, of three different kinds. The energy may be used to increase the temperature of the gas, meaning that it may be returned as translational, vibrational and rotational energy. Indeed, when equilibrium has been reached, the electron temperature and the gas temperature are the same in a high-pressure gas discharge. This is normally known as local thermal equilibrium (LTE). A second energy-transfer mode is the one in which electrons in the outer valence band of the gas molecule are excited to higher quantum levels. Such a higher state may be unstable and, as a result, an excited electron will return to the lower state, possibly through a succession of intermediate states. During this process photons are emitted. As a crude approximation these may be thought to belong to

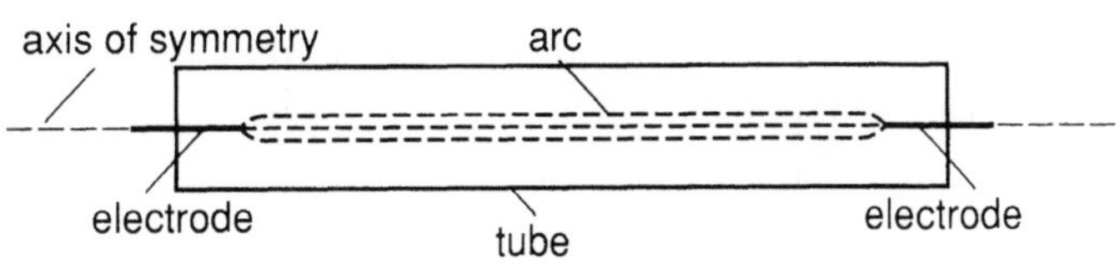

Fig. 1. Sketch of a gas-discharge tube.

two different categories: a) ones that leave the tube without interference; these constitute the lamp's radiation, both visible and invisible; b) ones that are absorbed within the plasma by the gas molecules; this constitutes an added heat-transfer mechanism. The former is called thin radiation and the second is thick radiation. A third electron-gas energy-transfer mode is that in which a gas molecule is ionized. In this process an electron is knocked loose, thereby becoming a free electron.

A knowledge of the temperature field is extremely important for a proper understanding of how a lamp operates. We shall consider the temperature distribution away from the electrodes where it may be thought to depend solely upon the radial coordinate r, assuming stationary conditions. We shall also assume that convection effects are absent, although in larger tubes, particularly in those which are in a vertical position, convection effects may dominate [1, 2, 3]. The one-dimensional rotationally symmetric model we wish to investigate was first described by Elenbaas. A summary of his work can be found in his book on the high-pressure gas discharge [4]. More recent accounts can be found in [5, 6]. Recently, we have shown [7, 8, 9] that this classical model can be treated by asymptotic methods. Since asymptotic methods are capable of giving a much clearer insight into the structure of solutions than could be achieved by any direct computational approach, we decided that the field was worth revisiting.

What we intend to study in this paper is a particular question left open in [9]. It concerns the structure of the temperature field when the maximum radiation efficiency is approached. In that case some important quantities characterizing the arc become singular. It is important to know how.

The model

As explained in [9], the temperature distribution in the middle section of an elongated tube containing a high-pressure gas-discharge arc is governed by the equation

$$\frac{1}{r}\frac{\mathrm{d}}{\mathrm{d}r}\, r\lambda(t)\,\frac{\mathrm{d}t}{\mathrm{d}r} + \sigma(t)E^2 - u(t) = 0\,, \tag{1}$$

where t is the temperature, r the distance to the axis symmetry, E the electric field and $\lambda(t)$, $\sigma(t)$ and $u(t)$ the temperature-dependent thermal conductivity, electrical conductivity and radiation output terms, respectively. The electric field E, which is assumed to be uniform, and the total arc current I are related as follows:

$$E = I\bigg/\int_0^a 2\pi r\sigma(t)\,\mathrm{d}r\,, \tag{2}$$

where a is the inner tube radius. The thermal conductivity is assumed to represent ordinary thermal conduction and short-range radiation effects. The latter, which is usually called thick radiation, is to be attributed to photons emitted and absorbed by nearby molecules. It has been shown [9] that the analysis is independent of the particular functional description of $\lambda(t)$. It may even be given in tabulated form.

The electrical conductivity is given by

$$\sigma(t) = \gamma t^{3/4}\exp(-t_i/t)\,, \tag{3}$$

where t_i is equal to half the ionization temperature. If V_i is the ionization potential, i.e. the potential that has to be overcome to dislodge totally an outer electron from the valence band of a gas molecule, then $t_i = eV_i/2k$, where e is the elementary charge and k is Boltzmann's constant. Further, γ is some constant.

Next we have to consider the radiation term $u(t)$ which represents the so-called thin radiation which leaves the tube without hindrance. In [9] we modelled this as follows:

$$u(t) = \omega t^p \exp(-t_*/t) , \tag{4}$$

where the temperature $t_* = eV_*/k$ represents an effective quantized state from which radiation occurs. Since ionization represents the highest energy level attainable, we always have

$$t_* < 2t_i . \tag{5}$$

Further, ω is a constant. The exponent p is usually taken as -1. Boundary conditions are defined as follows:

$$\frac{dt}{dr} = 0 \quad \text{at } r = 0 \text{ (symmetry)} , \tag{6}$$

$$t = t_w \quad \text{at } r = a , \tag{7}$$

where t_w is the (observed) temperature at the inner tube wall.

It has been emphasized before [7, 8, 9] that an effective treatment of a complicated nonlinear system such as the one defined by equations (1–6) can only be done when the system is properly nondimensionalized. Clearly, r can be rendered dimensionless by means of the tube's inner radius a. Arguments were presented in [7, 8, 9] that the temperature is best rendered dimensionless by referring to the axis temperature t_r, which is the maximum temperature in the system. Typical values of t_r are 3000 K–4000 K for sodium arcs and 5000 K–6000 K for mercury arcs. These values are considerably higher than t_w which can be in the range of 1000 K–1500 K, but much lower than either t_i or t_* which are in the range of 50000 K–100000 K. Therefore, neither t_w nor t_i or t_* are suitable reference temperatures. In view of the above, we introduce dimensionless variables as follows:

$$R = r/a , \qquad T = t/t_r . \tag{8}$$

Since t_r is now assumed to have a known value, another parameter has to be turned into an unknown instead. The most natural choice will be the arc current I. By doing this, we invert the problem definition. Instead of asking which maximum temperature t_r results from a given, i.e. observed, arc current, we ask ourselves what arc current is needed to achieve a required axis temperature.

We shall not repeat here the complete derivation of the final equation we wish to investigate. This has already been done in [9], to which we refer for the details. It suffices to note that by means of a Kirchhoff transformation an auxiliary function Q replacing T is defined as follows:

$$Q = T_i \int_T^1 \frac{\lambda(t_r \tilde{T})}{\lambda(t_r)} \, d\tilde{T} , \tag{9}$$

where

$$T_i = \frac{t_i}{t_r} \tag{10}$$

is a large parameter. Because of this, small variations in T result in large variations in Q. As a result, all the interesting phenomena occur in a region where T is close to unity. Asymptotics for $T_i \gg 1$ lead to the following equation for Q:

$$\frac{1}{Z}\frac{\mathrm{d}}{\mathrm{d}Z} Z \frac{\mathrm{d}Q}{\mathrm{d}Z} = \mathrm{e}^{-Q} - \xi\,\mathrm{e}^{-\eta Q}\,, \tag{11}$$

where Z is the stretched radial coordinate

$$Z = R(HT_i)^{1/2}\,. \tag{12}$$

For the definition of H we refer to either [8] or [9]. This parameter is related to the current I and is therefore unknown. It grows exponentially with T_i. It is emphasized that for $Z = O(1)$ the coordinate R is much less than unity. The parameter ξ is related to the radiation efficiency of the arc. It can assume values in the interval

$$0 \leqslant \xi < 1\,. \tag{13}$$

Further, we have

$$\eta = \frac{t_*}{t_i}\,, \tag{14}$$

so that in view of (5) η is always less than 2. In this paper we shall only consider values of η in the interval

$$1 < \eta < 2\,. \tag{15}$$

Boundary conditions at $Z = 0$ are easily obtained from (6) and the requirement that $t = t_r$ at $r = 0$:

$$Q = 0\,, \qquad \frac{\mathrm{d}Q}{\mathrm{d}Z} = 0 \quad \text{at } Z = 0\,. \tag{16}$$

The condition at the inner tube wall now reads

$$Q = T_i \int_{T_w}^{1} \frac{\lambda(t,T)}{\lambda(t_r)}\,\mathrm{d}T \quad \text{at } Z = (HT_i)^{1/2}\,. \tag{17}$$

We shall not concern ourselves here with condition (17). Its role in the determination of H is one of the main subjects of ref. [9]. Hence, we refer to that paper for the details. Instead, we consider the structure of solutions of the system consisting of equations (11) and (16) for $\xi \uparrow 1$. We shall show that the solution reveals some singular characteristics in that limit. Indeed, if $\xi = 1$, the system admits of the solution $Q \equiv 0$. Numerical solutions for values of ξ close to unity show Q-plots which are not at all as simple as that, or even close to it. These numerical calculations [9] also indicate that high-performance lamps operate at values of ξ close to unity. This is why a special analysis of the structure of solutions in this range seems warranted.

Asymptotics for $\xi\uparrow 1$ (or $\varepsilon\downarrow 0$)

Let us write

$$\xi = 1 - \varepsilon \tag{18}$$

and consider asymptotics for $\varepsilon\downarrow 0$. Equation (11) now reads

$$\frac{1}{Z}\frac{d}{dZ} Z \frac{dQ}{dZ} = e^{-Q} - e^{-\eta Q} + \varepsilon\, e^{-\eta Q}\,. \tag{19}$$

Since $Q \equiv 0$ is a solution for $\varepsilon = 0$, we consider small values of Q first. By introducing

$$Q = \varepsilon P\,, \tag{20}$$

and expanding the exponential functions for small values of the arguments, we find to first order in ε:

$$\frac{1}{Z}\frac{d}{dZ} Z \frac{dP}{dZ} - (\eta - 1)P = 1\,. \tag{21}$$

Since P must satisfy the conditions imposed upon Q by equation (16), we arrive at

$$Q = \frac{\varepsilon}{\eta - 1}\{I_0((\eta-1)^{1/2}Z) - 1\}\,, \tag{22}$$

where I_0 is a modified Bessel function. Since (see Eq. 9.7.1 of [10])

$$I_0(x) \sim e^{x}(2\pi x)^{-1/2} \quad \text{for } x \to \infty\,, \tag{23}$$

we conclude that eventually Q is no longer small. For large values of Z we have

$$Z \sim (\eta - 1)^{-1/2}\{\Omega(\varepsilon) + c_1 + \log(Q)\}\,, \tag{24}$$

where

$$c_1 = \tfrac{1}{2}\log(2\pi) + \log(\eta - 1) \tag{25}$$

and

$$\Omega(\varepsilon) = \log\left\{\frac{1}{\varepsilon}\left(\log\frac{1}{\varepsilon}\right)^{1/2}\right\}. \tag{26}$$

Clearly, the function Q becomes of order unity when $Z - (\eta-1)^{-1/2}\Omega(\varepsilon)$ is of order unity. To describe the function Q in this region we introduce the translated coordinate z as follows:

$$Z = (\eta - 1)^{-1/2}\Omega + z\,, \tag{27}$$

where z is $O(1)$ in the limit $\varepsilon\downarrow 0$. Since Q is now assumed to be of order unity, no further scaling is necessary. Introducing (27) in (19), assuming that Q and its derivatives with respect to z are all of order unity, we obtain the equation

$$\frac{\mathrm{d}^2Q}{\mathrm{d}z^2} = \mathrm{e}^{-Q} - \mathrm{e}^{-\eta Q}\,. \tag{28}$$

The solution to this equation must match with the previous one in accordance with the rule (24) for $z \to \infty$, i.e.

$$z \sim (\eta - 1)^{-1/2}\{c_1 + \log(Q)\} \quad \text{for } z \to -\infty \quad (Q\downarrow 0)\,. \tag{29}$$

On integrating equation (28) once, we obtain

$$\frac{1}{2}\left(\frac{\mathrm{d}z}{\mathrm{d}Q}\right)^{-2} = \text{constant} - \mathrm{e}^{-Q} + \frac{1}{\eta}\,\mathrm{e}^{-\eta Q}\,. \tag{30}$$

If the solution to this differential equation is to behave in accordance with the matching condition (29), the constant must be equal to $1 - \eta^{-1}$. Integrating once more, we find

$$z = c_2 + 2^{-1/2}\int_1^Q \frac{\mathrm{d}q}{\{1 - \mathrm{e}^{-q} - (1 - \mathrm{e}^{-\eta q})/\eta\}^{1/2}}\,, \tag{31}$$

where c_2 is a constant of integration. It can easily be shown that the behaviour of (31) for $Q\downarrow 0$ is

$$z \sim c_2 + c_3 + (\eta - 1)^{-1/2}\log(Q)\,, \tag{32}$$

where

$$c_3 = \int_0^1 \left\{\frac{1}{(\eta - 1)^{1/2}Q} - \frac{1}{2^{1/2}\{1 - \mathrm{e}^{-Q} - (1 - \mathrm{e}^{-\eta Q})/\eta\}^{1/2}}\right\}\mathrm{d}Q\,. \tag{33}$$

By comparing (29) and (32), we conclude that the integration constant c_2 is given by

$$c_2 = c_1(\eta - 1)^{-1/2} - c_3\,. \tag{34}$$

On the other hand, the behaviour of equation (31) for $z \gg 1$ is given by

$$z \sim c_2 + c_4 + \left\{\frac{\eta}{2(\eta - 1)}\right\}^{1/2} Q\,, \qquad (Q \to \infty)\,, \tag{35}$$

where

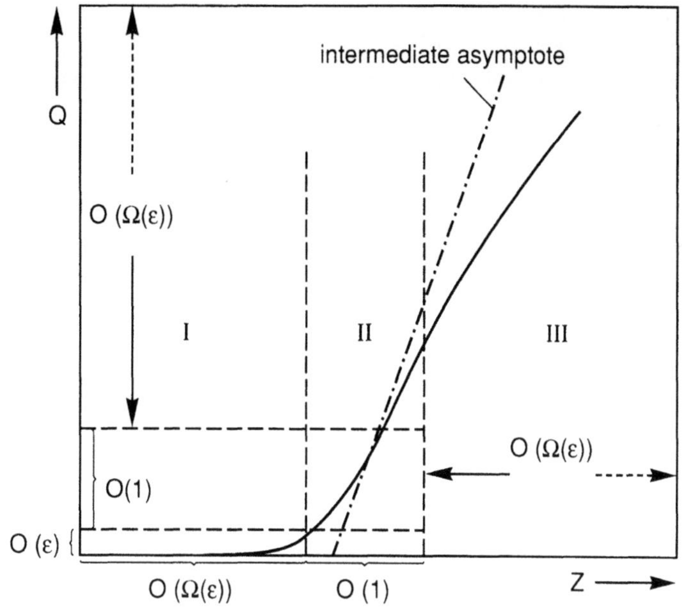

Fig. 2. Schematic picture of the three-layer structure of the temperature field.

$$c_4 = 2^{-1/2} \int_1^\infty \left\{ \frac{1}{\{1 - e^{-Q} - (1 - e^{-\eta Q})/\eta\}^{1/2}} - \left(\frac{\eta}{\eta - 1}\right)^{1/2} \right\} dQ - \left(\frac{\eta}{2(\eta - 1)}\right)^{1/2} . \tag{36}$$

Apparently, this solution approaches an oblique asymptote with an intercept on the Z-axis at $z = c_2 + c_4$ or, using (27), at

$$Z = Z_0 = (\eta - 1)^{-1/2}\Omega(\varepsilon) + c_2 + c_4 \tag{37}$$

and a slope equal to $\{\eta/2(\eta - 1)\}^{1/2}$.

In Fig. 2 we present a schematic picture of the two regions which emerged from the analysis so far. First we have region I in which Q remains $O(\varepsilon)$. The Z-range, in which this part of the solution defined by (22) is valid, extends up to distances from the origin which are $O(\Omega(\varepsilon))$. This is followed by a region II in which Q and its derivatives rise rapidly to values that are of order unity. The width of this region is of order unity. This part of the solution is defined by equations (27) and (31). The differential equation which governs the solution in region II is given by equation (28). From this equation the term $Z^{-1}\,dQ/dZ$ is absent. Indeed, with Q and its derivatives being of order unity here, the term Z^{-1} renders this particular term $O(\Omega^{-1})$. We shall see later that this term regains first-order importance in a third region in which both Z and Q are allowed to tend to infinity.

Region III

The solution we found in region II, the transition region, applies as long as z is much smaller than $\Omega(\varepsilon)$. Suppose $z = \delta\Omega(\varepsilon)$, where δ is a small but fixed parameter ($0 < \delta \ll 1$). Letting $\varepsilon\downarrow 0$, we conclude that the solution in region II is given approximately by the asymptote of equation (35). The term $Z^{-1}\,dQ/dZ$ is still of the order of Ω^{-1} and the term $\exp(-Q)$ is of the order $\exp(-\tau\Omega)$, where τ is some constant larger than zero and independent of ε.

Therefore, as soon as we are well into the tail of the solution in region II, the exponentials are negligibly small in comparison with the other terms in the equation. This leads to the conclusion that the solution in region III is governed by the differential equation

$$\frac{d^2Q}{dZ^2} + \frac{1}{Z}\frac{dQ}{dZ} = 0 . \tag{38}$$

The solution must approach the asymptote (35) from the other side, i.e.

$$Q = 0 , \quad \frac{dQ}{dZ} = \left(\frac{2(\eta - 1)}{\eta}\right)^{1/2} \quad \text{at } Z = Z_0 . \tag{39}$$

The solution to this problem is

$$Q = \left(\frac{2(\eta - 1)}{\eta}\right)^{1/2} Z_0 \log(Z/Z_0) . \tag{40}$$

This expression shows that in the outer region Z_0 is the natural scaling factor for both Q and Z. (See also Fig. 2.)

Composite solutions

We can define composite solutions on the basis of the partial solutions that are valid in regions I, II and III, respectively. Because of the way in which the solution is given in region II, we shall define composite solutions with Q as the independent variable. The common (matched) part of the solutions in regions I and II is

$$\text{CP(I, II)} = (\eta - 1)^{-1/2}\Omega(\varepsilon) + c_2 + c_3 + (\eta - 1)^{-1/2} \log(Q) . \tag{41}$$

This follows from (27) and (32). Therefore, the composite solution which comprises the regions I and II is

$$Z_{\text{comp}}(\text{I, II}) = (\eta - 1)^{-1/2} \operatorname{inv} I_0(Q(\eta - 1)/\varepsilon) + Z_{\text{II}} - \text{CP(I, II)} , \tag{42}$$

where inv I_0 denotes the inverse Bessel function I_0. Further, Z_{II} is defined by (27) and (31). This solution applies in the interval

$$0 \leq Z \leq (1 + \delta)Z_0 , \tag{43}$$

where δ is a small, but fixed, positive constant $(0 < \delta \ll 1)$.

When Q is of the order of ε, the inv I_0 function represents a non-trivial contribution to (42). It can easily be shown that Z_{II} is then approximately equal to the common part. On the other hand, when Q becomes much larger than ε, the common part cancels approximately the first term on the right of (42). The latter result will be used in a later section.

It is also possible to make a composite solution on the basis of the solutions that are valid in the regions II and III. The part these two solutions have in common is defined by (27) and (35). Therefore,

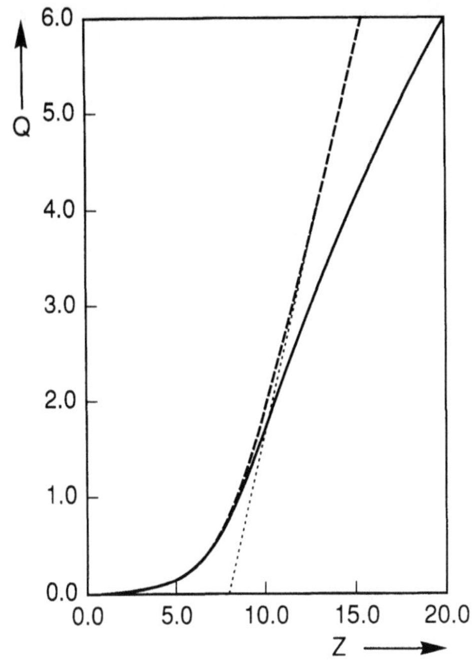

Fig. 3.

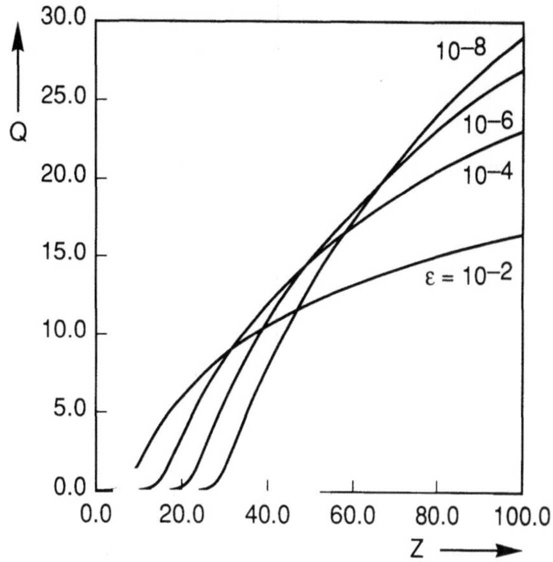

Fig. 4.

Fig. 3. $Z_{\text{comp}}(\text{I, II})$ given by dashed curve and $Z_{\text{comp}}(\text{II, III})$ given by fully drawn curve for $\varepsilon = 0.01$ and $\eta = 1.5$. The dotted line represents the intermediate asymptote.

Fig. 4. The temperature function Q for four different values of ε. $(\eta = 1.5)$

$$Z_{\text{comp}}(\text{II, III}) = Z_{\text{II}} + Z_0 \exp\left\{\left(\frac{\eta}{(2\eta-1)}\right)^{1/2} \frac{Q}{Z_0}\right\} - Z_0 - \left(\frac{\eta}{(2\eta-1)}\right)^{1/2} Q \,, \tag{44}$$

where Z_{II} is defined by the sum of equations (27) and (31). This composite solution applies in the regions II and III. This renders it at once more practical than equation (42). Indeed, since the values of Q in region I are only $O(\varepsilon)$, the overall picture of the temperature profile is well represented by (44). A graph showing the two composite expansions is presented in Fig. 3, together with the intermediate asymptote.

Figure 4 shows temperature profiles for various values of ε and $\eta = 1.5$. These profiles indicate that the transition layer moves further and further away from the origin, as ε becomes smaller and smaller. The changeover from region I to region III seems to become more and more rapid. It is also interesting to note that two different profiles have a point in common outside the origin. The envelope of the family of curves can easily be calculated from Eq. (40), namely by varying the parameter Z_0. This envelope is the straight line $Q = (2(\eta - 1)/\eta)^{1/2} e^{-1} Z$.

Characteristic functions

In the analysis of reference [9] two functions $g_1(\xi, \eta)$ and $g_2(\xi, \eta)$ are of particular importance. They are defined by the asymptotic behaviour of Q for $Z \to \infty$:

$$Q \to g_1(\xi, \eta) \log(Z) - g_2(\xi, \eta) \,. \tag{45}$$

From the analysis of this paper we conclude the following (see equations (40), (37) and (18)):

$$g_1(\xi,\eta)=(2/\eta)^{1/2}\{\Omega(1-\xi)+O(1)\}\,, \tag{46}$$

$$g_2(\xi,\eta)=g_1(\xi,\eta)\log\{(\eta-1)^{-1/2}\Omega(1-\xi)+O(1)\} \tag{47}$$

for $\xi\uparrow 1$. Another useful quantity defined in [9] is the lamp's efficiency:

$$W=\xi\int_0^\infty Z\,\mathrm{e}^{-\eta Q}\,\mathrm{d}Z\Big/\int_0^\infty Z\,\mathrm{e}^{-Q}\,\mathrm{d}Z\,. \tag{48}$$

To be able to calculate this function we must evaluate asymptotically the integral

$$I(\nu)=2\int_0^\infty Z\,\mathrm{e}^{-\nu Q(Z)}\,\mathrm{d}Z\,. \tag{49}$$

Since Q is a monotonically increasing function of Z we obtain from integration by parts:

$$I(\nu)=\tfrac{1}{2}\nu\int_0^\infty Z^2\,\mathrm{e}^{-\nu Q}\,\mathrm{d}Q\,. \tag{50}$$

In view of the availability of the two composite solutions (42) and (44), we shall write

$$I(\nu)=\tfrac{1}{2}\nu\int_0^{Q_0} Z^2_{\mathrm{comp}}(\mathrm{I},\mathrm{II})\,\mathrm{e}^{-\nu Q}\,\mathrm{d}Q+\tfrac{1}{2}\nu\int_{Q_0}^\infty Z^2_{\mathrm{comp}}(\mathrm{II},\mathrm{III})\,\mathrm{e}^{-\nu Q}\,\mathrm{d}Q\,, \tag{51}$$

where

$$Q_0=Q((1+\delta)Z_0)\,, \tag{52}$$

with $0<\delta\ll 1$ fixed. It can easily be seen that Q_0 is well within the asymptotic range defined by (35). Therefore, the upper bound of the first integral of (51) can be replaced by ∞. The very first part of the integration range, where Q is of the order of ε, can be disregarded, since its contribution to the integral is asymptotically zero, namely $O(\varepsilon)$, with respect to the terms retained. Therefore, the first integral reads approximately

$$\tfrac{1}{2}\nu\int_0^\infty Z^2_{\mathrm{II}}\,\mathrm{e}^{-\nu Q}\,\mathrm{d}Q\sim\frac{\Omega^2(\varepsilon)}{2(\eta-1)}+\frac{\Omega(\varepsilon)}{(\eta-1)^{1/2}}\left\{c_2+2^{-1/2}\int_1^\infty\frac{\mathrm{e}^{-\nu Q}\,\mathrm{d}Q}{\{1-\mathrm{e}^{-Q}-(1-\mathrm{e}^{-\eta Q})/\eta\}^{1/2}}\right.$$
$$\left.-2^{-1/2}\int_0^1\frac{(1-\mathrm{e}^{-\nu Q})\,\mathrm{d}Q}{\{1-\mathrm{e}^{-Q}-(1-\mathrm{e}^{-\eta Q})/\eta\}^{1/2}}\right\}+o(\Omega)\,. \tag{53}$$

The second integral of (51) can be shown to be asymptotically smaller than any of the terms retained in (53). Equation (48) can now be evaluated for $\xi\uparrow 1$, i.e. for $\varepsilon\downarrow 0$. After some manipulations we find

$$W\sim 1-2^{3/2}(\eta-1)\eta^{-1/2}\Omega^{-1}(\varepsilon)+o(\Omega^{-1})\,, \tag{54}$$

showing that the radiation-efficiency parameter tends to unity when $\xi\uparrow 1$ or $\varepsilon\downarrow 0$.

In [9] we derived

$$g_1(\xi,\eta)=\int_0^\infty Z(e^{-Q}-\xi\, e^{-\eta Q})\,dZ=I(1)-I(\eta)\,. \tag{55}$$

Using the result of Eq. (53), we find for $\xi\uparrow 1$

$$\begin{aligned} g_1(\xi,\eta) &\sim 2^{-1/2}(\eta-1)^{-1/2}\Omega(1-\xi)\int_0^\infty \frac{e^{-Q}-e^{-\eta Q}}{\{1-e^{-Q}-(1-e^{-\eta Q})/\eta\}^{1/2}}\,dQ \\ &= 2^{1/2}\eta^{-1/2}\Omega(1-\xi)\,. \end{aligned} \tag{56}$$

By comparing this with (46), we conclude that the leading-order terms match. This may serve as a useful check on the correctness of the present analysis.

Concluding remarks

In this paper we have described the asymptotic form of the temperature profile in a wall-stabilized high-pressure gas-discharge arc, when the radiation-efficiency parameter approaches unity. This case applies when the heat-input term and the radiation-loss term become almost equally important. In this limit the temperature remains virtually constant ($t\sim t_r$) in a central portion of the arc. This is followed by a thin transition region in which the temperature profile rapidly bends downwards. Both the heat-input term and the radiation-loss term play a first-order role in these two regions. Next to the transition region, away from the axis, the temperature profile follows a simple logarithmic rule. In this region, which extends all the way to the inner tube wall, neither the energy source nor the loss term play a first-order role. For the practical implementation of the results of this paper and for a more complete discussion of these, we refer to the companion paper [9].

In this communication we have limited ourselves to presenting the leading-order terms of the three expansion solutions. Clearly, these suffice for bringing out the structure of the temperature profile. It would seem to be possible to derive a few more terms in each of the expansions. However, we postpone that to a future paper [11]. To illustrate that, apart from bringing forth the structure of the asymptotic temperature profiles, the leading terms of the asymptotic expansions are also capable of giving useful numerical results, we present Table 1. This table gives asymptotic and numerically evaluated values of the radiation-efficiency

Table 1. Radiation efficiency W for various values of ε and for $\eta=1.5$

$\log_{10}(\varepsilon)$	Equation (54)	Exact (numerical)
−1	0.575	0.594
−2	0.785	0.768
−3	0.853	0.842
−4	0.888	0.880
−5	0.9093	0.9036
−6	0.9237	0.9194
−7	0.9341	0.9308
−8	0.9419	0.9393
−9	0.9481	0.9460
−10	0.9531	0.9513

parameter W. The numerical values were obtained with the software discussed in ref [9]. Clearly, even for $\varepsilon = 0.1$ or $\xi = 0.9$ the asymptotic result is quite useful. It is interesting to note that ε must become extremely small for W to approach the limiting value of unity. Since radiation efficiencies larger than 0.5 are not uncommon for many lamps, the asymptotic treatment would seem to be quite useful.

To conclude, we remark that the problem area studied here shows great similarities with certain fields in combustion theory. Expressions such as (3) or (4), with rapidly varying exponential terms, are also characteristic of problems dealing with thermal explosions. Singular perturbation techniques are frequently needed to describe the solutions of such problems adequately. We refer to [12, 13] for further details. What seems to distinguish our problem from combustion problems is that we have to deal with two competing exponential functions, as shown by equation (11). In this paper we studied what happens when this competition is at its extreme.

References

1. R.J. Zollweg, Convection in vertical high-pressure mercury arcs. *J. Appl. Phys.* 49 (1978) 1077–1091.
2. J.J. Lowke, Calculated properties of vertical arcs stabilized by natural convection. *J. Appl. Phys.* 50 (1979) 147–157.
3. H.K. Kuiken, A boundary-layer model for a plane free-burning high-pressure gas-discharge arc. *J. Appl. Phys.* 69 (1991) 2896–2903.
4. W. Elenbaas, *The High Pressure Mercury Vapour Discharge*. Amsterdam: North Holland Publishers (1951) 173 pp.
5. J.F. Waymouth, *Electric Discharge Lamps*. Cambridge: M.I.T. Press (1971) 353 pp.
6. J.J. de Groot and J.A.J.M. van Vliet, *The High-Pressure Sodium Lamp*. Deventer: Kluwer Technische Boeken B.V. (1986) 328 pp.
7. H.K. Kuiken, Thermal behaviour of a high-pressure gas-discharge lamp. To appear in: Lecture Notes in Mathematics, *Mathematical Modelling of Industrial Processes*. Heidelberg: Springer (1991).
8. H.K. Kuiken, An asymptotic treatment of the Elenbaas–Heller equation. *Appl. Phys. Letters* 58 (1991) 1833–1835.
9. H.K. Kuiken, An asymptotic treatment of the Elenbaas–Heller equation for a radiating wall-stabilized high-pressure gas-discharge arc. *J. Appl. Phys.* 70 (1991) 5282–5291.
10. M. Abramowitz and I.A. Stegun, *Handbook of Mathematical Functions*. Seventh Edition. Washington: Nat'l Bur. Stands. (1968) 1046 pp.
11. H.K. Kuiken, Higher approximations to the solution of a problem concerning a high-pressure gas-discharge arc. To appear in *Appl. Math. Letters*, 1992.
12. J.D. Buckmaster and G.S.S. Ludford, *Theory of Laminar Flames*. Cambridge: Cambridge University Press (1982) 266 pp.
13. A. Kapila, *Asymptotic Treatment of Chemically Reacting Systems*. Boston: Pitman (1983) 119 pp.

Journal of Engineering Mathematics **26**: 51–59, 1992.
H.K. Kuiken and S.W. Rienstra (eds), Problems in Applied, Industrial and Engineering Mathematics.

The Green function for potential flow in a rectangular channel

J.N. NEWMAN

Department of Ocean Engineering, Massachusetts Institute of Technology, Cambridge, MA 02139, USA

Abstract. The evaluation of the Green function is considered for the three-dimensional Laplace equation, in the interior of a rectangular channel subject to homogeneous Neumann conditions on the boundaries. To complement the Fourier eigenfunction expansion which is effective in the far-field, a near-field algorithm is developed based on the simpler Green function for a channel of infinite width, using images to account for the channel sides. Examples are given of numerical applications including the added mass of a sphere in a square channel, and the interaction force between a ship and an adjacent canal wall.

1. Introduction

Three-dimensional potential flows inside rectangular channels are of interest in various applications. These include wall corrections in water- and wind-tunnels, and problems associated with ships moving in canals. In the latter case, if the Froude number is sufficiently small, wave effects are negligible and the free-surface condition can be replaced by a homogeneous Neumann boundary condition. In applications where Dirichlet boundary conditions are applicable the corresponding Green functions can be derived by superposition of appropriate pairs of the functions considered here, with opposite signs.

The Green function for a rectangular channel can be constructed simply from the free-space singularity $1/r$ and a doubly-periodic array of images. This representation is useful conceptually, but the very slow convergence of the doubly-infinite series makes it unsuitable for routine computations. Alternative representations have been derived by Breit [1]. One, reproduced below as equation (7), is an eigenfunction expansion involving trigonometric functions of the coordinates transverse to the channel axis, and exponential functions of the axial coordinate. This expansion is simple and effective if the axial distance between the source and field points is moderate or large compared to the *larger* of the two transverse dimensions of the channel, the width w or height h. To complement this representation Breit derives an elegant Taylor series expansion of the images about the source point. Since the radius of convergence is limited by the distance to the nearest image, one expects this series to be useful computationally in a spherical domain with radius of the order of the *smaller* dimension of the channel. Thus for channels where h and w are substantially different, the two complementary expansions given in [1] do not suffice.

A simpler analysis applies in the limits $w/h \rightarrow \infty$ or $w/h \rightarrow 0$, which are fundamentally identical except for scaling and definition of the coordinates. For definiteness we consider the former case, corresponding to the domain between two parallel horizontal planes which are unbounded. The Green function can be constructed from a single periodic array of images, situated on the vertical axis above and below the source point. The potential is axisymmetric about this axis. The eigenfunction expansion involves trigonometric functions of the axial coordinate, and modified Hankel functions K_0 in the radial coordinate R. This expansion is efficient for large or moderate values of R, relative to the height h between the two boundaries. For smaller values of R/h a Taylor series has been derived by Breit [1], and Newman [2] has developed Chebyshev expansions and economized polynomials. The latter representation permits computation of the singly-periodic Green function with single-precision accuracy for $R<h$ with only 21 polynomial coefficients. In the complementary domain $R>h$ the eigenfunction expansion requires only six terms for single-precision accuracy, and thus the computation of the singly-periodic Green function is effectively achieved by these two complementary algorithms. An important feature of this Green function is its logarithmic behaviour for $R \gg h$, a consequence of the fact that at large radial distances the array is equivalent asymptotically to a two-dimensional line source.

In the present work the Green function for an arbitrary rectangular channel is constructed by periodic superposition of the simpler infinite-width Green functions, along the transverse axis normal to the channel sides. In its original form this representation is slowly convergent, due to the logarithmic component noted above, but this can be summed separately with a closed-form potential equivalent to an array of two-dimensional sources. The remainder, associated with the modified Bessel functions in the eigenfunction expansion, converges exponentially. Examples are given of computations based on this approach including the added mass of a sphere in a square channel, and the analysis of the bank-interaction force for a ship moving in a canal.

2. Analysis

A rectangular channel of width w and height h is considered, with unbounded length. Without loss of generality it can be assumed that $w \geqslant h$. Nondimensional coordinates are defined in terms of the height h, which is assumed hereafter to be equal to one. Thus the interior of the channel is the three-dimensional space $(-\infty<x<\infty,\ 0<y<w,\ 0<z<1)$, where $w \geqslant 1$. The desired Green function $G(x-\xi, y, \eta, z, \zeta)$ is the potential of a source, situated at the point ξ, η, ζ inside the channel and subject to homogeneous Neumann boundary conditions on these boundaries. The governing equation is

$$\nabla^2(G-1/r)=0\,, \tag{1}$$

where $1/r=[(x-\xi)^2+(y-\eta)^2+(z-\zeta)^2]^{-1/2}$ is the free-space Green function in the absence of the channel boundaries. The boundary conditions take the form

$$\partial G/\partial y=0\,, \quad \text{on } y=0, w\,, \tag{2}$$

$$\partial G/\partial z=0\,, \quad \text{on } z=0,1\,. \tag{3}$$

Any linear function of x with constant coefficients is a homogeneous solution of (1–3), corresponding physically to a uniform streaming flow through the channel. To eliminate such a component it is appropriate to impose the far-field condition

$$G=U|x|+o(1)\,, \quad \text{for } |x|\to\infty\,. \tag{4}$$

This states that the flux from the source is equally divided between $\pm\infty$, with limiting velocities $\pm U$. Since the singularity $1/r$ corresponds to a source with a flux -4π, the constant $U = -2\pi/w$ is specified by continuity. (The singularity corresponds more precisely to a *sink*, but the more general term *source* is retained here with the understanding that its strength is negative.) The far-field condition (4) also fixes the value of the arbitrary constant in the solution of (1–3).

A formal solution can be constructed by superposition of four simpler functions,

$$G(x-\xi, y, \eta, z, \zeta) = G_0(x-\xi, y-\eta, z-\zeta) + G_0(x-\xi, y+\eta, z-\zeta) + G_0(x-\xi, y-\eta, z+\zeta) + G_0(x-\xi, y+\eta, z+\zeta), \tag{5}$$

where $G_0(x, y, z)$ is defined by (1–4) with the source point at the origin ($\xi = 0$, $\eta = 0$, $\zeta = 0$). The function G_0 corresponds to a doubly-periodic array of free-space Green functions, situated at the points $x = 0$, $y = 2mw$, $z = 2n$, where the integers m, n take all positive and negative values as well as zero. The solution of (1–4) can thus be constructed from (5) with a basic potential of the form

$$G_0 = \frac{1}{r} + \sum_{m=-\infty}^{\infty} \sum_{\substack{n=-\infty \\ |m|+|n|>0}}^{\infty} \{[x^2 + (y+2mw)^2 + (z+2n)^2]^{-1/2} - [(2mw)^2 + (2n)^2]^{-1/2}\} + C. \tag{6}$$

Here the extra inverse square-root is a constant, subtracted from each term to secure convergence. The additional constant C is required to satisfy (4). (In [1] the unknown constant is included in (4), instead of (6). Since we do not use (6) directly in our analysis the present convention is more convenient.)

The series (6) is unsuitable for numerical applications, due to its very slow convergence. A more useful computational form can be derived by Fourier techniques, as in [1], with the result

$$G_0 = -\frac{\pi|x|}{2w} + \frac{\pi}{2w} \sum_{m=0}^{\infty} \sum_{\substack{n=0 \\ m+n>0}}^{\infty} \varepsilon_m \varepsilon_n \frac{\cos u_m y \cos v_n z}{(u_m^2 + v_n^2)^{1/2}} e^{-\sqrt{u_m^2+v_n^2}|x|}, \tag{7}$$

where

$$\varepsilon_0 = 1, \qquad \varepsilon_m = 2 \quad \text{for } m \geq 1,$$

$$u_m = m\pi/w, \qquad v_n = n\pi.$$

Note that the far-field condition (4) is satisfied without introducing an additional constant. This expansion is simple to evaluate and rapidly convergent, provided $|x| \geq O(w)$.

As the basis for effective numerical algorithms in the complementary domain $|x| \leq O(w)$, we consider a single periodic array, equivalent to the contribution with $m = 0$ in (6) except for a constant, and defined by the series expansion

$$g(R, z) = [R^2 + z^2]^{-1/2} + \gamma - \log 4 + \sum_{\substack{n=-\infty \\ n\neq 0}}^{\infty} \left\{[(R^2 + (z+2n)^2]^{-1/2} - \frac{1}{2|n|}\right\}. \tag{8}$$

Here $R = (x^2 + y^2)^{1/2}$ is the radial distance from the array axis, and $\gamma = 0.577\ldots$ is Euler's constant. Equation (8) defines a periodic function of z, which can be expanded in a Fourier series, as in Gradshteyn & Ryzhik [3, §8.526]. The resulting expression is

$$g(R, z) = -\log R + 2 \sum_{n=1}^{\infty} \cos(n\pi z) K_0(n\pi R) , \tag{9}$$

where K_0 is the modified Hankel function. Since K_0 is exponentially small for large values of its argument the series in (9) converges rapidly if $R \geqslant O(1)$. For large values of R

$$g(R, z) = -\log R + O(R^{-1/2} \mathrm{e}^{-\pi R}) . \tag{10}$$

The first term in (9) is a two-dimensional source; the asymptotic approximation (10) is equivalent to the statement that the array of three-dimensional point sources appears in its far field as a simple line source with the same total flux.

Returning now to the case of the doubly-periodic array (6), our plan is to construct this function by summation of the single array g and its images. Except for the term $\log R$, the asymptotic approximation (10) confirms the convergence of such a procedure. We therefore consider the modified series

$$G_0(x, y, z) = G_{2D}(x, y) + \sum_{m=-\infty}^{\infty} [g(R_m, z) + \log R_m] , \tag{11}$$

where $R_m = [x^2 + (y - 2mw)^2]^{1/2}$, and

$$G_{2D}(x, y) = -\log\left(\frac{\pi R_0}{w}\right) - \sum_{\substack{m=-\infty \\ m \neq 0}}^{\infty} \log\left(\frac{R_m}{2|m|w}\right)$$

$$= -\mathrm{Re} \log 2 \sinh \frac{\pi(x + \mathrm{i}y)}{2w} = -\frac{1}{2} \log\left(2 \cosh \frac{\pi x}{w} - 2 \cos \frac{\pi y}{w}\right) . \tag{12}$$

The function G_{2D} is the potential of a two-dimensional periodic array of sources. Except for a constant, the evaluation of the series in (12) can be confirmed by differentiation with respect to the complex variable $(x + \mathrm{i}y)$. The constant is evaluated by considering the limit of (12) for $x + \mathrm{i}y \to 0$. Finally, from the last expression in (12), it follows for large $|x|$ that

$$G_{2D} = -\frac{\pi |x|}{2w} + O(\mathrm{e}^{-\pi|x|/w}) . \tag{13}$$

Since each term in the infinite series of (11) is exponentially small, in accordance with (10), it follows that the asymptotic approximation (13) also applies to G_0, confirming the far-field form of the latter function and the constants in (11) and (12).

The validity of (11) can be confirmed directly. The function G_{2D} satisfies the boundary conditions (2), and each term satisfies (3). The boundary conditions (2) are satisfied by the sum in (11) as a result of periodicity and symmetry. Since the logarithmic singularities in the first and last terms cancel, the only remaining singularity within the channel domain is the contribution from the first term on the right-hand-side of (8), in the term $m = 0$ of (11), which is precisely the three-dimensional source singularity $1/r$ associated with the Green function. Thus all of the prescribed conditions (1–4) are satisfied by (11).

3. Numerical procedure

Three algorithms are effective collectively in the evaluation of the Green function G. The simplest, based on (7), is straightforward to implement. A maximum of eight terms in each series is sufficient to give 6D accuracy in the domain $|x|/w > 0.6$. In the complementary domain the series (11) can be utilized, with the single array g evaluated in the manner to be described below.

For arbitrary points within the domain of the channel, the radius R_m to the mth image in (11) is greater than or equal to the lower bound $2|m|w - 1$. For large values of $|m|$ the terms in (11) are exponentially small, in proportion to the factors $\exp(-2\pi|m|w)$. Since $w \geq 1$ the series in (11) can be summed numerically with rapid convergence. Eight terms ($-3 \leq m \leq 4$) are required for 6D accuracy. With the exception of the term $m = 0$, $R_m \geq w \geq 1$, and the summands in (11) can be evaluated effectively from (9).

A complementary algorithm can be employed to evaluate the term $m = 0$ if $R_0 < 1$, based on the two-dimensional Chebyshev expansions and economized polynomials presented in [2], with the result

$$g(R, z) \simeq [R^2 + z^2]^{-1/2} + [R^2 + (z+2)^2]^{-1/2} + [R^2 + (z-2)^2]^{-1/2} + \sum_{m,n} a_{mn} R^{2m} z^{2n} . \quad (14)$$

Using the coefficients in Table 1, this economized polynomial approximation is accurate in $(0 \leq R \leq 1, 0 \leq z \leq 1)$, to about 8 decimals. (Except for the constant a_{00}, the entries in Table 1 are identical to Table 2 of [2].)

When $R_0 < 1$ it is advisable to cancel the term $\log R_0$ in the series (11) with the corresponding singular term of (12). Note that $G_{2D} + \log R_0$ is a regular function of (x, y) in $0 \leq y \leq w$, but care is required to evaluate this sum robustly when both x and y tend to zero.

When one or more of the square-root singularities in (14) is large, it is advisable to subtract these terms from G and evaluate their contributions in the usual manner of treating the free-space Green function in a boundary integral equation.

4. Applications

The Green function described in the preceding sections has been combined with a three-dimensional panel code based on Green's theorem. This code, a derivative of the free-surface radiation/diffraction program WAMIT, assumes flat quadrilateral panels and constant values of the unknown potential, or source strength, on each panel. An iterative solver is used for the linear system to permit relatively large numbers of panels, and corresponding

Table 1. Coefficients a_{mn} in the economized polynomial approximation (14)

n	$m = 0$	$m = 1$	$m = 2$	$m = 3$	$m = 4$
0	−1.80907870	−0.02525711	0.00086546	−0.00004063	0.00000193
1	0.05051418	−0.00692380	0.00073292	−0.00006636	0.00000398
2	0.00230838	−0.00097875	0.00020597	−0.00003333	0.00000524
3	0.00012934	−0.00010879	0.00003965	−0.00000891	
4	0.00000913	−0.00001270	0.00000466		

unknowns, on the body surface. Two simple applications are described here to illustrate the utility of the channel Green function. More complicated applications are described in [4], involving the interactions between two ships where one is moving past the other in a canal.

In the first application we consider the added mass of a sphere, of diameter D, centered in a square channel of width w and height $h = w$. The added-mass coefficients are defined by the integrals

$$a_{ij} = \rho \int\int \phi_j n_i \, \mathrm{d}S \, . \tag{15}$$

Here ρ is the fluid density, ϕ_j is the velocity potential for motion of the body in the direction j with unit velocity, and n_i is the component of the unit normal vector in the direction i. (Rotational modes and moments can be considered by straightforward extensions of these definitions.) The integral in (15) is over the body surface. Due to the spherical geometry of the body, with its center on the axis of the channel, a_{11}, a_{22} and a_{33} are the only non-zero elements of the matrix a_{ij}. Since the channel is square, $a_{22} = a_{33}$. In an unbounded fluid, corresponding to the limit $D/w = 0$, these coefficients are equal to one-half of the displaced fluid volume. In the presence of the channel walls a_{11} and $a_{22} = a_{33}$ have different values which depend on D/w.

To test for convergence three discretizations of the sphere are used, with a total of $N = 128, 512, 2048$ panels. (In the computations one plane of symmetry is imposed, reducing the total number of unknowns to half of these numbers. Two or three planes of symmetry could be utilized with additional programming effort.) Computations have been performed with $D/w = 0(.05)1.0$ for each of the three discretizations. In the limit $D/w = 0$ only the $1/r$ free-space Green function is considered. The complementary limit $D/w = 1$ is non-physical, corresponding to the case where the sphere is tangent to the channel boundaries.

The results are listed in abbreviated form in Table 2, which also includes extrapolated values for $N = \infty$ based on Richardson extrapolation. The effectiveness of this extrapolation is confirmed by noting that in all cases, including the singular limit $D/w = 1$, the results of the first and second extrapolants agree to three or more significant figures. The extrapolated

Table 2. Added-mass coefficients a_{11} (upper table) and a_{22} (lower table) for a sphere of diameter D, centered in a square channel of width w

D/w	$N = 128$	$N = 512$	$N = 2048$	$N = \infty$	$N = \infty$
0.0	0.47968	0.49492	0.49874	0.50001	0.50001
0.2	0.48583	0.50157	0.50554	0.50686	0.50687
0.4	0.53041	0.54973	0.55466	0.56630	0.55631
0.6	0.66681	0.69802	0.70604	0.70871	0.70873
0.8	1.02544	1.09381	1.11177	1.11775	1.11783
1.0	2.34813	2.65608	2.74970	2.78091	2.78239
0.0	0.49018	0.49775	0.49947	0.50004	0.50002
0.2	0.49431	0.50218	0.50401	0.50462	0.50461
0.4	0.52403	0.53403	0.53637	0.53714	0.53713
0.6	0.61135	0.62785	0.63193	0.63330	0.63329
0.8	0.81811	0.85229	0.86100	0.86391	0.86392
1.0	1.39278	1.52964	1.57288	1.58730	1.58810

N denotes the total number of panels used to represent the sphere. The last two columns are based on Richardson extrapolation of the entries in the preceding two and three columns, respectively. All coefficients are normalized by the displaced mass of fluid, $\pi\rho D^3/6$.

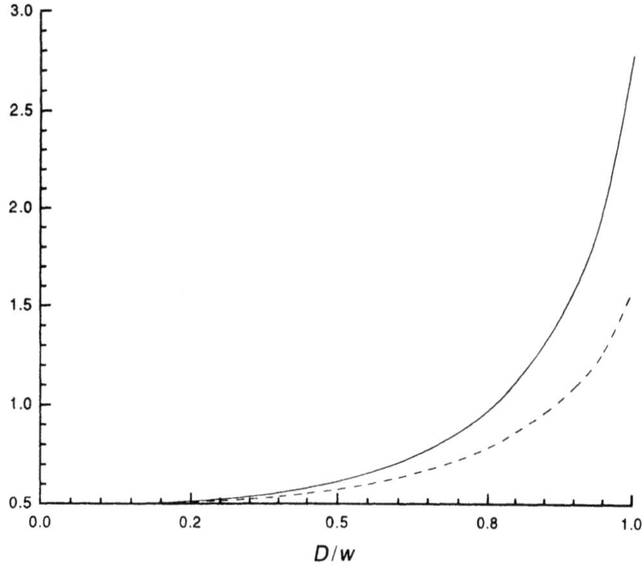

Fig. 1. Added mass of a sphere, of diameter D, in a square channel of width w, normalized by the mass of fluid displaced by the sphere. The solid curve represents the longitudinal added mass, for accelaration along the channel axis, and the dashed curve is for transverse acceleration, normal to one of the channel sides.

values are plotted in Fig. 1, as ratios of the added-mass coefficients in an unbounded fluid. The effect of the channel is to increase the added mass, particularly the longitudinal coefficient a_{11}, but the increase is small for $D/w \leqslant 0.5$.

As a more practical example, a ship hull is considered to move with constant velocity along rectangular canal and the same panel program is used to calculate the 'bank suction force' which acts in the transverse direction, forcing the ship toward the nearest side of the canal. The free surface is considered to be rigid and represented by a simple image, as is

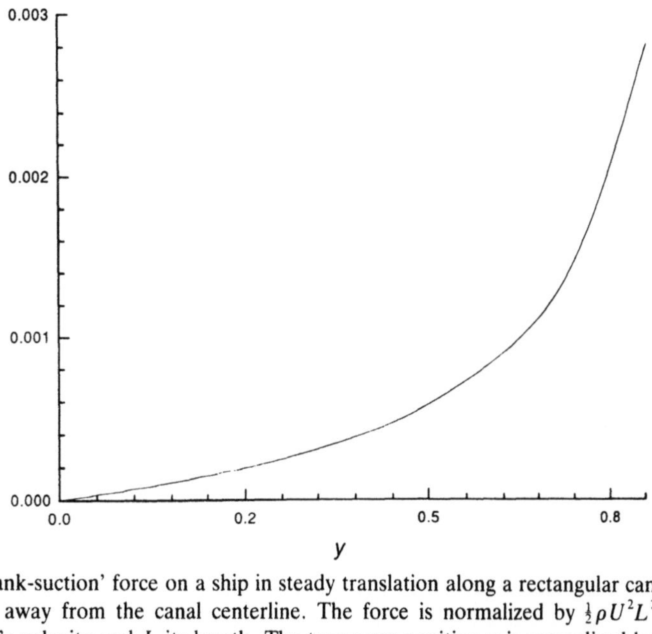

Fig. 2. Transverse 'bank-suction' force on a ship in steady translation along a rectangular canal, as a function of its transverse position y away from the canal centerline. The force is normalized by $\frac{1}{2}\rho U^2 L^2$ where ρ is the fluid density, U is the ship's velocity and L its length. The transverse position y is normalized by the half-width of the canal.

appropriate when the Froude number is small. The ship considered is a 'Mariner' hull form, represented by a total of 400 panels. We take the ship length to be one, giving a normalized draft of 0.057 and a half-beam of 0.073. The canal is assumed to have a unit width and a depth of 0.068. Figure 2 shows the resulting computed values of the transverse force coefficient, as a function of lateral position in the canal. The force vanishes when the ship is on the centerline, from symmetry, and increases rapidly as the ship moves toward one side of the canal.

5. Discussion

The Green function for Laplace's equation in the domain interior to a rectangular channel, of infinite length and constant width and height, can be represented directly in terms of the doubly-periodic array of free-space Green functions $1/r$, as in (6). In the limiting case where the width (or alternatively the height) is infinite, the problem reduces to that of a single source between two parallel infinite planes, or equivalently to the singly-periodic array (8). In their direct representations (6) and (8), neither the doubly-periodic or singly-periodic arrays are sufficiently convergent to permit effective numerical evaluation.

The principal result of the present analysis is the representation (11) for the potential of the doubly-periodic array, in terms of a sum of the simpler singly-periodic array and its images. The practical value of this construction results from the analytic simplicity of the singly-periodic array. Most significantly, the logarithmic behavior of the singly-periodic array can be analysed separately and summed over the image system in closed form, corresponding in effect to a two-dimensional far-field solution in planes normal to the array axis. After subtracting this two-dimensional part, the remaining components of the image system are exponentially convergent, and can be summed directly. Another advantage of the construction in terms of the singly-periodic array is the fact that it is amenable to efficient computation, using complementary algorithms such as the eigenfunction expansion (9) and economized polynomial approximation (14).

This procedure has been tested extensively in applications to problems of ship hydrodynamics in rectangular canals, using discretized boundary integral equations to solve for the velocity potential. The computational domain can be reduced to the submerged surface of the ship(s) if a Green function is used which satisfies the appropriate boundary conditions on the canal walls and bottom (and on the free surface, which is simplified in this context to a rigid flat boundary). This is in fact the motivation for the present work. In a typical application involving the interactions between two ships, each discretized with 400 panels, a linear system of dimension 800 must be set up and solved. The set-up requires the evaluation of the Green function and its gradient, for each of the $(800)^2$ combinations of the source and field point, corresponding to each element of the coefficient matrix in the linear system. These computations must be repeated for a sequence of different relative positions of the two ships, to fully describe their interactions as a function of time if one ship is passing the other. The total number of Green function evaluations is in excess of 10^7 in this case, and efficiency is clearly important. Further details of this application are presented in a separate paper [4].

In applications where Dirichlet boundary conditions are applicable the corresponding Green functions can be derived by superposition of appropriate pairs of the functions considered here, with opposite signs. Alternatively, the singly-periodic array with one

Neumann and one Dirichlet boundary condition can be evaluated from the corresponding results in [2]. (In this context a correction is required in equation (7) of [2], replacing the plus sign before the last term by $\pm$.)

Acknowledgement

This work has been supported by the National Science Foundation, Grant CTS-8921011.

References

1. S.R. Breit, The potential of a Rankine source between parallel planes and in a rectangular cylinder, *J. Engg. Math.* 25 (1991) 151–163.
2. J.N. Newman, The approximation of free-surface Green functions, Meeting in honour of Professor Fritz Ursell, University of Manchester, in *Wave Asymptotics*, Cambridge University Press (1991).
3. I.S. Gradshteyn and I.M. Ryzhik, *Tables of Integrals, Series and Products*, Academic Press, New York (1965).
4. F.T. Korsmeyer, C.-H. Lee and J.N. Newman, The computation of ship-interaction forces in restricted waters, submitted to *J. Ship Research*.

Journal of Engineering Mathematics **26**: 61–69, 1992.
H.K. Kuiken and S.W. Rienstra (eds), Problems in Applied, Industrial and Engineering Mathematics.

A note on the Kutta condition in Glauert's solution of the thin aerofoil problem

S.W. RIENSTRA

Dept. of Mathematics and Computing Science, Eindhoven University of Technology, P.O. Box 513, 5600 MB Eindhoven, The Netherlands

Abstract. Glauert's classical solution of the thin aerofoil problem (a coordinate transformation, and splitting the solution into a sum of a singular part and an assumed regular part written as a Fourier sine series) is usually presented in textbooks on aerodynamics without a great deal of attention being paid to the rôle of the Kutta condition. Sometimes the solution is merely stated, apparently satisfying the Kutta condition automatically. Quite often, however, it is misleadingly suggested that it is by the choice of a *sine* series that the Kutta condition is satisfied.

It is shown here that if Glauert's approach is interpreted in the context of generalised functions, (1) the whole solution, i.e. both the singular part and any non-Kutta condition solution, can be written as a sine-series, and (2) it is really the coordinate transformation which compels the Kutta condition to be satisfied, as it enhances the edge singularities from integrable to non-integrable, and so sifts out solutions not normally representable by a Fourier series.

Furthermore, the present method provides a very direct way to construct other, more singular solutions.

A practical consequence is that (at least, in principle) in numerical solutions based on Glauert's method, more is needed for the Kutta condition than a sine series expansion.

Introduction

Incompressible inviscid stationary two-dimensional aerodynamics is a well-established discipline of fluid mechanics ([1, . . . , 19]). The basic problem of a solid body in a uniformly moving medium is described mathematically by the much studied Laplace's equation together with suitable boundary conditions. A great variety of solutions and methods of solution are known, among which the most important are those based on the construction of an equivalent flow pattern via a distribution of elementary sources such as monopoles, dipoles, and vortices.

A classic example is the thin aerofoil where the perturbation to an otherwise uniform mean flow is described by a vortex sheet (of strength to be determined) positioned along the aerofoil camberline (thickness is ignored). Via the law of Biot and Savart the induced velocity field is determined. The condition of a vanishing normal velocity component at the aerofoil surface, together with a linearisation (the planar wing approximation) using a small mean camber and small angle of attack yields an integral equation for the vorticity distribution, known as the aerofoil equation.

The physically acceptable solutions to this equation have at most a square root singularity at the edges. As the effect of viscosity is excluded from the model the solution will not be unique without an additional condition. This condition is usually taken to be the level of singularity at the trailing edge. The most common assumption then is the flow being smooth near the trailing edge (Kutta condition), which amounts to a vanishing vorticity distribution.

The aerofoil equation is studied thoroughly (Tricomi [20]) and analytical solutions are known. A well-known elegant analytical solution, cited in many textbooks on aerodynamics, is Glauert's Fourier sine series expansion in a transformed variable (Glauert [1]). The steps taken in this approach are clearly inspired by physical intuition, and motivated by the fact that they lead in an ingenious way to a construction of the solution.

Considering the method in more detail, however, it appears that the Kutta condition is nowhere explicitly applied, so this condition must somehow be inherent in one of the steps taken. Although not stated by Glauert himself, many authors suggest in their presentation of the method that the Kutta condition is applied via the choice of the sine series ([10, . . . , 19]). At the origin (~ the trailing edge) the sine series vanishes irrespective of the coefficients, so the Kutta condition is "satisfied". This is, however, only true pointwise. The series tends continuously to zero at the origin only if it converges uniformly. This is a property which, in general, can only be verified after having obtained the solution.

The real reason that Glauert's series solution indeed satisfies the Kutta condition is that the other singular solutions have no expansion of the proposed type: the coordinate transformation changes the square root singularity into a non-integrable singularity. This suggests at the same time that solutions represented by divergent series in the context of generalised functions, will include these other solutions.

In the present note we will try to show how (i) Glauert's method in the context of generalised functions solves the aerofoil equation very directly and completely, and that (ii) Glauert's original solution satisfies the Kutta condition because of the (tacit) assumption of allowing only normally converging series expansions. A direct practical consequence is that, at least in principle, great care is necessary when applying numerical methods based on Glauert's series expansion in related but more complex lifting surface problems. It is *not* sufficient for the Kutta condition merely to pick a series which converges only pointwise to zero at the trailing edge, ([21, . . . , 29]).

The Kutta condition in Glauert's method

Glauert's approach consists of the following steps (Batchelor's notation [2]): (i) a new variable θ is introduced according to $x = \frac{1}{2}c(1 - \cos\theta)$, (ii) an anticipated singular part, $\tan(\frac{1}{2}\theta)$, of the vorticity distribution Γ is taken apart (the solution of the flat plate problem), (iii) the rest is expanded into a Fourier sine series:

$$\Gamma = A_0 W \tan(\tfrac{1}{2}\theta) + W \sum_{n=1}^{\infty} A_n \sin(n\theta)\,, \tag{1}$$

which is (iv), after integration, equated term by term to the respective series expansion of the aerofoil shape function. For smooth shape functions the Kutta condition is automatically satisfied, and the eigensolution, related to the generated circulation, does not appear.

It appears to be widely believed that step (i) and (ii) are just for convenience, while step (iii) is employed to satisfy the Kutta condition $\Gamma = 0$ at $\theta = 0$, by the argument that "the sum is zero because every term is zero". Now this may be true pointwise, but for the Kutta condition this is meaningless if Γ is not continuous at and near $\theta = 0$. A representation of a function by a Fourier series is not necessarily pointwise, and we cannot prescribe the behaviour of Γ *near* $\theta = 0$ by its value *at* $\theta = 0$. This is obviously illustrated by the example

$$\sum_{n=1}^{\infty} \frac{\sin(n\theta)}{\sqrt{n}}\,, \tag{2}$$

which is zero at $\theta = 0$ but behaves like $\sim\theta^{-1/2}$ for $\theta \to 0$. A Fourier series is continuous only if it converges uniformly. Therefore, the choice of a sine series is immaterial for a Kutta condition.

The real reason why the Kutta conditions is, indeed, satisfied, is step (i). This step transforms an integrable singularity (of order $x^{-1/2}$) into a non-integrable singularity (of order θ^{-1}). Thus the (tacit) assumption that the resulting Fourier series should converge only in ordinary sense excludes the divergent series representations of θ^{-1}, which would otherwise have appeared. (Indeed, under this assumption it is necessary to separate the leading edge singularity by means of the $\tan(\frac{1}{2}\theta)$-term.)

If a purely analytical solution is possible it will be produced by the above procedure. However, if this procedure is part of a numerical approach, complications may arise. Suppose we have a complicated lifting surface problem which has to be solved numerically, and we apply the above transformation, isolate the $\tan(\frac{1}{2}\theta)$ term, and expand the rest into a sine series, together with (for example) a collocation procedure. In this case the series expansion is necessarily finite, and, without further precautions, there is no reason why a divergent part would not be generated, for example, the eigensolution, or, more generally, any more singular non-physical solutions. So the solution is still not unique, and may become dependent on details of the numerical scheme. See for example [30].

It may be noted that there is an analogy with the expansion in duct modes of acoustic or electromagnetic waves in wave guides. Suppose we try to solve the bifurcated waveguide problem by means of the technique of mode matching ([31], p. 30). Continuity conditions for the field at the interfaces lead to an infinite set of simultaneous equations for the unknown modal amplitudes. For a physically acceptable solution, the field must remain integrable near the edges. This *edge condition* is reflected in the convergence rate of the modal amplitudes ([31], p. 34). If the infinite set of equations is truncated, it appears that any of the mathematically possible solutions may be selected. Depending on the way of truncation the solutions converge to different sets of amplitudes with different asymptotic behaviour, corresponding to different edge behaviour. The correct way of truncation is determined by the required edge behaviour.

In the next chapter we adapt the techniques of Glauert's transformation and series expansion to the setting of generalised functions. From this we derive the general solution of the aerofoil equation and show that both the term $\tan(\frac{1}{2}\theta)$ and the eigensolution appear naturally. Furthermore, the same method will produce every other more singular (and hence non-physical) solution to the original problem without any difficulty.

Analysis

Consider a thin cambered aerofoil (Fig. 1), described by its camberline $y = \eta(x)$, $0 \leq x \leq c$. This aerofoil perturbs a uniform stream $\mathbf{W} = (-W \cos\alpha,\ W \sin\alpha)$ by an amount $\mathbf{v}$. Assume the angle of incidence α, the aerofoil shape η and its derivative $d\eta/dx$ are small enough for linearisation. Then the aerofoil can be represented by a vortex sheet along $0 \leq x \leq c$ of strength $\Gamma(x)$. This gives

$$\mathbf{v} = \frac{1}{2\pi} \int_0^c \frac{(-y,\ x - \xi)}{(x - \xi)^2 + y^2}\, \Gamma(\xi)\, d\xi\ . \tag{3}$$

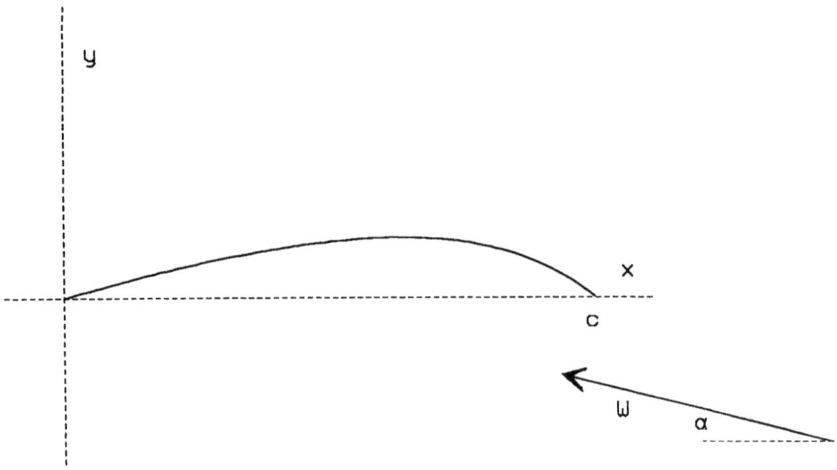

Fig. 1. Camberline $\eta(x)$ of thin aerofoil in stream W.

The boundary condition of the normal component of velocity $\mathbf{W}+\mathbf{v}$ vanishing at the aerofoil surface yields, after linearisation, the aerofoil equation

$$\frac{1}{2\pi W} ⨍_0^c \frac{\Gamma(\xi)}{x-\xi}\,\mathrm{d}\xi = -\alpha - \frac{\mathrm{d}\eta(x)}{\mathrm{d}x}\,, \quad 0<x<c\,, \tag{4}$$

where $⨍$ is to be interpreted as a Cauchy principal value. The unknown function here is Γ. This is to be determined under the additional conditions that it is integrable near $\xi=0$ and $\xi=c$ (finite energy) and furthermore that it vanishes at $\xi=0$ (Kutta condition).

The transformation

$$x=\tfrac{1}{2}c(1-\cos\theta)\,, \qquad \xi=\tfrac{1}{2}c(1-\cos\varphi) \tag{5}$$

yields

$$\frac{1}{2\pi} ⨍_0^\pi \frac{\gamma(\varphi)\sin\varphi}{\cos\theta-\cos\varphi}\,\mathrm{d}\varphi = v(\theta)\,, \quad 0<\theta<\pi\,, \tag{6}$$

where $\gamma(\theta)=\Gamma(x)/W$ and $v(\theta)=\alpha+\mathrm{d}\eta/\mathrm{d}x$. If Γ acts like $x^{-1/2}$ near the edges this is transformed into θ^{-1} behaviour of γ, but of course in the integration this is compensated by the factor $\sin\varphi$.

Up to this point we have closely followed Glauert [1]. We now deviate slightly from the way the method is usually presented. We introduce Fourier sine and cosine series for γ and v, but now without taking apart any foreseen singularities.

Continue γ antisymmetrically and v symmetrically on $(-\pi,\pi)$, and extend both 2π-periodically. Assume the Fourier expansions of γ and v are

$$\gamma(\theta)=\sum_{n=1}^{\infty}\gamma_n\sin(n\theta)\,, \tag{7}$$

$$v(\theta)=\sum_{n=0}^{\infty}v_n\cos(n\theta)\,. \tag{8}$$

Of course, this continuation is only for convenience. It is possible to write any solution for which a Fourier series can be defined in this way. The sine series is certainly not chosen in order that the Kutta condition be fulfilled.

Substitution of (7) and (8) into (6), assuming that integration and summation may be exchanged (which is consistent with the assumption of the existence of a Fourier series of γ), and then using (one of) the famous Glauert integrals ([1])

$$\int_0^\pi \frac{\sin(n\varphi)\sin\varphi}{\cos\varphi-\cos\theta}\,\mathrm{d}\varphi = -\pi\cos(n\theta)\,, \qquad \int_0^\pi \frac{\cos(n\varphi)}{\cos\varphi-\cos\theta}\,\mathrm{d}\varphi = \pi\,\frac{\sin(n\theta)}{\sin\theta} \tag{9}$$

yields

$$\tfrac{1}{2}\sum_{n=1}^{\infty}\gamma_n\cos(n\theta) = \sum_{n=0}^{\infty} v_n\cos(n\theta)\,, \quad 0<\theta<\pi\,. \tag{10}$$

Since in general $v_0 \neq 0$, equation (10) cannot be solved by any sequence (γ_n) in the ordinary sense. So it would seem that γ cannot be expanded into a Fourier series. As we will see, this is due to the non-integrable behaviour $\sim\theta^{-1}$ and $\sim(\pi-\theta)^{-1}$ near the edges. So the transformation (5) together with the assumption of a Fourier series acts like a sieve allowing no solutions at all. The usual way to remedy this is to take the singularities apart (the $\tan(\frac{1}{2}\theta)$ in eq. (1)), however this is not necessary. A more direct way is to identify solutions with generalised functions ([32, 33]), thus allowing a much wider class of Fourier series expansions.

The question now is how to make use of this greater freedom. We are free to add or subtract generalised functions provided we still satisfy equation (10). We then examine their effect on the Fourier series. Adding zero does not change anything since even in generalised sense the Fourier coefficients of the zero function are zero ([33], p. 58). However, since equation (10) need only be satisfied on the *open* intervals $0<\theta \bmod \pi<\pi$ (i.e. excluding the end points), we may add generalised functions whose support is concentrated at the integer multiples of π without affecting the general solution. Now, it is a standard theorem in generalised functions (Jones [32], p. 89) that a generalised function has support concentrated at the origin only if it is a finite linear combination of derivatives of delta functions. More explicitly, if

$$f(x)=0 \quad \text{for } x>0 \text{ and } x<0$$

then

$$f(x) = \sum_{n=0}^{N} a_n\delta^{(n)}(x) \quad \text{for any } a_n \text{ and finite } N\,. \tag{11}$$

This means that we can add multiples of the 2π-periodic δ-functions of θ and $\theta-\pi$, and its derivatives, to the right-hand side of (10). Their generalised Fourier expansions ([32], p. 153) are given by

$$2\pi\sum_{m=-\infty}^{\infty}\delta^{(k)}(\theta-2\pi m) = \sum_{n=-\infty}^{\infty}(\mathrm{i}n)^k\,\mathrm{e}^{\mathrm{i}n\theta} \tag{12}$$

(similarly for $\theta =: \theta - \pi$). For the moment, however, we will restrict ourselves to the δ-function itself, since the derivatives lead to solutions too singular at the edges. So to equation (10) we add

$$2\pi \sum_{m=-\infty}^{\infty} \delta(\theta - 2\pi m) = 1 + 2 \sum_{n=1}^{\infty} \cos(n\theta)$$

A times, and

$$2\pi \sum_{m=-\infty}^{\infty} \delta(\theta - \pi - 2\pi m) = 1 + 2 \sum_{n=1}^{\infty} (-1)^n \cos(n\theta)$$

B times to obtain

$$\tfrac{1}{2} \sum_{n=1}^{\infty} \gamma_n \cos(n\theta) = v_0 + A + B + \sum_{n=1}^{\infty} (v_n + 2A + 2(-1)^n B) \cos(n\theta) .$$

This equation has solution

$$\begin{aligned} B &= -v_0 - A , \\ \gamma_n &= 2v_n - 4(-1)^n v_0 + 4(1 - (-1)^n) A . \end{aligned} \tag{13}$$

Using the generalised identities ([32], p. 155)

$$\sum_{n=1}^{\infty} \sin(n\theta) = \tfrac{1}{2} \operatorname{cotan}(\tfrac{1}{2}\theta) ,$$

$$\sum_{n=1}^{\infty} (-1)^n \sin(n\theta) = -\tfrac{1}{2} \tan(\tfrac{1}{2}\theta) ,$$

we obtain the usual form

$$\gamma(\theta) = 2v_0 \tan(\tfrac{1}{2}\theta) + 2 \sum_{n=1}^{\infty} v_n \sin(n\theta) + 4A/\sin\theta . \tag{14}$$

This is in physical coordinates

$$\begin{aligned} \Gamma(x)/W = {} & 2v_0 \frac{\sqrt{x/c}}{\sqrt{1 - x/c}} + 4\sqrt{x/c}\,\sqrt{1 - x/c} \sum_{n=1}^{\infty} v_n U_{n-1}(1 - 2x/c) \\ & + 2A \frac{1}{\sqrt{x/c}\,\sqrt{1 - x/c}} , \end{aligned} \tag{15}$$

where U_n denotes the Chebyshev polynomial of second kind and degree n, satisfying $\sin(n\theta) = \sin\theta\, U_{n-1}(\cos\theta)$ ([34]).

The factor of A in (15) is indeed the eigenfunction of the problem without Kutta condition (see [20]). For η (and so: v) smooth enough the Kutta condition is satisfied by taking $A = 0$.

Further generalisations

Motivated by the physics of the problem we have restricted ourselves until now to integrable singular behaviour of Γ. It is clear that once we have admitted generalised function solutions there is, other than physically, no restriction to the singular behaviour. We then have, in fact, an infinite number of eigensolutions. Although these eigensolutions may be of little importance practically, it is possible that they could play an unfavourable rôle in numerical solutions. We therefore include them here for the sake of completeness.

It is sufficient to consider equation (10) with zero right-hand side. With equation (12) we then have in general

$$\sum_{n=1}^{\infty} \gamma_n \cos(n\theta) = A_0 + C_0 + 2 \sum_{k=0}^{K} \sum_{n=1}^{\infty} (-1)^k n^{2k} [(A_k + (-1)^n C_k) \cos(n\theta) - n(B_k + (-1)^n D_k) \sin(n\theta)]$$

which has solutions

$$\gamma_n = 2 \sum_{k=0}^{K} (-1)^k n^{2k} (A_k + (-1)^n C_k) \tag{16}$$

if

$$A_0 + C_0 = B_k = D_k = 0 .$$

Clearly, the solution consists of linear combinations of *even* derivatives of $\tan(\frac{1}{2}\theta)$ and $\operatorname{cotan}(\frac{1}{2}\theta)$. To see this, consider for $k \geqslant 0$

$$\gamma_n = 2(-1)^k n^{2k} ,$$

then

$$\gamma(\theta) = 2 \frac{d^{2k}}{d\theta^{2k}} \sum_{n=1}^{\infty} \sin(n\theta) = \frac{d^{2k}}{d\theta^{2k}} \operatorname{cotan}(\tfrac{1}{2}\theta). \tag{17}$$

Similarly, if

$$\gamma_n = -2(-1)^{k+n} n^{2k} ,$$

then

$$\gamma(\theta) = -2 \frac{d^{2k}}{d\theta^{2k}} \sum_{n=1}^{\infty} (-1)^n \sin(n\theta) = \frac{d^{2k}}{d\theta^{2k}} \tan(\tfrac{1}{2}\theta) . \tag{18}$$

Conclusions

The classical series solution of the thin aerofoil problem developed by Glauert is cited throughout the literature. However, the rôle of the Kutta condition, together with the matter of uniqueness of the solution, sometimes seems to have escaped attention. It is either not mentioned (since, in general, the Kutta condition is satisfied), or it is claimed to be satisfied by the pointwise behaviour of the chosen sine-series (which is not correct). This incorrect interpretation has not only relevance to the original incompressible $2-D$ problem, but also to the application in more complex flow problems, to the finite wing problem where a very similar technique is used, and to numerical methods based on this type of series.

In the present note we have tried to rectify the incorrect interpretation that the choice of a Fourier-sine series is sufficient for the Kutta condition, and we have shown that in general the Kutta condition is only indirectly satisfied by excluding divergent series. By posing the problem in the context of generalised functions it is possible to handle divergent series, and the general, non-unique, solution can be found directly without further assumptions.

References

1. H. Glauert, *The Elements of Aerofoil and Airscrew Theory*. Cambridge: Cambridge University Press (1926) p. 87.
2. G.K. Batchelor, *An Introduction to Fluid Mechanics*. Cambridge: Cambridge University Press (1967) p. 467.
3. L.M. Milne-Thomson, *Theoretical Aerodynamics*, 2nd edn. London: MacMillan & Co. (1952) p. 138.
4. B. Thwaites (ed.), *Incompressible Aerodynamics*. Oxford: Clarendon Press (1960) p. 123.
5. N.E. Kochin, I.A. Kibel' and N.V. Roze, *Theoretical Hydrodynamics*. New York: Interscience Publishers (1964).
6. K. Karamcheti, *Principles of Ideal-Fluid Aerodynamics*. New York: John Wiley & Sons (1966) p. 506.
7. H. Schlichting and E. Truckenbrodt, *Aerodynamics of the Airplane*. New York: McGraw-Hill (1979).
8. L.J. Clancy, *Aerodynamics*. London: Pitman (1975) p. 172.
9. I.H. Abbott and A.E. Von Doenhoff, *Theory of Wing Sections*. New York: Dover Publications (1959) p. 66.
10. A.M. Kuethe and C.-Y. Chow, *Foundations of Aerodynamics*, 4th edn. New York: Wiley & Sons (1986) p. 116.
11. H.W. Försching, *Grundlagen der Aeroelastik*. Berlin: Springer-Verlag (1974) p. 255.
12. H. Ashley and M. Landahl, *Aerodynamics of Wings and Bodies*. Addison-Wesley (1965) p. 150.
13. N.A.V. Piercy, *Aerodynamics*, 2nd edn. London: The English University Press (1947) p. 241.
14. E.L. Houghton and A.E. Brock, *Aerodynamics for Engineering Students*, 2nd edn. London: Edward Arnold Publishers (1970) p. 317.
15. B.K. Shivamoggi, *Theoretical Fluid Dynamics*. Dordrecht: Martinus Nijhoff Publishers (1985) p. 139.
16. A.M. Kuethe and J.D. Schetzer, *Foundations of Aerodynamics*, New York: John Wiley (1950) p. 78.
17. W.J. Duncan, A.S. Thom and A.D. Young, *The Mechanics of Fluids*. London: Edw. Arnold (1960) p. 589.
18. J.D. Anderson, *Fundamentals of Aerodynamics*. London: McGraw-Hill (1984) p. 275.
19. J.H. Spurk, *Strömungslehre*. Berlin: Springer-Verlag (1987) p. 350.
20. F.G. Tricomi, *Integral Equations*. New York: Interscience Publishers (1957) p. 173.
21. M.E. Goldstein, *Aeroacoustics*. New York: McGraw-Hill (1976) p. 230.
22. G.F. Homicz and J.A. Lordi, Three-dimensional theory for the steady loading on an annular blade row. *AIAA Journal* 19 (4) (1981) 492–495.
23. S. Kaji and T. Okazaki, Generation of sound by rotor-stator interaction. *Journal of Sound and Vibration* 13 (3) (1970) 281–307.
24. N. Namba, Lifting surface theory for a rotating subsonic or transonic blade row, Aeronautical Research Council Reports and Memoranda No. 3740 (1972).
25. N. Namba, Three-dimensional analysis of blade force and sound generation for an annular cascade in distorted flows. *Journal of Sound and Vibration* 50 (4) (1977) 479–508.
26. S.M. Ramachandra, Acoustic pressures emanating from a turbo-machine stage, AIAA paper 84–2325 (9th Aeroacoustics Conference) (1984).
27. J.B.H.M. Schulten, Sound generated by rotor wakes interacting with a leaned vane stator. *AIAA Journal* 20 (10) (1982) 1352–1358.
28. F. Lane and M. Friedman, Theoretical Investigations of Subsonic Oscillating Blade-Row Aerodynamics NACA TN 4136 (1958).
29. D.S. Whitehead, Vibration and sound generation in a cascade of flat plates in subsonic flow, Aeronautical Research Council Reports and Memoranda No. 3685 (1972).

30. R.M. James, On the remarkable accuracy of the vortex lattice method. *Computer Methods in Applied Mechanics and Engineering* 1 (1972) 59–79.
31. R. Mittra and S.W. Lee, *Analytical Techniques in the Theory of Guided Waves*. New York: The MacMillan Company (1971).
32. D.S. Jones, *The Theory of Generalised Functions*, 2nd edn. Cambridge: Cambridge University Press (1982).
33. M.J. Lighthill, *Introduction to Fourier Analysis and Generalised Functions*. Cambridge: Cambridge University Press (1958).
34. M. Abramowitz and I.A. Stegun (ed.), *Handbook of Mathematical Functions*. National Bureau of Standards (1964) (22.3.16).

Journal of Engineering Mathematics **26**: 71–79, 1992.
H.K. Kuiken and S.W. Rienstra (eds), Problems in Applied, Industrial and Engineering Mathematics.

Free convection from a vertical cooling fibre

N. RILEY

School of Mathematics, University of East Anglia, Norwich, NR4 7TJ, UK

Abstract. We study the free-convective boundary-layer flow that is induced when a slender circular cylinder emerges from an orifice and moves vertically downwards. We demonstrate, by numerical solution of the equations, that the boundary-layer solution develops a singularity at a finite point, where the limiting form of solution is as predicted by Kuiken [3] for an analogous two-dimensional flow.

1. Introduction

There are many industrial processes which involve the cooling of cylinders, fibres or sheets of material. An example is provided by the manufacture of glass fibre. The fibre issues from an orifice at the bottom of a crucible containing liquid glass. The fibre cools as it moves downwards, before being rolled onto a drum. It is the free-convective flow associated with such a process that we are concerned with in this paper.

Kuiken [1, 2, 3] has been much concerned with cooling in such extrusion processes. In [1, 2] he considered the cooling of thin sheets and slender cylinders in the absence of free-convective effects. However, in a study [3] of the cooling of a hot thin sheet moving vertically downwards he included free convection due to gravity and discovered a most interesting singular solution, of self-similar form, of the governing equations. This solution of the boundary-layer equations has the property that it decays algebraically, rather than exponentially, far from the sheet. On that account, and bearing in mind the work of Goldstein [4], and Brown and Stewartson [5] on boundary layers with algebraic decay, the role of Kuiken's solution might be thought to be that of a limit solution as the singular point is approached. However Khan and Stewartson [6], from consideration of a full numerical solution of the governing parabolic partial differential equations, show that for this problem the similarity solution gives a remarkably accurate estimate of flow properties at the sheet over a much greater extent of it than might have been presupposed.

In the present paper we study the free-convective boundary layer on a heated, downward moving circular cylinder which is sufficiently slender to model a fibre. Although there is no analogue of Kuiken's self-similar solution [3] for this problem, we demonstrate that it is the leading term in a series solution that may be developed about a singular point. A full numerical solution of the governing boundary-layer equations identifies the point at which the solution breaks down, and at the same time confirms Kuiken's result as the limiting form of the solution at that point.

2. Problem formulation

We consider the axisymmetric, laminar, free-convective flow about a thin, hot vertical cylinder or fibre of circular cross-section and diameter $2a$. The cylinder moves with speed u_s

in the direction of the gravity vector **g**, emerging from an orifice with excess temperature T_e over the ambient temperature T_∞, and disappearing into another orifice with an excess temperature $T_1 < T_e$ which is unknown *a priori*. The exposed length of the cylinder is L. With reference to Fig. 1 we choose cylindrical polar co-ordinates (x', r'), with origin at the lower orifice, and corresponding velocity components (u', v'). We write the excess temperature as $T' + T_\infty$. If we assume that the thickness of the moving region of fluid adjacent to the cylinder is small compared to L, then the boundary-layer equations expressing conservation of mass, momentum and energy, for the problem we have defined, are

$$\frac{\partial u'}{\partial x'} + \frac{\partial v'}{\partial y'} + \frac{v'}{a+y'} = 0\,, \tag{2.1}$$

$$u'\frac{\partial u'}{\partial x'} + v'\frac{\partial u'}{\partial y'} = \nu\left(\frac{\partial^2 u'}{\partial y'^2} + \frac{1}{a+y'}\frac{\partial u'}{\partial y'}\right) + g\beta T'\,, \tag{2.2}$$

$$u'\frac{\partial T'}{\partial x'} + v'\frac{\partial T'}{\partial y'} = \frac{\nu}{\sigma}\left(\frac{\partial^2 T'}{\partial y'^2} + \frac{1}{a+y'}\frac{\partial T'}{\partial y'}\right) \tag{2.3}$$

where $y' = r' - a$, ν is the kinematic viscosity of the fluid and β its coefficient of thermal expansion; σ is the Prandtl number. We suppose that the temperature of the moving cylinder, or fibre, is constant across its cross-section with excess value $T_s(x')$. The boundary

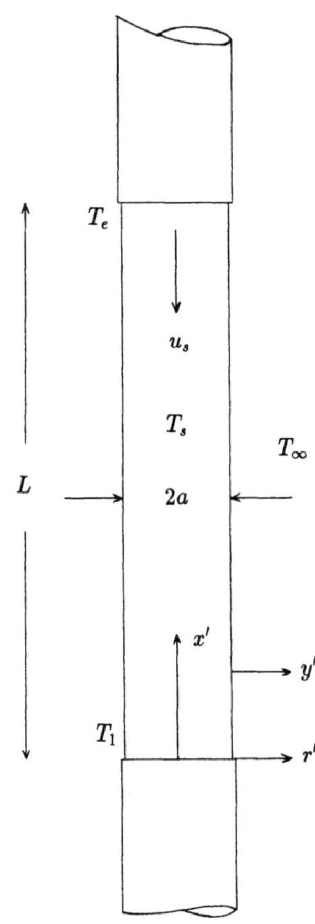

Fig. 1. Definition sketch.

conditions for our equations (2.1) to (2.3) are, then,

$$\left.\begin{array}{l} u' = -u_s\,, \quad v' = 0\,, \quad T' = T_s(x')\,; \quad y' = 0\,, \quad x' \geqslant 0\,, \\ u',\, T' \to 0 \text{ as } y' \to \infty\,, \quad x' \geqslant 0\,, \\ u',\, T' = 0\,; \quad x' = 0\,, \quad y' > 0\,. \end{array}\right\} \tag{2.4}$$

Since the cylinder temperature $T_s(x')$ is not known *a priori* an additional condition is required. This is determined from a consideration of the heat balance in the cylinder itself. Between two fixed stations the net flux of heat carried into the section, by the cylinder's own motion, is balanced by a transport of heat across the curved surface of the cylinder by conduction. This may be expressed mathematically as

$$\frac{\partial T'}{\partial x'} + \lambda \frac{\partial T'}{\partial y'} = 0 \quad \text{at } y' = 0\,, \tag{2.5}$$

with $\lambda = 2\rho c \nu / \sigma a \rho_s c_s u_s$, where ρ is the density, c the specific heat and a suffix s refers to the solid material. In (2.5) $\partial T'/\partial y'|_{y'} = 0$ is calculated from the solution of (2.3).

We now introduce dimensionless variables by writing $x' = lx$, $y' = ay$, $u' = U'u$, $v' = \varepsilon U'v$, $T' = T_1 T$. Here $l = O(L)$ is an axial length scale, $\varepsilon = a/l \ll 1$, and for the velocity scale we choose $U' = (g\beta T_1 l)^{1/2}$ for this free-convection dominated problem. The equations (2.1) to (2.3) and the condition (2.5) may then be written in non-dimensional form, respectively, as

$$\frac{\partial u}{\partial x} + \frac{\partial v}{\partial y} + \frac{v}{1+y} = 0\,,$$

$$u\frac{\partial u}{\partial x} + v\frac{\partial u}{\partial y} = \frac{1}{\varepsilon\, \mathrm{Gr}^{1/2}}\left(\frac{\partial^2 u}{\partial y^2} + \frac{1}{1+y}\frac{\partial u}{\partial y}\right) + T\,,$$

$$u\frac{\partial u}{\partial x} + v\frac{\partial T}{\partial y} = \frac{1}{\varepsilon\, \mathrm{Gr}^{1/2}\sigma}\left(\frac{\partial^2 T}{\partial y^2} + \frac{1}{1+y}\frac{\partial T}{\partial y}\right),$$

$$\frac{\partial T}{\partial x} + \frac{\lambda}{\varepsilon}\frac{\partial T}{\partial y} = 0\,,$$

where $\mathrm{Gr} = g\beta T_1 l a^2/\nu^2$ is a Grashof number, with $\mathrm{Gr} \gg 1$ for consistency with our framework of boundary-layer theory. Unlike the two-dimensional situation considered by Khan and Stewartson [6], and the two-dimensional equations may be derived from (2.1) to (2.3) by formally allowing $a \to \infty$, our non-dimensional equations do not appear to be reducible to a suitable canonical form. For the purposes of the present paper, as outlined in Section 1, we proceed as follows. We make the approximation of unit Prandtl number, which is acceptable for air. Also, since in our formulation $\varepsilon \ll 1$ and $\mathrm{Gr} \gg 1$ we take $\varepsilon\, \mathrm{Gr}^{1/2} = 1$. Further since $\lambda \ll 1$, and indeed is required to be so if our boundary-layer theory is to apply as we see from (2.5), we also take $\lambda/\varepsilon = 1$. The values we have taken are representative of a practical situation, and the results we obtain below will only be modified in detail if other values for these parameters are chosen. We make one further simplification which is that $u_s \ll U'$; that is we assume the fibre's downward velocity to be small compared with the induced free-convective velocity in the surrounding medium. This assumption will be justified in practice except close to the origin $x = 0$.

With the above assumptions the problem we address is as follows:

$$\frac{\partial u}{\partial x} + \frac{\partial v}{\partial y} + \frac{v}{1+y} = 0\,, \tag{2.6}$$

$$u\,\frac{\partial u}{\partial x} + v\,\frac{\partial u}{\partial y} = \frac{\partial^2 u}{\partial y^2} + \frac{1}{1+y}\,\frac{\partial u}{\partial y} + T\,, \tag{2.7}$$

$$u\,\frac{\partial T}{\partial x} + v\,\frac{\partial T}{\partial y} = \frac{\partial^2 T}{\partial y^2} + \frac{1}{1+y}\,\frac{\partial T}{\partial y}\,, \tag{2.8}$$

together with

$$\left.\begin{aligned} &u = v = 0\,, \qquad \frac{\partial T}{\partial x} + \frac{\partial T}{\partial y} = 0 \quad \text{at } y = 0\,, \quad x > 0\,,\\ &u \to 0\,, \quad T \to 0 \quad \text{as } y \to \infty \text{ for } x > 0\,,\\ &u = 0\,, \quad T = 1 \quad \text{at } x = 0\,, \quad y = 0\,,\\ &u = T = 0 \quad \text{at } x = 0\,, \quad y > 0\,. \end{aligned}\right. \tag{2.9}$$

We remark finally that, following the solution of (2.6) to (2.9), the temperature T_1 of the fibre at $x = 0$ is determined in terms of the exit temperature T_e at $x_e = L/l$ as $T_1 = T_e/T(x_e, 0)$.

3. Numerical results

For the analogous problem of the cooling of a vertical two-dimensional thin film Kuiken [3] has obtained a remarkable similarity solution of the equations analogous to (2.6) to (2.9). The solution is remarkable in that it is singular at a finite point $x = x_0 > 0$ on the film. As that point is approached the temperature and vertical velocity increase without bound, with the thickness of the boundary layer diminishing to zero. For the problem posed by equations (2.6) to (2.9) there appears to be no similarity solution. However, the solution of Kuiken can be used as the leading term in the development of a singular solution close to the point $x = x_0$. Thus, we write

$$\left.\begin{aligned} u &= \frac{\mu^{1/2}}{x_0 - x} \sum_{n=0}^{\infty} (x_0 - x)^n f_n'(\eta)\,,\\ v &= \frac{\mu^{1/4}}{x_0 - x} \sum_{n=0}^{\infty} (x_0 - x)^n (n f_n - \eta f_n')\,,\\ T &= \frac{\mu}{(x_0 - x)^3} \sum_{n=0}^{\infty} (x_0 - x)^n \theta_n(\eta)\,, \end{aligned}\right\} \tag{3.1}$$

where

$$\eta = \frac{\mu^{1/4} y}{x_0 - x}\,, \tag{3.2}$$

and μ is a constant to be determined. The leading terms of (3.1) provide the Kuiken similarity solution, with f_0, θ_0 satisfying

$$f_0''' - f_0'^2 + \theta_0 = 0\,, \qquad \theta_0'' - 3\sigma f_0'\theta_0 = 0\,, \tag{3.3}$$

where a prime denotes differentiation with respect to η. Kuiken solved these equations numerically subject to the conditions $f_0(0) = f_0'(0) = 0$, $\theta_0(0) = 1$, $\theta_0(\infty) = f_0'(\infty) = 0$ and obtained, in particular, for $\sigma = 1$ the results

$$f_0''(0) = 0.69321\,, \qquad \theta_0'(0) = -0.76986\,, \qquad f_0(\infty) = 2.43998\,. \tag{3.4}$$

The constant μ in (3.1), (3.2) is then obtained from the heat flow condition in (2.9) as

$$\mu^{1/4} = \{-\tfrac{1}{3}\theta'(0)\}^{-1} \quad \text{or} \quad \mu = 2.30589 \times 10^2\,. \tag{3.5}$$

A feature of the solution of (3.3), noted by Kuiken is that the decay of f_0', θ_0 to zero as $\eta \to \infty$ is algebraic rather than exponential. Such behaviour of solutions of the boundary-layer equations had been observed by Goldstein [4] who cast doubt upon their value, since solutions which exhibit such behaviour over a finite length cannot be satisfactorily matched to an inviscid potential flow. However Brown and Stewartson [5] subsequently showed that algebraic decay is usually only encountered at one streamwise point, and is associated with non-commutative limits. As a consequence Kuiken conjectured that his solution should perhaps be interpreted as a limit solution of the boundary-layer equations as $x \to x_0$. This conjecture provided the starting point for the investigation by Khan and Stewartson [6]. They not only verified Kuiken's conjecture, with $x_0 = 6.01252$, but in addition showed that the similarity solution predicts conditions close to the boundary very accurately for $x \geqslant 2$. Less good is a comparison with the displacement thickness which represents the overall structure of the boundary layer and its previous history. We may conclude that the similarity solution is indeed a limit solution in the inner part of the boundary layer, as originally conceived by Kuiken, but that it has a wider range of validity than might have been supposed, since the situation in the inner part of the layer tends to be dominated by local conditions.

For the axisymmetric problem under consideration there is no comparable similarity solution but, as we have already indicated in (3.1), (3.2), the Kuiken solution appears to dominate the flow as a singular point x_0 is approached. The principal aim of the present paper is to demonstrate that this is indeed the case. To achieve this aim we integrate our governing equations (2.6) to (2.9) from $x = 0$ in the direction of x increasing, until a singular behaviour is encountered. To enable this we introduce new co-ordinates which are sympathetic, not only to the nature of the singularity as $x \to x_0$ but also to the essential singularity at $x = 0$. Close to $x = 0$ the free-convective boundary layer behaves like that on a heated semi-infinite plate with a thickness $O(x^{1/4})$, and streamwise velocity $O(x^{1/2})$. Thus we introduce co-ordinates X, Y with

$$x = \frac{3}{4} X^{4/3}\,, \qquad y = \frac{X^{1/3}(X_0 - X)}{X_0}\, Y\,, \tag{3.6}$$

where X_0, corresponding to x_0, remains to be determined. The velocity components and temperature are then transformed as

$$u=\frac{X_0X^{2/3}}{X_0-X}\,U(X,Y)\,,\qquad V=\frac{X_0}{X^{1/3}(X_0-X)}\,V(X,Y)\,,\qquad T=\frac{X_0^3}{(X_0-X)^3}\,\Theta(X,Y)\,. \tag{3.7a,b,c}$$

From (2.6) to (2.9), (3.6) and (3.7) we then have, with

$$\left.\begin{aligned}&F_1(X)=\frac{2X_0+X}{3X_0}\,,\qquad F_2(X)=\frac{X(X_0-X)}{X_0}\,,\qquad F_3(X,Y)=\left(\frac{4X-X_0}{3X_0}\right)Y\,,\\&F_4(X,Y)=\frac{X^{1/3}(X_0-X)}{X_0+X^{1/3}(X_0-X)Y}\,,\qquad F_5(X)=\frac{3X}{X_0}\,,\end{aligned}\right\} \tag{3.8}$$

$$F_2\,\frac{\partial U}{\partial X}+F_3\,\frac{\partial U}{\partial Y}+F_1U+\frac{\partial V}{\partial Y}+F_4V=0\,, \tag{3.9}$$

$$\frac{\partial^2U}{\partial Y^2}-F_2U\,\frac{\partial U}{\partial X}+(F_4-F_3U-V)\,\frac{\partial U}{\partial Y}-F_1U^2+\Theta=0\,, \tag{3.10}$$

$$\frac{\partial^2\Theta}{\partial Y^2}-F_2U\,\frac{\partial\Theta}{\partial X}+(F_4-F_3U-V)\,\frac{\partial\Theta}{\partial Y}-F_5U\Theta=0\,, \tag{3.11}$$

together with

$$\left.\begin{aligned}&U=V=0\,,\quad 3\Theta+(X_0-X)\,\frac{\partial\Theta}{\partial X}+X_0\,\frac{\partial\Theta}{\partial Y}=0\,;\quad Y=0\,,\;X>0\,,\\&U\to0\,,\;\Theta\to0\,;\;Y\to\infty\,,\;X>0\,,\\&U=0\,,\;\Theta=1\,;\;X=Y=0\,,\\&U=\Theta=0\,;\;X=0\,,\;Y>0\,.\end{aligned}\right\} \tag{3.12}$$

If we set $X\equiv0$ then equations (3.9) to (3.12) become identical with those for flow over a semi-infinite heated flat plate, and their numerical solution provides an initial solution for step-by-step integration of the full equations. To advance the solution beyond $X=0$ in such a manner we proceed as follows. First we quasi-linearise (3.10). Then, with the solution known at $X=X_i$, X_{i-1} etc., we provide an estimate of U, Θ at $X=X_{i+1}$ by extrapolation from previous stations, and gain a first estimate of V from the solution of (3.9). The quasi-linearised equation (3.10) then provides an updated solution for U. In obtaining this we do not seek a fully converged, but only a partially converged, solution of (3.10). Finally we update our estimate of Θ by solving (3.11). This iterative scheme, through equations (3.9) to (3.11), is repeated until overall convergence according to some pre-set criterion is achieved. The method of solution adopted is a finite-difference method in which all derivatives in (3.9) to (3.11), and the heat-flow condition in (3.12), are approximated by central differences. Thus, second-order accuracy is achieved, and the solution of (3.10), (3.11) is simply an adaptation of the well-known Crank–Nicolson method. For the results presented in this paper we have set step lengths $\delta X=0.05$, $\delta Y=0.1$, and placed the outer boundary of the computational domain at $Y=Y_\infty=60$. These values provide the necessary level of accuracy for our purposes.

A feature which complicates the procedure described above is the introduction, in the transformations (3.6) and (3.7), of the parameter X_0 which is unknown *a priori*. In order to implement our step-by-step procedure an estimate, X_0^e say, of this quantity must be made. If our estimate $X_0^e > X_0$ then, assuming the Kuiken solution is indeed appropriate, the solution will clearly exhibit a singular behaviour with, *inter alia*, $\Theta(X,0) \to \infty$ as X approaches the unknown value X_0. On the other hand, if $X_0^e < X_0$, T remains finite as $X \to X_0^e$ and as a consequence, as we see from (3.7c), Θ and indeed $\partial^2\Theta/\partial X^2$ both approach zero. When X_0 is identified correctly Θ remains finite and smooth as X approaches X_0. With these considerations in mind it is not difficult to estimate X_0, and for the step length $\delta X = 0.05$ we readily find that $3.55 < X_0 < 3.60$. Proceeding in this way, with smaller step lengths δX, would enable us to refine this estimate. However, a consequence of the procedure described above is that Kuiken's solution is indeed emerging as a candidate for the limiting solution as $X \to X_0$. We therefore exploit that as follows. If we define a skin friction coefficient $C_f = \partial u/\partial y|_{y=0}$, then from (3.1), (3.2), (3.4) and (3.5) we have, close to $x = x_0$,

$$\left.\begin{aligned} T^{-1/3}(x,0) &= 0.163075(x_0 - x) + O\{(x_0 - x)^2\}\,, \\ C_f^{-1/2}(x) &= 0.156136(x_0 - x) + O\{(x_0 - x)^2\}\,. \end{aligned}\right\} \qquad (3.13)$$

If now, in the course of our numerical solution of (3.9) to (3.11), we tabulate the quantities shown in, and make a comparison with, equations (3.13) we can without difficulty estimate X_0 and hence x_0 to a high degree of accuracy. In this way we find

$$X_0 = 3.58634 \quad \text{and} \quad x_0 = 4.11715\,. \qquad (3.14)$$

In Figs 2a and 2b we compare (3.13) with the corresponding quantities obtained from our numerical solution, with x_0 as in (3.14). The agreement is excellent and provides striking confirmation that Kuiken's solution is indeed the appropriate limiting solution as $x \to x_0$. To substantiate this further we shown in Figs 3a and 3b temperature and velocity profiles to demonstrate that these do indeed approach the limiting forms appropriate to Kuiken's solution.

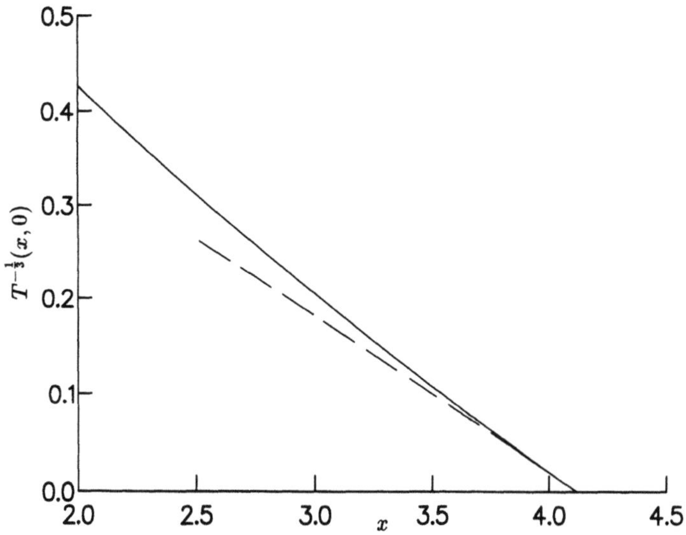

Fig. 2(a). A comparison between the wall temperature obtained from the full solution and the asymptotic result (3.13), shown as a broken line.

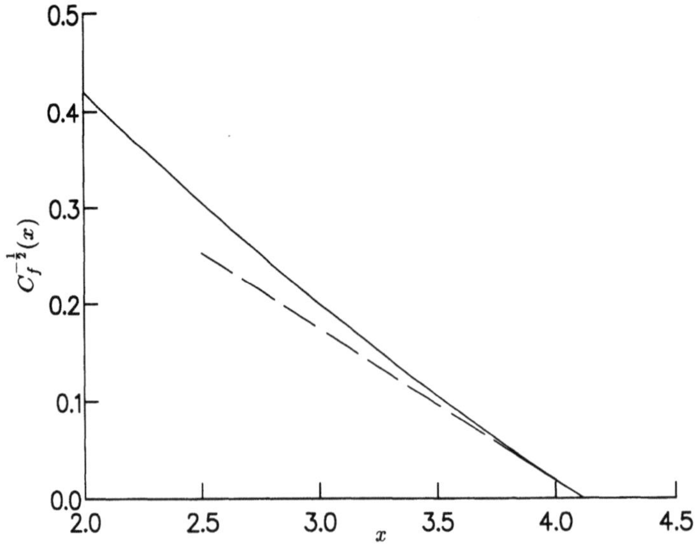

Fig. 2(b). A comparison between the skin friction coefficient C_f from the full solution and the asymptotic result (3.13), shown as a broken line.

4. Discussion

In the foregoing we have demonstrated that the singular, self-similar solution introduced by Kuiken [3], in association with the cooling of a vertically moving thin film has an important role to play in the solution of the axisymmetric boundary-layer equations for a hot, vertically downward moving fibre that is cooling. We have modelled the fibre as a long, thin cylinder of circular cross-section and shown that Kuiken's solution provides the leading term in the development of a singular solution in the neighborhood of the point at which the solution, of the boundary-layer equations, breaks down. A physical explanation for this breakdown is as

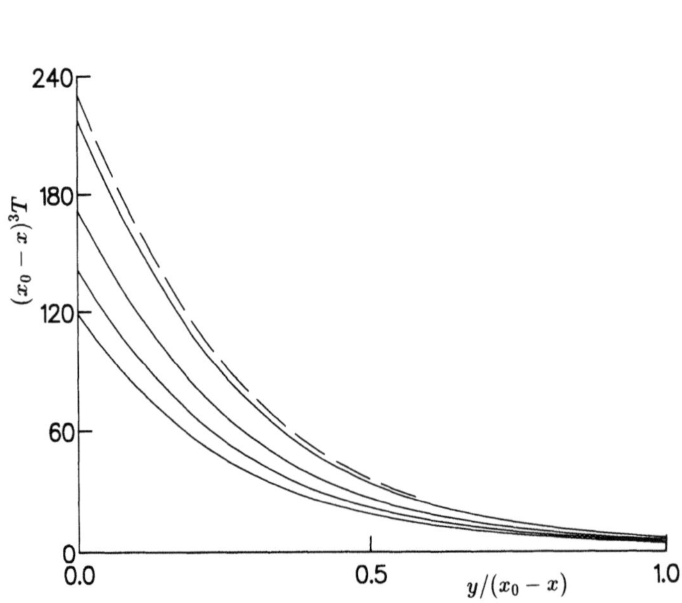

Fig. 3(a). Temperature profiles, successively as we move up, for $X = 2, 2.5, 3.0, 3.5$. The limit solution of Kuiken is shown as a broken line.

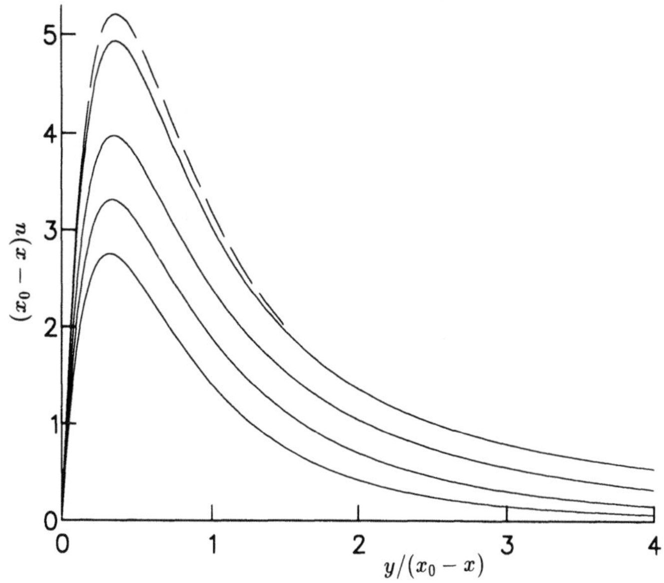

Fig. 3(b). Velocity profiles, successively as we move up, for $X = 2, 2.5, 3.0, 3.5$. The limit solution of Kuiken is shown as a broken line.

follows. Heat is carried down in the fibre from $x' = L$, and simultaneously free convective effects transport heat upwards. There is therefore an accumulation of heat which results in an acceleration of the free-convective process to the extent that the boundary-layer equations fail at some point $x' = x_0'$. As we see from equation (3.1), not only do the streamwise velocity and temperature become unbounded but so, also, does streamwise diffusion of vorticity and heat. In that case the boundary-layer equations become inappropriate to describe the flow. Whether or not such a breakdown does occur, leading to enhanced convection and therefore augmented cooling of the fibre, depends upon the length L of the exposed fibre. It is self-evident that the shorter the length of the exposed fibre the less effective is the cooling process. But what has been shown in the present investigation is that for a sufficient length of exposed fibre the free-convective, and hence cooling, process can be significantly enhanced.

References

1. H.K. Kuiken, The cooling of a low-heat-resistance sheet moving through a fluid. *Proc. Roy. Soc. Lond.* A341 (1974) 233–252.
2. H.K. Kuiken, The cooling of a low-heat-resistance cylinder moving through a fluid. *Proc. Roy. Soc. Lond.* A346 (1975) 23–35.
3. H.K. Kuiken, A backward free-convective boundary layer. *Quart. Jl Mech. appl. Math.* 34 (1981) 397–413.
4. S. Goldstein, On backward boundary layers and flow in converging passages. *J. Fluid Mech.* 21 (1965) 33–45.
5. S.N. Brown and K. Stewartson, On similarity solutions of the boundary-layer equations with algebraic decay. *J. Fluid Mech.* 23 (1965) 673–687.
6. M.A. Khan and K. Stewartson, On natural convection from a cooling vertical sheet. *Quart. Jl Mech. appl. Math.* 37 (1984) 325–338.

Journal of Engineering Mathematics **26**: 81–86, 1992.
H.K. Kuiken and S.W. Rienstra (eds), Problems in Applied, Industrial and Engineering Mathematics.

Effective slip in numerical calculations of moving-contact-line problems

J.A. MORIARTY and L.W. SCHWARTZ*

*Department of Mechanical Engineering, University of Delaware, Newark, DE 19716, USA (*author for correspondence)*

Abstract. For many coating flows, the profile thickness h, near the front of the coating film, is governed by a third-order ordinary differential equation of the form $h''' = f(h)$, for some given $f(h)$. We consider here the case of dry wall coating which allows for slip in the vicinity of the moving contact-line. For this case, one such model equation, due to Greenspan, is $f(h) = -1 + (1+\alpha)/(h^2+\alpha)$, where α is the slip coefficient. The equation is solved using a finite difference scheme, with a contact angle boundary condition prescribed at the moving contact-line. Using the maximum thickness of the profile as the control parameter, we show that there is a direct relationship between the effective Greenspan slip coefficient and the grid-spacing of the numerical scheme used to solve the model equation. In doing so, we show that slip is implicitly built into the numerical scheme through the finite grid-spacing. We also show why converged results with finite film thickness cannot be obtained if slip is ignored.

1. Introduction

Many researchers have shown (e.g., Levich 1962; see Tuck and Schwartz 1990 for a recent survey) that a study of coating flow problems, where surface tension forces are significant, very often involves solving third-order ordinary differential equations for the film coating thickness in a region local to the moving front of the coating film. For example, in the case of a thin film draining under gravity down a dry vertical wall, as shown in Fig. 1, the equation governing the profile thickness near the front of the film is given by

$$h''' = -1 + \frac{1}{h^2}\,, \tag{1.1}$$

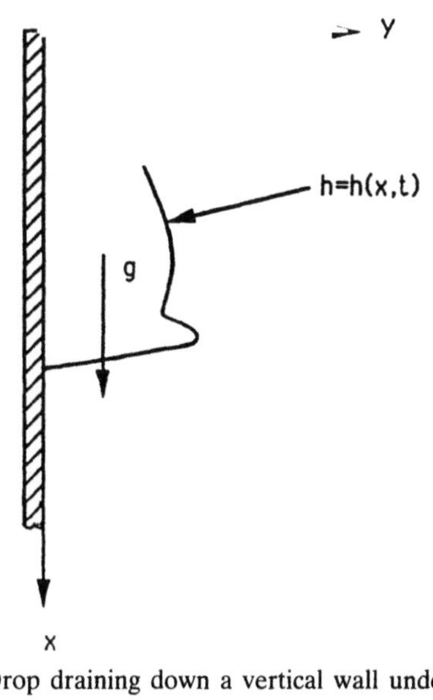

Fig. 1. Drop draining down a vertical wall under gravity.

where $h(x)$ is the non-dimensinal profile thickness, measured in multiples of the profile thickness upstream of the front, h_F. The non-dimensional spatial variable x, measured along the wall in the direction of propagation of the film, is defined in multiples of $(\sigma h_F/\rho g)^{1/3}$ where σ is the surface tension of the fluid, ρ is the fluid density and g is acceleration due to gravity.

In a recent study of wet-wall coating, where the solid substrate is pre-wetted with a thin film of non-dimensional thickness δ, Moriarty et al. (1991) have shown that (1.1), which is the limiting case as $\delta \rightarrow 0$, is also the model equation for two other coating flows, these being spin-coating and spray-coating. The distinction between each coating mechanism on the meniscus length scale at the front of the film enters only through the non-dimensionalisation of the spatial variable x. Thus study of (1.1) is relevant to many coating flows and is not restricted to just one coating process alone.

Despite the seeming simplicity of (1.1), there are some fundamental difficulties in its solution, since it becomes singular at the contact-line, where $h = 0$. The genesis of this singularity is entirely physical, and lies in the so-called 'contact-line paradox'; the classical no-slip boundary condition being in direct conflict with the requirement of contact-line movement.

One way of circumventing the contact-line singularity is to introduce a slip coefficient into the governing equation which relaxes the no-slip condition in a small region local to the contact-line. This idea was used by Greenspan (1978), when he postulated that there was some small amount of slip in the vicinity of the contact-line. The slip he proposed was of Navier-type (see, e.g., Panton 1984), and so was proportional to the local velocity gradient at the contact-line; the factor of proportionality being $\alpha/3h$ where α is a non-dimensional slip coefficient and is generally a small number. Hocking (1981) also added a form of slip in to the equation; his slip formulation weakened the singularity, rather than removed it.

Despite the formidable nature of (1.1) when $h = 0$, it is nevertheless possible to compute free surface profiles by ignoring the singularity altogether. Infinite derivatives at the contact-line, which would normally be a result of the singularity there, are avoided because there is always some amount of numerical slip implicit to any numerical scheme used to solve (1.1). The singularity only becomes apparent through the fact that convergence under spatial refinement cannot be established. This numerical manifestation arises due to the fact that the governing equation is solved at discrete points on the flow domain, i.e. equation (1.1) is satisfied at a point close to the contact-line but not at the contact-line itself.

It has been suggested to us (Davis 1988) that this implicit numerical slip is a function of the numerical grid-spacing, Δx. One way of thinking about it is that the no-slip condition can only be enforced at discrete nodes in the flow domain. Thus, in the space *between* the nodes, the no-slip condition is not enforced. It is in this space, of length Δx, that the contact-line is free to move, so that the smaller the Δx, the smaller the amount of movement – or slip. In the present work, we show that there is a direct relationship between effective slip and numerical grid-spacing.

2. Procedure

Following Greenspan's formulation, the governing equation for coating profiles, with the addition of slip, is given by

$$h''' = -1 + \frac{1+\alpha}{h^2+\alpha}, \tag{2.1}$$

which is a more general form of (1.1), reducing to (1.1) when $\alpha = 0$. The finite difference scheme used to approximate the derivative in (2.1) is $O(\Delta x)$ accurate, so that (2.1) reduces to the singular problem of (1.1) if $\alpha < O(\Delta x) \ll 1$. In other words, for all non-zero α, with $\alpha > O(\Delta x)$ so that the numerical scheme can 'see' the slip, the stress singularity at the contact-line is alleviated.

In order to ascertain the relationship between the slip coefficient α and the numerical grid-spacing Δx, we use the maximum thickness of the profile (called the overshoot) as the control parameter. Proceeding in two steps, we firstly solve (2.1) for a given grid-spacing and no slip (i.e. $\alpha = 0$), to determine the overshoot. We then solve (2.1) with some finite amount of slip, and a grid-spacing $\Delta x \ll \alpha$ so that all results are converged, iterating on α to determine the degree of slip required to produce the same amount of overshoot as for the zero slip case.

Ordinarily, (2.1) would be solved as an initial value problem using a Runge–Kutta formulation, with the initial condition being a perturbation from the linearized problem far upstream of the contact-line. However, this method requires input of the small perturbation parameter; the contact angle at the contact-line comes out as part of the solution and thus cannot be specified *a priori*. Solution of a general class of these problems using a Runge–Kutta method is discussed in detail by Tuck and Schwartz (1990).

Since the contact angle is a monotonically increasing function of the overshoot (Tuck and Schwartz 1990), in order to obtain meaningful results, all computations must be done for a constant contact angle ϕ. This requires us to solve (2.1) as a boundary value problem using a finite-difference scheme, so that the contact angle can be prescribed.

Two of the boundary conditions are imposed at the contact line, and are

$$h = 0$$

and

$$h' = -\tan\phi .$$

Note here that ϕ is the contact angle for the nondimensional problem. The correspondence between this and the physical contact angle ϕ_p is $\tan\phi = (\sigma/\rho g h_F^2)^{1/3}\tan\phi_p$.

The third boundary condition, derived by Goodwin and Homsy (1990), is an asymptotic boundary condition far upstream of the contact line,

$$h'' - \left(\frac{2}{1+\alpha}\right)^{1/3} h' + \left(\frac{2}{1+\alpha}\right)^{2/3} (h-1) = 0 .$$

This boundary condition is in agreement with Tuck and Schwartz's initial condition, but does not require the input of a small parameter.

3. The numerical scheme and results

Equation (2.1) is solved by dividing the flow domain into n discrete points, and using low-order central differences. The profile thickness $h(x)$ is evaluated at the midpoint of the nodes, so that (2.1) is not actually solved at the contact-line itself. Thus, in discrete form (2.1) can be written

$$h(i+2)^{k+1} - 3h(i+1)^{k+1} + 3h(i)^{k+1} - h(i-1)^{k+1} - \Delta x^3\left[-1 + \frac{4(1+\alpha)}{(h(i)^k + h(i+1)^k)^2 + 4\alpha}\right] = 0 \tag{3.1}$$

where the k superscript refers to the kth iteration, and i is the nodal reference point, with $i = n$ corresponding to the nodal point at the contact-line. A schemata is illustrated in Fig. 2.

In the above equation, the h^2 term is evaluated at the previous iteration level; thus we have a system of n linear equations in n unknowns to be solved at each iteration. The resultant coefficient matrix is banded with a bandwidth of four, and a pentadiagonal solver is used to calculate $h(i)$ at each iteration. A Newton–Raphson scheme is used to perform the iterations, and convergence is typically established in five iterations. Computations were performed on an IBM 3090, using between 10^3 and 10^4 nodal points.

All calculations were done for a contact angle of 45 degrees, with respect to the nondimensional variables. This would correspond to a physical contact angle such that $\tan\phi_p = (\sigma/\rho g h_F^2)^{-1/3}$.

Figure 3 shows a comparison between the profile calculated without slip, $\alpha = 0$, and grid-spacing $\Delta x = 0.05$, to that profile calculated with a finite degree of slip, $\alpha = 0.0064$, and a grid-spacing $\Delta x = 0.0002$. In the latter (finite-slip) calculation, the amount of slip was chosen to provide the same overshoot as in the zero-slip case, whilst the grid-spacing was chosen small enough to ensure converged results. Note that the zero-slip calculations are non-convergent solutions of the singular problem (2.1) with $\alpha = 0$. The finite-slip calculations, on the other hand, are converged solutions to the well-posed problem (2.1), when $\alpha \neq 0$. The two curves are indistinguishable, with agreement being to within three decimal places.

A graph of slip coefficient versus grid-spacing for a contact angle $\phi = 45°$ is denoted by the curve with symbols in Fig. 4. The solid line pertains to the polynomial,

$$\alpha = 0.0173\,\Delta x + 2.24(\Delta x)^2\,, \tag{3.2}$$

which is a good fit to the numerical data.

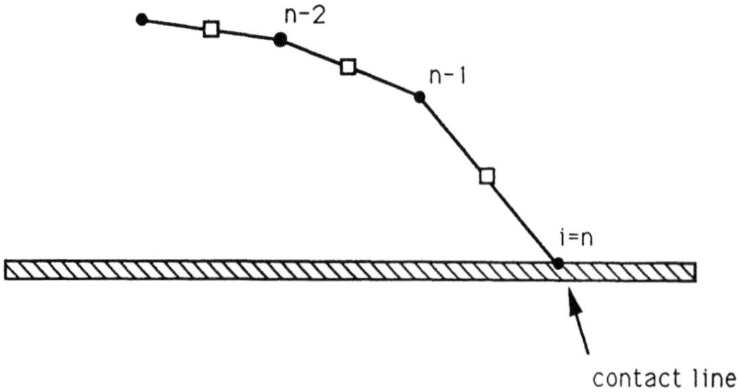

Fig. 2. Schematic of the numerical scheme. Circles represent element end points, and squares represent collocation points, where (3.1) is solved. Note that (3.1) is not solved at the contact-line itself.

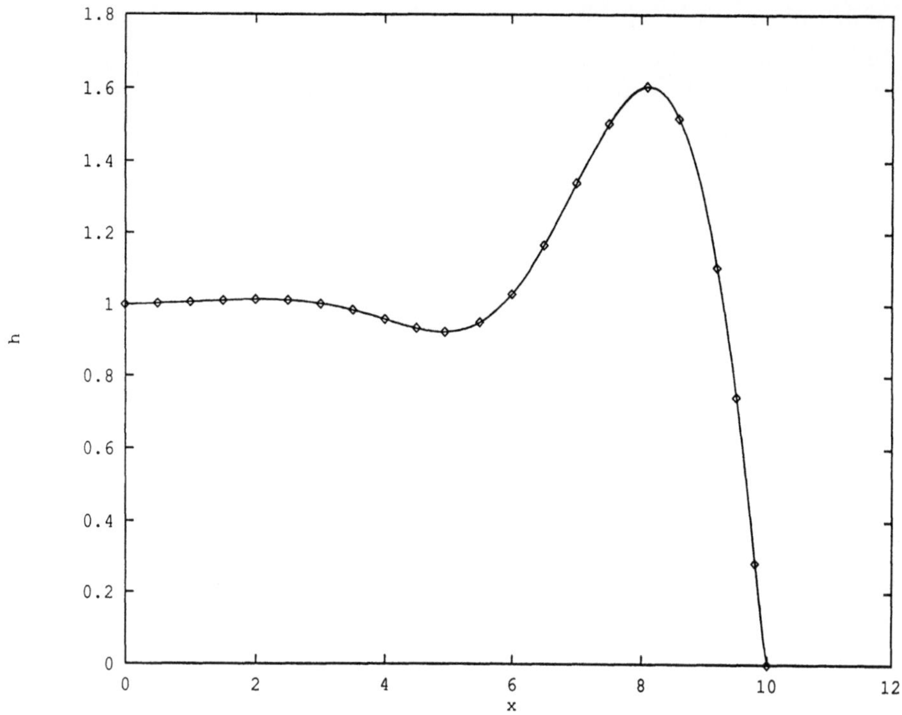

Fig. 3. Comparison of computed profiles with and without slip. The zero-slip model (symbols) is computed with a grid-spacing of $\Delta x = 0.05$. The finite-slip ($\alpha = 0.0064$) model, denoted by the solid line, is converged.

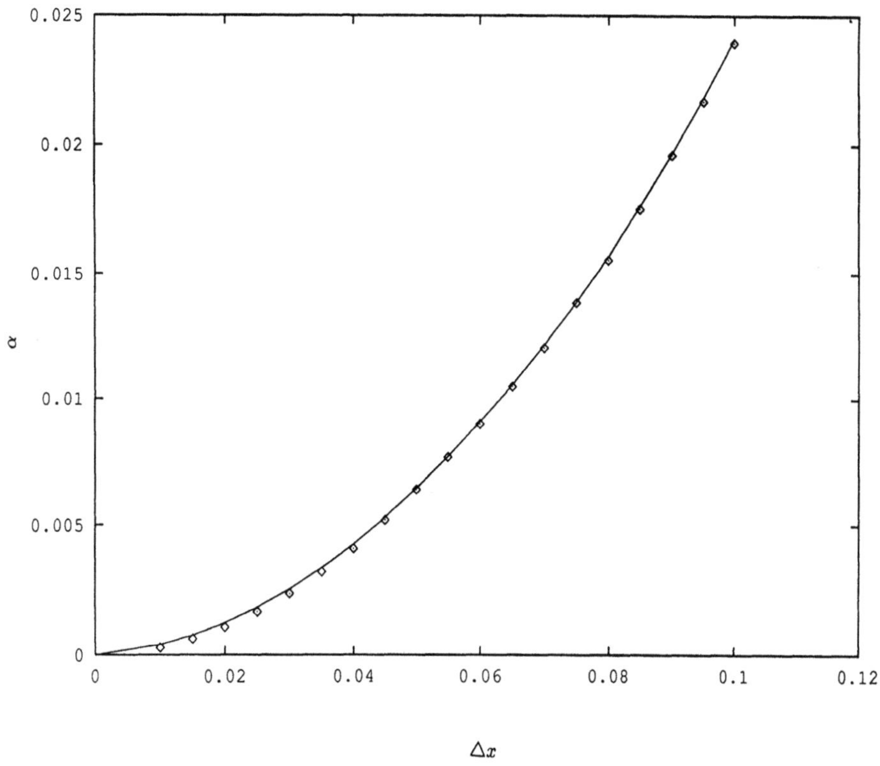

Fig. 4. Slip coefficient α vs grid-spacing Δx (symbols) for a contact angle of $\phi = 45°$. The solid line pertains to the polynomial (3.2).

The fact that the α vs Δx curve passes through the origin is important. It shows definitively that, in order to obtain finite overshoot and converged results with $\alpha = 0$, a grid-spacing $\Delta x = 0$ would be required. In other words, converged finite results, if slip is ignored, can never be obtained. This is the numerical manifestation of the non-integrable force singularity at a moving contact-line when slip is not permitted. If α is set equal to zero, whilst Δx is made small, the overshoot will increase monotonically to infinity.

Since the overshoot is a function of the contact angle, (3.2) would not universally describe a relationship begween Δx and α for all contact angles. For example, a plot of α vs Δx for the 60° case is steeper than (3.2), although it still demonstrates the same general trend.

Calculations to look at the relationship between grid-spacing and Hocking's slip coefficient would follow along similar lines to those described in the present work.

References

Davis, S.H., *Private Communication* (1988).

Goodwin, R. and Homsy, G.M., Viscous flow down a slope in the vicinity of a contact line. *Phys. Fluids* A3 (1991) 515–528.

Greenspan, H.P., On the motion of a small viscous droplet that wets a surface. *J. Fluid Mech.* 84 (1978) 125–143.

Hocking, L.M., Sliding and spreading of thin two-dimensional drops. *Quart. J. Mech. Appl. Math.* 34 (1981) 37–55.

Levich, V.G., *Physicochemical Hydrodynamics*. Englewood Cliffs, NJ: Prentice-Hall (1962).

Moriarty, J.A., Schwartz, L.W. and Tuck, E.O., Unsteady spreading of thin liquid films with small surface tension. *Phys. Fluids* A3 (1991) 733–742.

Panton R.L., *Incompressible Flow*. New York: Wiley (1984) pp. 194ff.

Tuck, E.O. and Schwartz, L.W., A numerical and asymptotic study of some third-order ordinary differential equations relevant to draining and coating flows. *SIAM Review* 32 (1990) 453–469.

Journal of Engineering Mathematics **26**: 87–106, 1992.
H.K. Kuiken and S.W. Rienstra (eds), Problems in Applied, Industrial and Engineering Mathematics.

Initial-value problems for spot disturbances in incompressible or compressible boundary layers

D.J. DOORLY[1] and F.T. SMITH*

*Department of Mathematics, University College London, Gower Street, London WC1E 6BT, UK (*author for correspondence); [1]current address: Department of Aeronautics, Imperial College, London SW7 2AZ, UK*

Abstract. This theoretical study of spot development in boundary layers is motivated by the need for some basic understanding of nonlinear spots, on which there appears to be little or no previous acceptable theory. The eventual aim is to be able to describe theoretically the transitional and/or turbulent spots which are often investigated experimentally. Here, as a starting point to guide possible nonlinear studies, we concentrate in particular on relatively long-scale inviscid spot disturbances in the context of the unsteady Euler equations, although short-scale with respect to Tollmien–Schlichting lengths for example. Both the incompressible and the compressible ranges are examined, for small disturbances, with the corresponding initial-value problems being treated computationally and analytically. The spatial spreading rate of the resulting spot is affected significantly by compressibility, and in fact tends to zero in the hypersonic extreme. The typical amplitudes provoked downstream and their decay lengths also vary considerably with the free-stream Mach number, producing two distinct structures in the transonic regime for example and an elongated structure in the hypersonic regime. The downstream behaviour found at comparatively large times, for finite Mach numbers, is used to provide guidance for nonlinear theory. There are also some potentially useful comparisons and links found with experiments, in both laminar and turbulent conditions.

1. Introduction

A 'spot disturbance' is the flow perturbation that develops from an initial disturbance, to a boundary layer in the present setting, with the typical development involving mostly downstream travel, some amplitude growth, and spatial spreading of the spot. The spot is a three-dimensional (3D) unsteady phenomenon, in general, and there would appear to be basically three types of spot, namely laminar, transitional and turbulent, depending on the amplitude and spectra of the initial disturbance. The three types have certain features in common, and others not. The two extremes of laminar and turbulent flow are studied experimentally and numerically in interesting works by Gaster & Grant (1975), Gaster (1975) and by Katz, Seifert & Wygnanski (1990), Riley & Gad-el-Hak (1985) (and references therein) for example, respectively, and many other experimental studies have also been made in the turbulent case, e.g. see additional references in the papers above and in Smith & Burggraf (1985) (SB), Smith, Doorly & Rothmayer (1990) (SDR), Gaster (1990). In the present work we start an attempt to theoretically understand the transitional and turbulent types of spot, in a compressible or incompressible boundary layer. Some simplification is necessary of course because such spots are very complex in practice, but the aim is to compare eventually with the experimental work above on such spots. While some previous studies of *ad hoc* linear theory have been made, little or no systematic theory/analysis has been done, as far as we know, especially on nonlinear initial-value problems relevant to transitional or turbulent spots, and on the scales and flow structures necessary for a clear physical understanding of the spots' behaviour. These aspects in fact provide the setting for the current study.

Concerning a systematic theory, and corresponding modelling, the 3D unsteady Euler system (acting within the boundary layer) is believed to be very relevant to the transitional-turbulent regime: see SB, SDR, Walker & Smith (1991) and references therein. Thus SDR find agreement between Euler-stage theory and experiments on certain of the established scales for fully turbulent flow, pointing to the use of the Euler system coupled with an allowance for viscous sublayer bursting. The present investigation is therefore based on applying the 3D unsteady Euler system to the study of spot disturbances, this being related more to the transitional-turbulent regime than to the laminar one, and indeed we concentrate below on the inviscid dynamics of the initial-value problem for such spots.

The context for the spot-disturbance analysis is described in Section 2 below, first for an *incompressible* boundary layer (cf. the compressible case below) with an originally inflexion-free velocity profile. The main concern is with the 3D Euler setting, for the reasons mentioned above, and attention is then focussed on an unsteady 3D thin-layer version appropriate to relatively long-scale disturbances. In the thin-layer version the precise details of the original boundary-layer profile have negligible effect, apart from the $O(1)$ skin-friction factor (λ) and the local external velocity. The same version applies for the 3D unsteady triple-deck setting, which governs nonlinear Tollmien–Schlichting transitions (Smith 1989), provided relatively high-frequency disturbances are considered as described also in Section 2. These disturbances are relatively fast and have possible connections with some forms of by-pass transition (see in references above), although it should be observed that the high-frequency range tends to subjugate, or at least delay, the nonlinear break-up and possible intermittency described by Smith (1988), Peridier et al. (1991), Hoyle et al. (1991) for the complete 3D triple-deck system. Nevertheless, the break-up can still arise, as Peridier et al. show. We should add also that the response of a single or finite number of waves provoked by a maintained disturbance (e.g. as in Smith 1984, Smith & Stewart 1987a,b) is distinct from the present initial-value setting since here in effect all waves are activated. In Section 3, a linearized version is examined, as a first step, permitting a relatively simple analysis which produces useful and 'universal' results at large times, this representing an advantage over previous linear-disturbance studies, e.g. see references above and Ryzhov (1987), Smith (1987a). In particular, concentrated wake activity is found far downstream inside a wedge of half-angle $\sin^{-1}(1/3) = 19.47°$ (in plan view) and the maximum amplitude growth occurs at the edges of this wake (cf. the experiments of Wygnanski's group, e.g. Glezer, Katz & Wygnanski, 1989). This aspect is the same as for the Kelvin ship-wake (Whitham 1974), and as studied in another context by Cheng & Johnson (1982). The *compressible* boundary layer is then addressed in Section 4, yielding some very interesting changes in the spot development with increasing Mach number. These include especially the *transonic* range, where two main kinds of spot-disturbance amplitude response appear spatially, and the *hypersonic* range, in which the typical amplitudes produced are much reduced but last longer. The change in the wake half-angle mentioned above is also considered, as the Mach number increases, with the half-angle being found to tend to zero in the hypersonic range. Further comments are provided in Section 5, including suggestions for the nonlinear regime.

It is believed that the nonlinear 3D problems posed in Sections 2 and 4 could be very relevant to transitional and turbulent spot behaviour in practice, and further computations and analysis are desirable on these. The present first step, on the linear 3D regime, reveals certain interesting features of the incompressible and compressible initial-value problems, nevertheless, and in fact there are some signs that the experiments tie in qualitatively with the overall picture that emerges below, especially regarding the enhanced waviness in the downstream wake. This tie-in is felt likely to increase once nonlinearity is accommodated. In the 2D case nonlinearity can lead to the Benjamin–Ono or Burgers equations (SB), after some simplification, and indeed the large-time scales found in Sections 3 and 4 agree with the scales associated with those equations, but in the more intriguing 3D case the nonlinear governing equations do not simplify so readily. The current work however enables estimates to be made (in Section 5) for the nonlinear 3D regime, studies of which are in progress (F.T.S.), while the results obtained below are also of potential practical value in gauging the spread of spots, following an initial disturbance in an incompressible or compressible boundary layer. There may be broader interest in addition with regard to the development of wavefronts.

The nondimensional velocity vector $\mathbf{u} \equiv (u, v, w)$ along with the corresponding Cartesian coordinates (x, y, z) (streamwise, normal, spanwise, respectively), the pressure p, and the time t, are used such that the typical free-stream velocity is $(1, 0, 0)$ and the undisturbed steady planar boundary layer has its x, y scales of order 1, $\mathrm{Re}^{-1/2}$ in turn, with a streamwise velocity profile written u_0. Here the global Reynolds number Re is taken to be large, for an incompressible (Sections 2 and 3) or compressible boundary layer (Section 4), and the 3D disturbance develops on length and time scales between $O(\mathrm{Re}^{-1/2})$ and $O(1)$. Some of the results from the present work are summarized and used in comparisons by Smith (1987a).

2. The two main contexts in 3D boundary-layer flow

There are two major areas in which the present work applies for boundary layers, (a), which is perhaps the more relevant to turbulent flow, and (b), which is more concerned with transition. These are addressed in turn below.

(a) The first concerns 3D unsteady Euler flow, in which the velocities and pressure are $O(1)$ but the spatial scales are all comparable with the boundary-layer thickness $O(\mathrm{Re}^{-1/2})$ and the corresponding time scale is then the convective one, $O(\mathrm{Re}^{-1/2})$. So the governing equations become

$$\nabla \cdot \mathbf{u} = 0 , \qquad \mathbf{u}_t + (\mathbf{u} \cdot \nabla)\mathbf{u} = -\nabla p , \qquad \nabla = (\partial_x, \partial_y, \partial_z) , \tag{2.1}$$

subject to appropriate boundedness conditions in the farfield including matching to the basic boundary-layer profile, $u = u_0(y)$ say upstream, and tangential flow $v = 0$ at $y = 0$. One solution is $\mathbf{u} \equiv (u_0, 0, 0)$ corresponding to undisturbed parallel motion, but linear and nonlinear waves are also possible and are of much interest. Nonlinear phenomena are especially important, in 2D or 3D (see SDR and references therein), giving possible connections with turbulence. Here we investigate long-scale 3D disturbances in particular, as suggested by Section 1, where a three-region structure arises analogous to that in (b) below. Thus $(x, z) = l(X, Z)$, say, with the length scale l being large, the main time scale has $t = l^2 T$, where X, Z, T are typically $O(1)$, and three regions I–III appear in the y-direction, of scales l^{-1}, 1, l in which $y = l^{-1}Y$, y, $l\bar{y}$ and

$$[u, v, w, p] = \begin{cases} [l^{-1}U, l^{-3}V, l^{-1}W, l^{-2}P] + \cdots & \text{(2.2a)} \\ [u_0 + l^{-1}Au_0', -l^{-2}A_x u_0(y), l^{-2}D_1/u_0, l^{-2}P] + \cdots & \text{(2.2b)} \\ [1, 0, 0, 0] + l^{-2}[\bar{u}_2, \bar{v}_2, \bar{w}_2, \bar{p}_2] + \cdots & \text{(2.2c)} \end{cases}$$

in turn. Here P (the pressure), A (the negative displacement), D_1 are unknown functions of X, Z, T, with $D_{1X} \equiv -P_Z$. The small-disturbance expansion (2.2b) holding across most of the boundary layer satisfies (2.1) to the required order, whereas the near-wall response (2.2a) is controlled by the nonlinear unsteady thin-layer system

$$\bar{\nabla}\cdot\mathbf{U}=0\,, \qquad \{\partial_T+\mathbf{U}\cdot\bar{\nabla}\}(U,W)=-(P_X,P_Z)\,, \quad \text{with } \bar{\nabla}\equiv(\partial_X,\partial_Y,\partial_Z) \tag{2.3a–c}$$

subject to

$$V=0 \quad \text{at } Y=0\,, \qquad U\sim\lambda(Y+A)\ \&\ W\to 0 \quad \text{as } Y\to\infty\,, \tag{2.3d}$$

from (2.1) and matching with (2.2b); and the outermost response (2.2c) leads from (2.1) to 3D quasi-steady potential-flow properties in the upper reaches of the boundary layer, so that $\bar{p}_2$ satisfies

$$(\partial_X^2+\partial_{\bar{y}}^2+\partial_Z^2)\bar{p}_2=0\,, \quad \text{with } \bar{p}\to P\,, \qquad \bar{p}_{2\bar{y}}\to A_{XX} \quad \text{as } \bar{y}\to 0\,, \tag{2.4a,b}$$

again to join to (2.2b), together with boundedness in the farfield. The solution of (2.4) therefore gives the relation

$$P(X,Z,T)=-\frac{1}{2\pi}\int_{-\infty}^{\infty}\int_{-\infty}^{\infty}\frac{A_{\xi\xi}(\xi,\eta,T)\,\mathrm{d}\xi\,\mathrm{d}\eta}{[(X-\xi)^2+(Z-\eta)^2]^{1/2}}\,. \tag{2.5}$$

Hence we are left with solving the nonlinear problem (2.3) allied with the pressure-displacement law (2.5). We note in passing that the constant λ may be normalised to unity, the decay into the large-Y asymptote is algebraic, and a self-consistent initial condition is assumed at time $T=0$.

(b) The second context is in 3D nonlinear triple-deck theory, which applies to the initial linear or nonlinear development of 3D TS waves and transition (Smith (1988, 1989) and references therein) and captures their viscous-inviscid nature. The governing equations here, mainly from the lower-deck flow where u, v, w, p, x, y, z, t are scaled with $\mathrm{Re}^{-m/8}$, $m=1, 3, 1, 2, 3, 5, 3, 2$ respectively, are the 3D nonlinear boundary-layer equations

$$\nabla\cdot\mathbf{u}=0\,, \qquad \{\partial_t+\mathbf{u}\cdot\nabla\}(u,w)=-(p_x,p_z)+\partial_y^2(u,w)\,, \tag{2.6a,b}$$

along with $p_y\equiv 0$, the constraints

$$\mathbf{u}=0 \quad \text{at } y=0\,, \qquad u\sim y+a \quad \text{and} \quad w\to 0 \quad \text{as } y\to\infty \tag{2.6c,d,e}$$

and the pressure-displacement law, which is given by (2.5) with the unknown functions $(p,a)(x,z,t)$ replacing (P, A) here. The basic inviscid problem (2.3), (2.5) therefore arises again if the viscous terms on the right in (2.6b) are simply neglected. More precisely, as in SB, for relatively high frequencies $|\partial_T|\sim\Omega\gg 1$ the scalings

$$\begin{aligned}&(u,v,w,p,a,x,y,z,t)\\&\quad\sim(\Omega^{1/2}U,\Omega V,\Omega^{1/2}W,\Omega P,\Omega^{1/2}A,\Omega^{-1/2}X,\Omega^{1/2}Y,\Omega^{-1/2}Z,\Omega^{-1}T)\end{aligned} \tag{2.7}$$

lead to (2.3) coupled with (2.5).

In both of the contexts (a), (b) it should be emphasized that, as in 2D flow (SB, SDR), vorticity bursting is possible from the viscous sublayer assumed near the wall, in the nonlinear setting, although on the other hand bursting is not relevant in the linear regime studied subsequently. Further, in the special 2D case a simple solution of (2.3) is available, namely $U \equiv Y + A$, and the coupling with (2.5) thereby leads to the Benjamin–Ono equation for $A(X, T)$. Unfortunately no such simplification seems available in general in the full 3D case (Rothmayer & Smith 1987), and so (2.3), (2.5) as they stand, i.e. in terms of X, Y, Z, T, are in effect the 3D counterparts of the nonlinear Benjamin–Ono equation. The system (2.3), (2.5) is of much interest, then, in both contexts (a), (b) and as a start we concentrate next on one aspect of the system.

3. Development from an initial spot disturbance

Aspects of the travel, growth and spreading of an initial localised disturbance are considered in (i)–(iii) below.

(i) Linearized features hold for relatively small disturbances where the basic flow $U = Y$ is slightly disturbed. There to leading orders $(U - Y, V, W, P, A) = h(\hat{U}, \hat{V}, \hat{W}, \hat{P}, \hat{A})$, say, with $h \ll 1$, and so (2.3a–c) become

$$\nabla \cdot \hat{\mathbf{U}} = 0\,, \qquad \hat{U}_T + Y\hat{U}_X + \hat{V} = -\hat{P}_X, \; \hat{W}_T + Y\hat{W}_X = -\hat{P}_Z \tag{3.1a–c}$$

with $\hat{V} = 0$ at $Y = 0$, $\hat{U} \to \hat{A}$ and $\hat{W} \to 0$ as $Y \to \infty$, and (2.5) holds with $(\hat{P}, \hat{A})$ instead of (P, A). An appropriate solution may be derived from adding together the X-, Z-derivatives of (3.1b,c) respectively, which yields a quasi-2D system for $\hat{U}_X + \hat{W}_Z$, $\hat{V}_X$ whose solution, subject to suitable initial conditions, has $\hat{U}_X + \hat{W}_Z$ independent of Y and hence identically equal to $\hat{A}_X$. That is consistent with the momentum balances provided

$$\hat{A}_{XT} = -(\partial_X^2 + \partial_Z^2)\hat{P}\,. \tag{3.2}$$

Hence, combining this with (2.5), we obtain the equation

$$2\pi\hat{A}_{XT} = (\partial_X^2 + \partial_Z^2)\int_{-\infty}^{\infty}\int_{-\infty}^{\infty} \frac{\hat{A}_{\xi\xi}(\xi, \eta, T)\,\mathrm{d}\xi\,\mathrm{d}\eta}{[(X-\xi)^2 + (Z-\eta)^2]^{1/2}} \tag{3.3}$$

for $\hat{A}(X, Z, T)$, with a similar equation governing $\hat{P}$. Here (3.3) is a 3D extension of the 2D linearised Benjamin–Ono equation. We should remark that a benign critical layer is present in the $\hat{W}$ solution; see also later.

The solution of (3.3) may be found via a double Fourier transform (denoted **) in X, Z, which converts (3.3) to $\mathrm{i}\hat{A}_T^{**} = \alpha(\alpha^2 + \beta^2)^{1/2}\hat{A}^{**}$ where α, β are the transform variables corresponding to X, Z respectively. Hence

$$4\pi^2\hat{A}(X, Z, T) = \int_{-\infty}^{\infty}\int_{-\infty}^{\infty} Q(\alpha, \beta)\exp\{\mathrm{i}\alpha X + \mathrm{i}\beta Z - \mathrm{i}\alpha(\alpha^2 + \beta^2)^{1/2}T\}\,\mathrm{d}\alpha\,\mathrm{d}\beta \tag{3.4}$$

where $Q(\alpha, \beta)$ is the initial distribution of the negative displacement transform, $A^{**}(\alpha, \beta, 0)$. Here (3.4) determines the development of the displacement in space and time, and hence the pressure from (2.5), for any prescribed initial distribution. We note that (3.4) agrees with the dispersion relation obtained in allied previous papers, although the flow structure shown above in the derivation of (3.4) is worth emphasising, as discussed later. It is apparent also that the dispersion relation involved yields only neutral waves in the present regime.

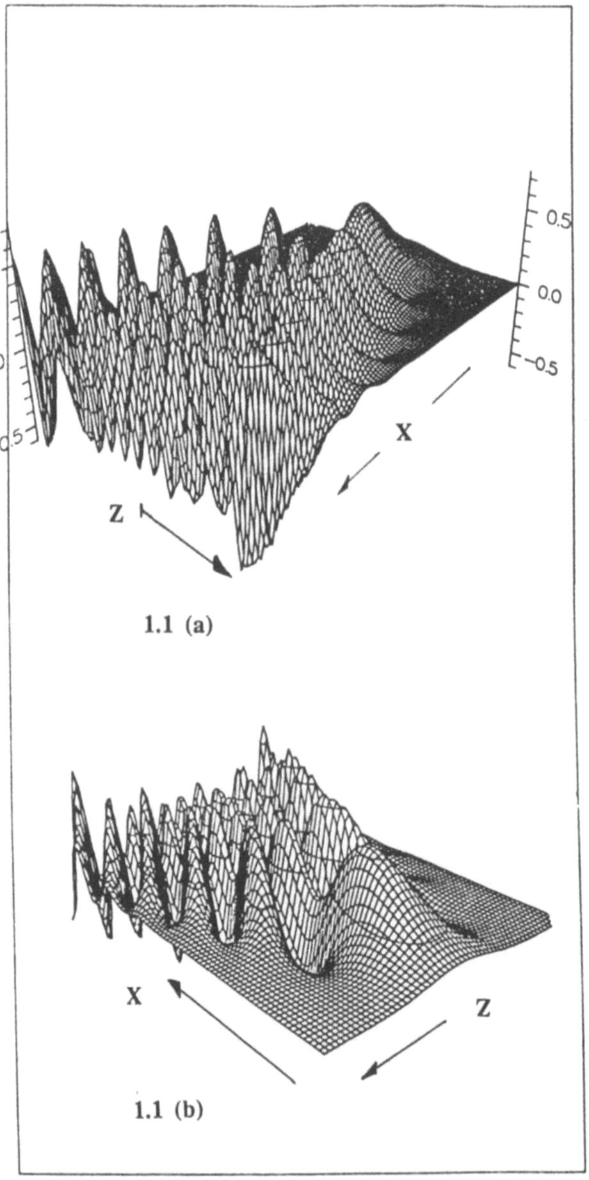

Fig. 1. Numerical results for the incompressible case. (See also final page.)
1.1. (a)(b) Surface plots of disturbance $\hat{A}(\bar{X}, \bar{Z}, T)$ for very large time T. ($T = 256$)
1.2. (a)(b) Surface plot of $\hat{A}(\bar{X}, \bar{Z}, T)$ for short time T. ($T = 2$)
1.3. (a) Contour plot of $\hat{A}(\bar{X}, \bar{Z}, T)$ for short (a) and large time (b). (Results symmetric about $\bar{Z} = 0$. At large times, the maximum amplitudes are clearly seen to lie along the caustic.)
1.4. (a) Centreline amplitudes for various values of time T.
(b) Amplitudes along caustic at various values of T.
1.5. Amplitude growth along caustic. Theoretical slope $= \frac{1}{3}$ (since $\hat{A} \sim O(r^{1/3})$), computational slope $= 0.3$.

(ii) Computations of (3.4) were performed for a number of different initial conditions, by use of a spectral scheme. Sample results are shown in Fig. 1.

(iii) Of special interest now is the behaviour of the flow solution at large times, for $O(1)$ or large distances from the start. There (3.4) suggests that significant effects arise mainly in two regions far downstream, the first of which has X, Z large and $O(T^{1/2})$, say $(X, Z) = T^{1/2}(\bar{X}, \bar{Z}) = T^{1/2}R(\cos\theta, \sin\theta)$. There, in (3.4), $(\alpha, \beta) = T^{-1/2}(\bar{\alpha}, \bar{\beta}) = T^{-1/2}\bar{r}(\cos\phi, \sin\phi)$ in polars, yielding

$$2\pi\hat{A} \sim T^{-1}Q(0,0)\int_0^\infty \bar{r}J_0(\Phi)\,\mathrm{d}\bar{r} \quad \text{with } \Phi^2 = \bar{r}^2R^2 - 2\bar{r}^3R\cos\theta + \bar{r}^4\,, \tag{3.5}$$

after a ϕ-integration, where J_0 is the standard Bessel function of zeroth order. Here (3.5) gives most of the large-time behaviour for $O(1)$ values of the scaled distance $R \equiv$

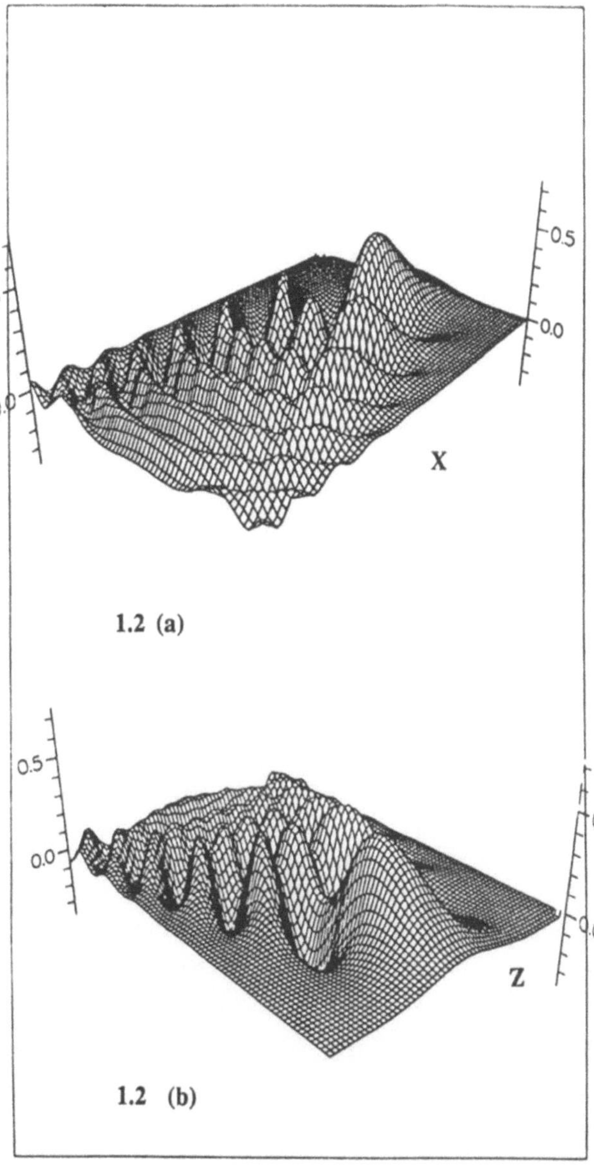

Fig. 1 (cont.).

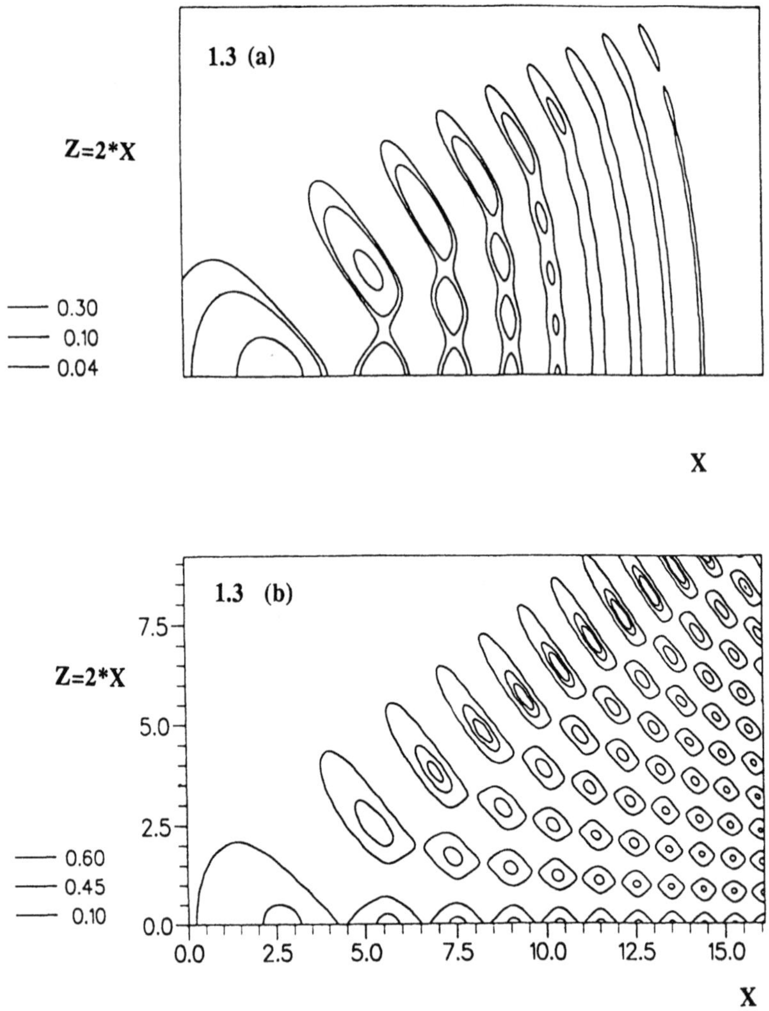

Fig. 1 (cont.).

$(X^2+Z^2)^{1/2}T^{-1/2}$ and angles θ. The constant $Q(0,0)$ is an integral property of the initial displacement, specifically the double integral with respect to X, Z of the initial disturbance $A(X, Z, 0)$. For *smaller* R values, the range $\bar{r} \sim 1$ dominates, giving

$$\hat{A} \sim T^{-1}Q(0,0)\{c_1 + c_2T^{-1/2}X + O(T^{-1})\} \text{ [for } R \ll 1] \tag{3.6}$$

where c_1, c_2 are constants, whereas for *larger* distances R the beginning of a confined 'wake' can be distinguished. For then the range $\bar{r} = \sigma R \gg 1$ matters, giving $\Phi^2 = R^4(\sigma^2 - 2\sigma^3c + \sigma^4) \gg 1$ [where $c = \cos\theta$] and so $J_0(\Phi) \sim (2/\pi\Phi)^{1/2}\cos(\Phi - \pi/4)$. The major effects therefore come from the maxima of $h(\sigma) \equiv (\sigma^2 - 2\sigma^3c + \sigma^4)$, since J_0 is predominantly oscillatory. But $h'(\sigma) = 2\sigma\{2(\sigma - 3c/4)^2 + (1 - 9c^2/8)\}$. So the maxima are possible only for $c^2 > 8/9$, i.e. at angles $|\theta| < \theta_s$ where

$$\theta_s = \sin^{-1}\left(\frac{1}{3}\right) = 19.47^\circ . \tag{3.7}$$

For such angles, (3.5) then gives the asymptote

1.4 (a)

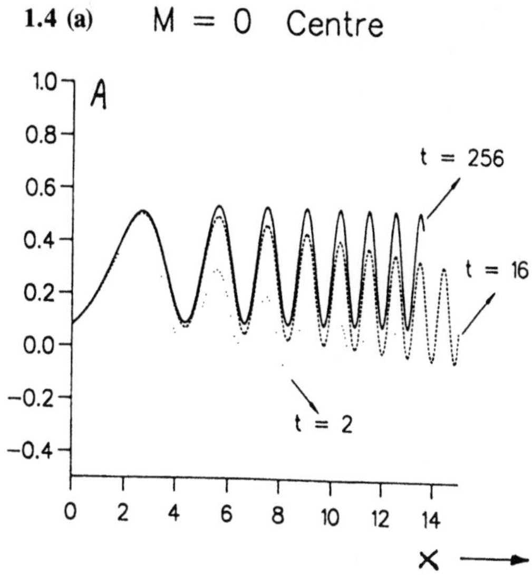

1.4 (b)

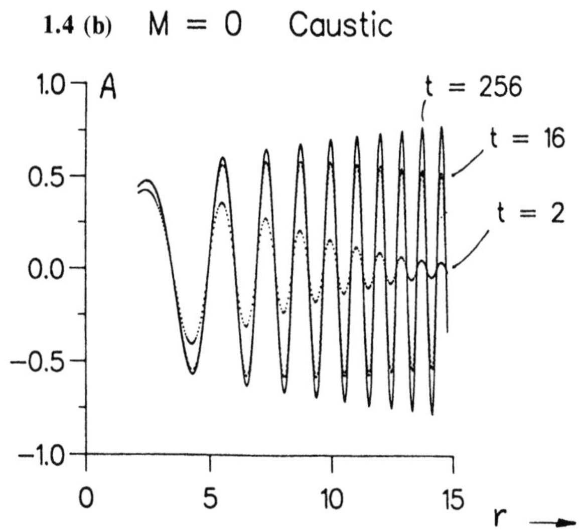

1.5

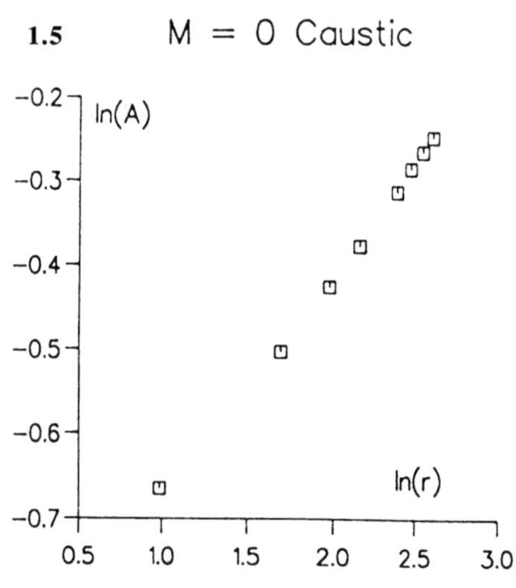

Fig. 1 (cont.).

$$2^{1/2}\pi\hat{A} \sim T^{-1}Q(0,0)\left[\frac{\sigma_1^+}{(B_1^+B_2^+)^{1/2}}\cos(R^2B_1^+) + \frac{\sigma_1^-}{(B_1^-B_2^-)^{1/2}}\cos(R^2B_1^-)\right],$$

$$\text{for } R \gg 1 \text{ with } -\theta_s < \theta < \theta_s\,, \tag{3.8a}$$

consisting of two wave forms in which

$$B_1(\theta) = (\sigma_1^2 - 2\sigma_1^3 c + \sigma_1^4)^{1/2}\,, \tag{3.8b}$$

$$B_2(\theta) = \frac{1}{2}(1 - 6\sigma_1 c + 6\sigma_1^2)B_1^{-1}\,, \qquad \sigma_1^{\pm}(\theta) = \frac{1}{4}\{3c \pm (9c^2 - 8)^{1/2}\}\,. \tag{3.8c,d}$$

From this it can be seen that increasingly concentrated (i.e. relatively short-wavelength) waves form in the wake far downstream, with their maximum amplitudes occurring at the *edges* $\theta \sim \pm\theta_s$, since $B_2 \to 0$, $B_1 \to O(1)$ as $|\theta| \to \theta_s^-$. Near each edge (or caustic), a relatively thin layer is present in which $\hat{A}$ acquires the form of an Airy function, matching with (3.8a) inside the wake and with the exponentially small response outside (for $|\theta| > \theta_s$). In detail, (3.5) gives

$$2\pi\hat{A} \sim \left(\frac{8\pi}{B_1}\right)^{1/2}\sigma_1\left(\frac{2}{a}\right)^{1/3}R^{1/3}\cos(B_1R^2)Ai(\eta)\,, \quad \text{for } R \gg 1 \text{ near } \theta = \theta_s\,, \tag{3.9}$$

where $\eta \equiv 2^{1/2}\bar{X}^{4/3}3^{-1/3}(\bar{Z}/\bar{X} - 8^{-1/2})$ is $O(1)$ in the edge layer and now $\sigma_1 = 3c/4$, $B_1 = 3^{-1/2}/2$, $c = 3^{-1}2^{3/2}$, $\bar{X} = cR$, $a = 2^{5/2}3^{1/2}$.

The second region, further downstream, occurs where the distances X, Z are increased to $O(T)$, cf. Figs 1.3, 1.4 and below. There, in view of (3.5)ff, the major contributions to the integral (3.4) come from $O(1)$ values of α, β. So the new feature emerging here is that the detailed distribution $Q(\alpha, \beta)$ now modifies the solution amplitude at leading order, i.e. this region feels the 'footprint' of the initial disturbance $\hat{A}(X, Z, 0)$. There are again two wave families present containing an infinite number of waves analogous with those in (3.8a) but the typical wavelength is reduced to $O(1)$. Moreover, the wake remains confined to $|\theta| < \theta_s$,

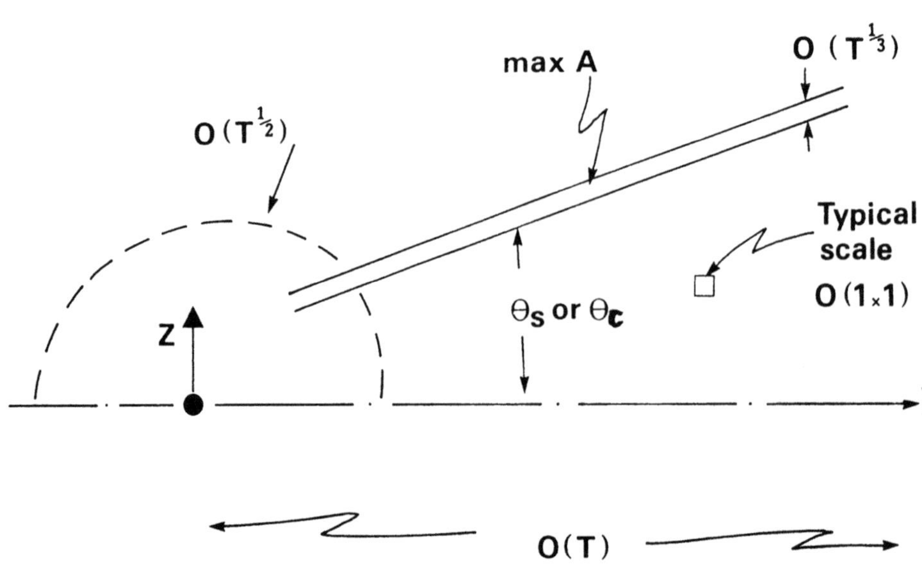

Fig. 2. Sketch of the structure for the large-time large-distance behaviour. Plan view: upper half only.

with a modified form of (3.9) applying in an edge region of thickness $O(T^{1/3})$ in terms of $Z - 8^{-1/2}X$. The final downstream decay of the solution is then obtained at distances greater than $O(T)$.

The analytical features here, summarized in Fig. 2, are also well brought out in the computed results of Fig. 1, and there is broad agreement in fact with the computational trends obtained in (ii) above. We refer in particular to Figs 1.4a, 1.4b concerning the centre-line and the near-caustic calculations, as time T increases, and note that the slope of the log–log plot in Fig. 1.5 at large distances is very close to $\frac{1}{3}$, cf. (3.9).

4. The compressible regime

Most of the reasoning in Section 2 for the 3D nonlinearly disturbed incompressible boundary layer can be adapted also for the compressible case studied next. Thus, from scaling of the unsteady compressible Euler equations, (2.3a–d) still hold, but the outer potential-flow part in (2.4a,b) or (2.5) must be replaced by

$$(M_\infty^2 - 1)(\bar{p}_{2XX} \pm \bar{p}_{2\bar{y}\bar{y}}) - \bar{p}_{2ZZ} = 0 \tag{4.1a,b}$$

(e.g. as in Smith 1989), along with (2.4b), for the subsonic ($0 \leq M_\infty < 1$) and supersonic ($M_\infty > 1$) ranges respectively, where M_∞ is the Mach number. [Shorter-scale contributions are omitted; see Section 5]. In consequence the linearized properties replacing those of Section 3 are as follows, for the development from an initial spot disturbance.

In the *subsonic* regime, (4.1a) leads to the result

$$4\pi^2\hat{A}(X, Z, T) = \int_{-\infty}^{\infty}\int_{-\infty}^{\infty} Q(\alpha, \beta)\exp\left\{\mathrm{i}\alpha X + \mathrm{i}\beta Z - \frac{\mathrm{i}\alpha(\alpha^2+\beta^2)T}{(\alpha^2+\kappa^2\beta^2)^{1/2}}\right\}\mathrm{d}\alpha\,\mathrm{d}\beta \tag{4.2}$$

for the negative displacement, with $\kappa \equiv (1 - M_\infty^2)^{-1/2}$ ranging from 1 to ∞. Spectral computations for (4.2) were performed as in Section 3(ii) at several subsonic Mach numbers, giving the results shown in Fig. 3. Analytically, rather than proceeding as in Section 3(iii) we may use the method of stationary phase to deduce the large-time behaviour, setting $(X, Z) = T^{1/2}R(\cos\theta, \sin\theta)$, $(\alpha, \beta) = T^{-1/2}\bar{r}(\cos\phi, \sin\phi)$ again and then $\bar{r} = \sigma R$, to convert (4.2) to the form

$$4\pi^2\hat{A} \sim Q(0,0)T^{-1}R^2 \int_0^{\infty}\int_0^{2\pi} \exp\left\{\mathrm{i}\left[\sigma\cos(\phi-\theta) - \sigma^2\frac{\cos\phi}{K(\phi)}\right]R^2\right\}\sigma\,\mathrm{d}\sigma\,\mathrm{d}\phi \tag{4.3a}$$

for large distances R, where

$$K(\phi) \equiv (\cos^2\phi + \kappa^2\sin^2\phi)^{1/2}. \tag{4.3b}$$

The maximum contributions to $\hat{A}$ at large R therefore come from the extrema of the expression in square brackets in (4.3a). With the derivatives of that expression, with respect to σ, ϕ, equated to zero, then, we obtain two equations to determine the extremal values of σ, ϕ, leading to the equation

$$\kappa^2\chi^4\tan\theta - \kappa^2\chi^3 + 3\kappa^2\chi^2\tan\theta + (\kappa^2-2)\chi + 2\tan\theta = 0 \tag{4.4}$$

3.1

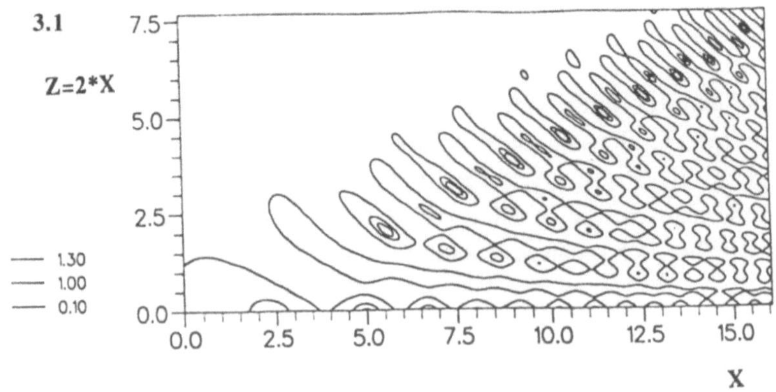

3.2

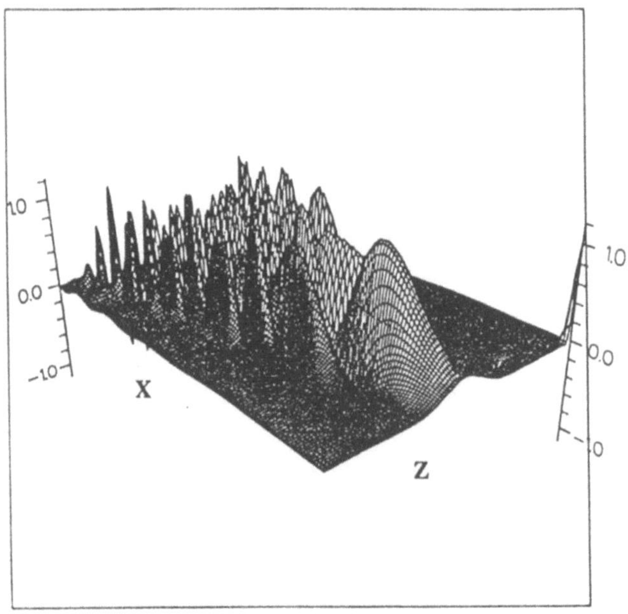

Fig. 3. Numerical results for subsonic range. (See also final page.)
3.1. Contours of disturbance amplitude at large time for $M = 0.9$.
3.2. Surface plot at $M = 0.9$ for large time.
3.3. Intermediate time ($T = 16$) results for $M = 0.9$.
3.4. Results for same time, but $M = 0.98$.
The nearly planar waves predicted by theory are clearly evident, as is their persistence after the centreline waves die out.
3.5. Surface plot (for $\bar{Z} > 0$) showing persistence of plane waves at $M = 0.98$, $T = 16$.
3.6. Comparison of centreline amplitude for $M = 0.9$, $M = 0.98$ at $T = 16$. Note very rapid initial growth at higher Mach number, but more rapid subsequent decay.

for $\chi \equiv \tan \phi$. Potentially there are four main waves possible downstream now rather than just two as in Section 3(iii); see also (4.15a–c) later. We observe also that the solutions near the edges of the wake(s) downstream are still described by an Airy function, as in Section 3(iii), and the ultimate downstream decay at large times T is associated with larger distances, of order T, similarly to Section 3(iii), at general Mach numbers in the subsonic or supersonic range.

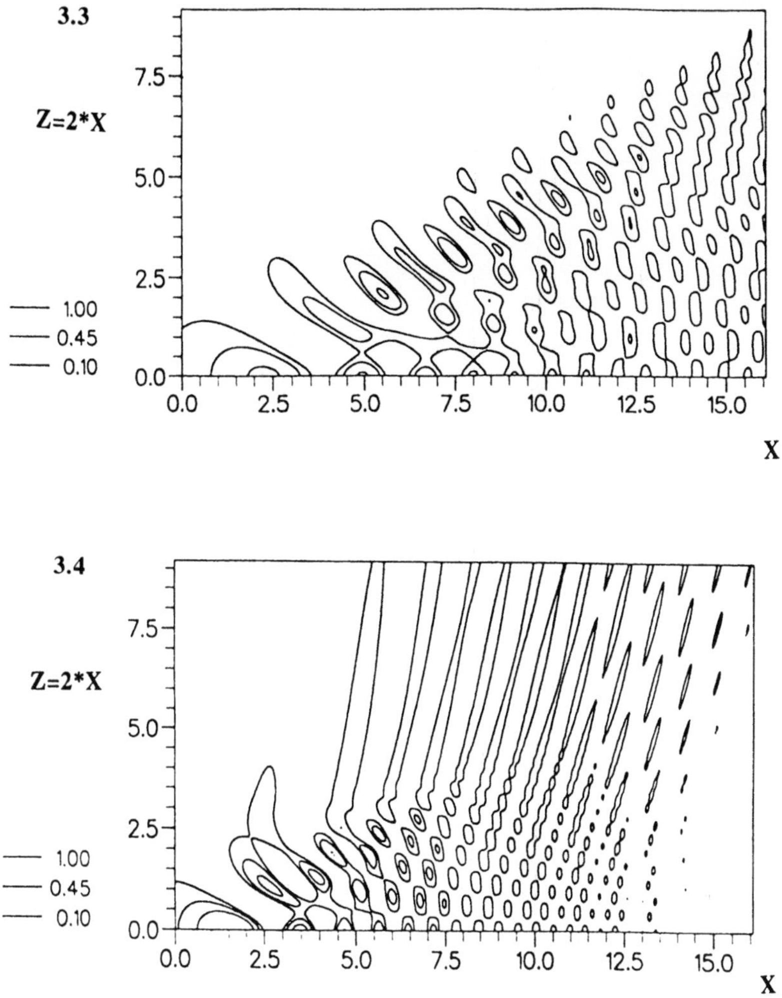

Fig. 3 (cont.).

Two special limits are of most interest next. First, in the *incompressible* limit $M_\infty \to 0$, $\kappa \to 1+$, only two waves remain since (4.4) then yields only two real roots $\chi = \{1 \pm (1 - 8\tan^2\theta)^{1/2}\}/(2\tan\theta)$, the other two roots being imaginary, $\chi = \pm i$. The two real roots here confirm the results in Section 3(iii) and in particular the half-wake angle of $\tan^{-1}(8^{-1/2}) = 19.47°$. Second, in the *transonic* limit $M_\infty \to 1-$, $\kappa \to \infty$, there can be four real roots, three of which are governed by

$$\chi^3 \tan\theta - \chi^2 + 3\chi\tan\theta + 1 = 0 \tag{4.5}$$

with χ of $O(1)$, when θ is $O(1)$, whereas the fourth root has $|\chi| \ll 1$, from (4.4). One root of (4.5) is negative and persists for all angles θ, starting at $\chi = -1$ for zero θ, then increasing monotonically with θ and tending to zero from below as $\theta \to \pi/2$. The other two, positive, roots of (4.5) start at $\chi = 1, \infty$ in effect for $\theta = 0$, then increase/decrease with θ respectively and coalesce at $\chi = 6\tan\theta_c/(1 - 9\tan^2\theta_c)$ for $\theta = \theta_c$, thereafter becoming irrelevant (complex) for $\theta > \theta_c$, where

$$\theta_c = \tan^{-1}[(9 + 6\sqrt{3})^{-1/2}] = 12.79° \tag{4.6}$$

3.5

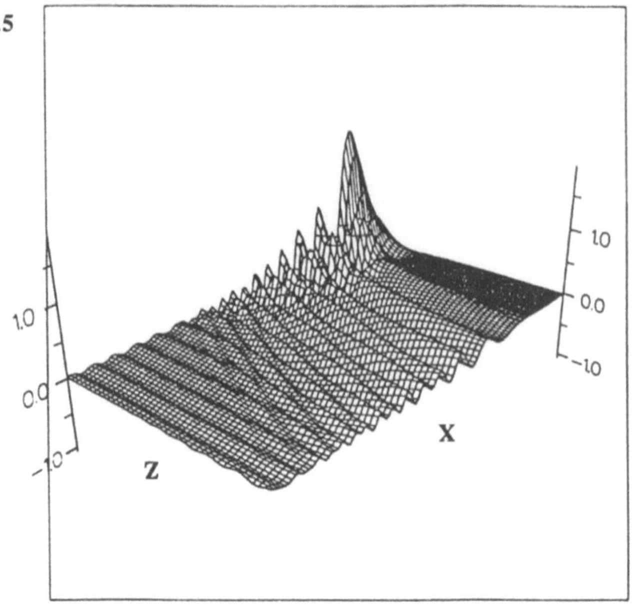

3.6

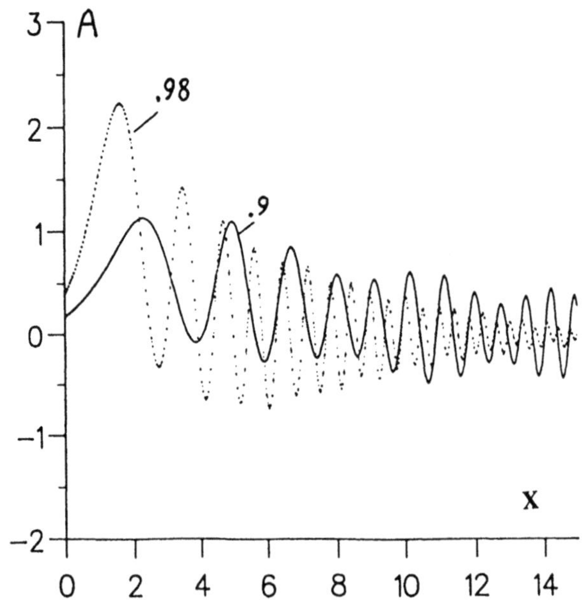

Fig. 3 (cont.).

defines a cut-off angle. Further, at sufficiently large angles $\theta \sim \pi/2 - O(\kappa^{-1})$ the negative root from (4.5) and the fourth root mentioned above become comparable and are associated with a distinct structure in which $\tan\theta \sim \kappa b$, $\chi \sim \kappa^{-1}\tilde{\chi}$ typically, with b, $\tilde{\chi}$ of order unity. Then (4.4) gives a quadratic equation for $\tilde{\chi}$, yielding the two roots

$$\tilde{\chi} = [-1 \pm \{1 - 24b^2\}^{1/2}]/(6b) \tag{4.7a}$$

and hence the second cut-off criterion

$$\theta \sim \frac{\pi}{2} - (24)^{1/2}\kappa^{-1} . \tag{4.7b}$$

Thus the roots (4.7a) correspond to wide waves, of quasi-planar form. An additional feature here, however, concerns the wave amplitudes. The roots in (4.5) correspond to the scalings

$$(\alpha, \beta) = \kappa^{1/2}(\alpha^*, \beta^*) , \qquad (X, Z) = \kappa^{-1/2}(X^*, Z^*) , \tag{4.8a,b}$$

effectively, thereby altering the right-hand side in (4.2) to the form

$$\kappa \iint \exp\{i\alpha^* X^* + i\beta^* Z^* - i\alpha^*(\alpha^{*2} + \beta^{*2})T/\beta^*\} \, d\alpha^* \, d\beta^* , \tag{4.8c}$$

apart from the Q factor. In consequence, although the associated downstream solution has increased amplitude of order κ, it decays fast spatially, in length scales reduced by $O(\kappa^{-1/2})$. By contrast, the roots in (4.7a) point to the scalings

$$(\alpha, \beta) = (\alpha, \kappa^{-1}\tilde{\beta}) , \qquad (X, Z) = (X, \kappa\tilde{Z}) , \tag{4.9a,b}$$

reducing the right-hand side of (4.2) to

$$\kappa^{-1} \iint \exp\left\{i\alpha X + i\tilde{\beta}\tilde{Z} - \frac{i\alpha^3 T}{(\alpha^2 + \tilde{\beta}^2)^{1/2}}\right\} d\alpha \, d\tilde{\beta} , \tag{4.9c}$$

except for effects due to Q. Hence this contribution to the downstream response starts with less amplitude, of order κ^{-1}, but then lasts longer since the X-scale stays at $O(1)$. Here (4.8c), (4.9c), when subjected to analysis as in (4.3a,b)ff, can be shown to produce the root equations (4.5), (4.7a), in turn, as expected. More significantly, all the above transonic-flow features, namely (4.6), (4.8a–c) for the *finite-angle waves*, and (4.7b), (4.9a–c) for the *wide quasi-planar waves*, are closely in keeping with the computational results of Fig. 3 at Mach numbers near unity. The relatively rapid spatial decay of the finite-angle contributions, compared with the slow spatial decay of the initially smaller quasi-planar contributions, is particularly noticeable.

In the *supersonic* regime, (4.1b) applies and so, in (4.2),

$$(\alpha^2 + \kappa^2\beta^2)^{1/2} \text{ is replaced by } (\kappa_1^2\beta^2 - \alpha^2)^{1/2} , \tag{4.10}$$

with $\kappa_1 \equiv (M_\infty^2 - 1)^{-1/2}$ ranging from 0 to ∞, and the integration domain is restricted to $|\alpha| < \kappa_1|\beta|$. Again, we obtained computational solutions of (4.2) with (4.10) at several supersonic Mach numbers, as shown in Fig. 4. The analytical working for large times and distances is similar to that for the subsonic case except that now

$$K(\phi) = (\kappa_1^2 \sin^2\phi - \cos^2\phi)^{1/2} \tag{4.11}$$

and hence (4.4) is replaced by

$$\kappa_1^2\chi^4 \tan\theta - \kappa_1^2\chi^3 + 3\kappa_1^2\chi^2 \tan\theta + (\kappa_1^2 + 2)\chi - 2\tan\theta = 0 . \tag{4.12}$$

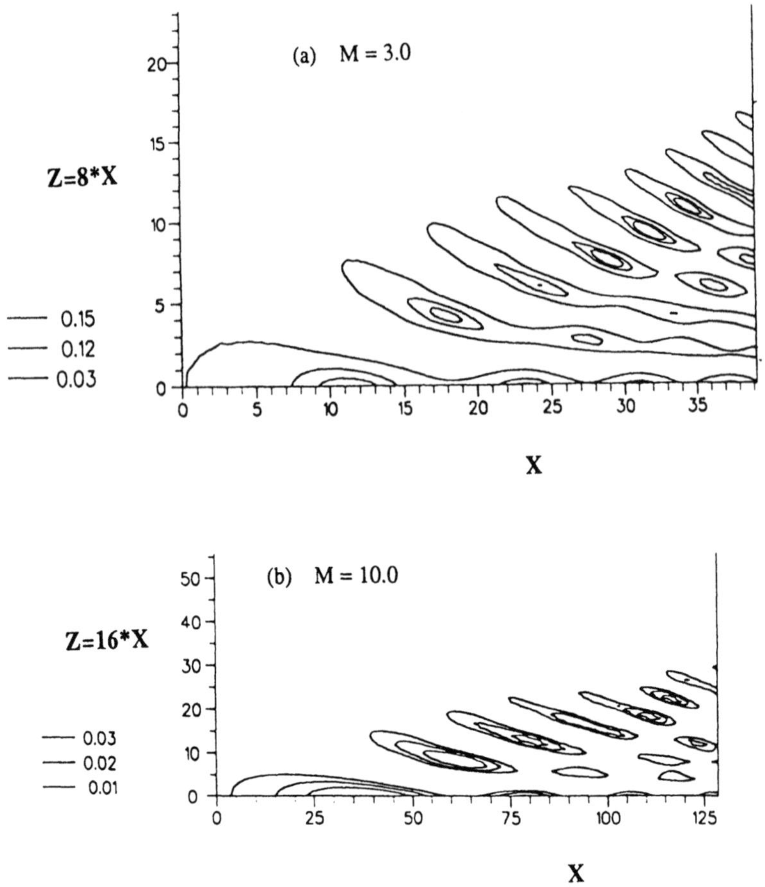

Fig. 4. Numerical results for supersonic range. (See also final page.) (a) $M = 3.0$. (b) $M = 10.0$. The results are similar to those at low Mach numbers but are confined in a very narrow wake.

The *transonic* limit, as $M_\infty \to 1+$, $\kappa_1 \to \infty$, is therefore broadly the same as for $M_\infty \to 1-$, in (4.5)–(4.9c). The *hypersonic limit*, on the other hand, is associated mainly with small angles of spreading downstream. Here $M_\infty \gg 1$, $\kappa_1 \to 0$, and the dominant effects are confined to the interval where $\theta = \kappa_1 \bar{\theta}$, $\chi = \kappa_1^{-1}\bar{\chi}$ with $\bar{\theta}$, $\bar{\chi}$ of $O(1)$. So from (4.12) the governing equation for the four roots present reduces to

$$\bar{\chi}^4\bar{\theta} - \bar{\chi}^3 + 2\bar{\chi} = 0 \,. \tag{4.13}$$

For small $\bar{\theta}$ the four roots are at $\bar{\chi} = \pm 2^{1/2}$, 0, $+\infty$ in effect. Their behaviour for increasing $\bar{\theta}$ is then analogous with that in the previous transonic-flow analysis, and in particular a double root and consequent cut-off arise at $\bar{\chi} = 6^{1/2}$ for $\bar{\theta} = 3^{-3/2}2^{1/2}$, i.e. for the small angle

$$\theta = \theta_c \sim 3^{-3/2}2^{1/2}M_\infty^{-1} \,. \tag{4.14a}$$

Concerning the hypersonic amplitudes provoked, the scalings involved now have

$$(\alpha, \beta, X, Z) = O(\kappa_1, 1, \kappa_1^{-1}, 1) \,. \tag{4.14b}$$

As a result, the typical amplitude is reduced by a factor $\kappa_1(\sim M_\infty^{-1})$, from (4.2), but the typical wake length of the spot increases by a factor $\kappa_1^{-1}(\sim M_\infty)$, at least in terms of the present coordinates. The analytical properties above again agree well with the computations (Fig. 4). Indeed, all the different decay rates summarized in (4.8b), (4.9b), (4.14b) are evident numerically in Figs 3 and 4 with the long scale (4.14b) for the hypersonic regime being particular clear in Fig. 4.

There are two final points of interest here, the first of which concerns the spot response near the line of symmetry, where $\theta \to 0$. In the subsonic range (4.4) gives the two roots

$$\chi = \pm(1 - 2\kappa^{-2})^{1/2}, \tag{4.15a}$$

apart from those at small and large χ. Hence the crossover from two to four waves downstream occurs at the subsonic Mach number

$$M_\infty = 2^{-1/2}. \tag{4.15b}$$

(Dr R.I. Bowles, private communication 1991, has kindly pointed out to us that at the crossover value (4.15b) the spanwise group velocity becomes negative, opposite in sign to the spanwise velocity.) The four-wave description also continues throughout the supersonic range, with the two finite-χ roots been given by

$$\chi = \pm(1 + 2\kappa_1^{-2})^{1/2}, \tag{4.15c}$$

from (4.12). These are continuous with the roots (4.15a) in the transonic limit, both sets then agreeing with the limiting result (4.5). Second, the scales quoted in (4.8), (4.9), (4.14b) hold outside the edge layers that arise near the caustics. In contrast, the amplitude maxima tend to occur inside these layers, cf. (3.9). For the transonic limit for instance the maximum amplitude is found to increase like $\kappa^{5/6}T^{1/3}$, being attained where X is $O(\kappa^{-1}T)$, for the contributions corresponding to (4.8), whereas the edge-layer maximum corresponding to (4.9) decreases like $\kappa^{-1}T^{1/3}$, being attained at $X = O(T)$. Likewise, the hypersonic limit can be analyzed for its edge-layer response, yielding a maximum amplitude which decreases as $\kappa_1^{4/3}T^{1/3}$ and is reached at distances X of order T. These results again are in good agreement with the computations of Figs 3 and 4.

5. Further comments

An investigation into initial-value problems for spots in incompressible or compressible boundary layers has been presented above. Figure 5 summarizes the complete Mach-number range, as regards the wake half-angles computed far downstream, and, as with other quantities, there is good agreement with the analysis. The results, throughout Sections 3 and 4, are felt to be of potential practical value in characterizing the downstream progress of spots after an initial disturbance to the boundary layer. The theory developed so far would not necessarily be expected to be applicable directly to previously computed or experimentally observed spots, at least not until nonlinearity is incorporated (see below), but there is some measure of agreement nevertheless. The main analytical prediction in the incompressible case concerns the concentrated wake effect inside a wedge of half-angle 19.47° downstream (as for Kelvin's ship wake, see Section 1), with the maximum disturbance

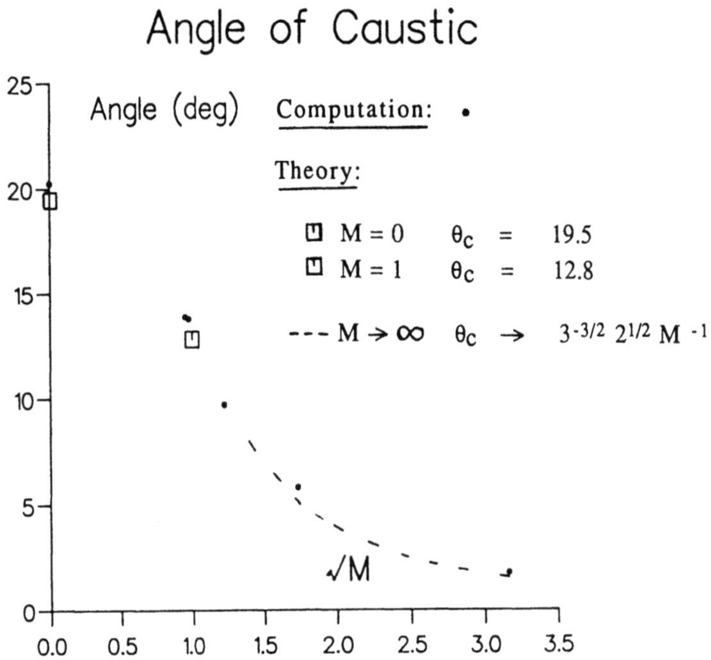

Fig. 5. Summary. Wake half-angle versus Mach number, including comparison between theory and computation.

amplitude occurring near the edges of the wake and with multiple waves in-between. Comparisons between this theoretical response and computational results for the 3D triple-deck problem (see Section 2) are presented by Smith (1987a) and are fairly affirmative; see also Duck (1987), Ryzhov (1987). Comparisons with Gaster & Grant's (1975) laminar-spot experiments are also quite encouraging (Smith 1987a), including the accentuated edge effects downstream at relatively large times, even though in these comparisons and those mentioned in the previous sentence the theory tends to over-estimate the half-angle downstream. Concerning transitional and turbulent spots, there is some additional encouragement both from the general 3D structure (Sections 3 and 4) with multiple waves present downstream, suggesting the study of nonlinear effects there, and from Glezer et al.'s (1989) experimental study focussing on effects near, and just trailing, the edges of spots. In the compressible case, the most interesting features found in the spot development are the changes in the flow scales, structure, and disturbance amplitude, as well as in the wake half-angle, as the Mach number increases [these changes are apart from other factors due to the scalings leading to (2.3) with (4.1), e.g. see Smith 1989]. This is especially so for the transonic and hypersonic ranges, as Section 4 above shows, e.g. with the transonic range producing two spatially distinct kinds of amplitude response downstream, while in the hypersonic range the wake-spreading angle tends to zero.

There are many other areas of possible applications. The influence of cross-flow in the original boundary layer can be incorporated similarly to compressibility effects in Section 4, i.e. by alteration of the pressure-displacement interaction law as in Smith & Stewart (1987b). Channel flows, jets, liquid-layer flows, in certain regimes, and other flows with other interaction laws all require study, as do shorter-scale Rayleigh wave contributions in the incompressible or compressible boundary layer (e.g. Mack 1984, Malik 1987, Brown & Smith 1990) and non-aerodynamic flows. In channel flows, for example, the corresponding wake half-angle is increased to 30°. Spot-disturbance experiments in channel flows, with reduced initial amplitudes, could be of much interest.

Several other applications and consequences suggest themselves, but the nonlinear aspects within the context of Section 2 seem most worthy of emphasis. Among these, there are the 3D nonlinear vortex structures pointed out by Smith (1987b); the wall-sublayer and micro-scales within the Euler stage (SDR); the possibility of small-scale vortex-wave interactions within the present scales; short- or long-scale effects in the spanwise direction; and the eventual occurrence of interactive break-up during sublayer eruption near the wall (Smith 1988, Peridier et al. 1991), as mentioned in Section 1. The most immediate impact of nonlinearity, however, can be seen in the downstream asymptotes of Section 3, for the incompressible regime, and in particular in the finding that the maximum disturbance amplitude occurs in comparatively thin edge-regions. One would therefore expect nonlinear effects to become important first in the edge-regions, rather than in the middle part of the far-wake. Here, as a guide for a subsequent nonlinear study, we note the following. In the edge region, if

$$A \propto T^{-1/2}(XT^{-1/2})^{N}\tilde{A}(\eta) + \cdots \quad \text{at large } X, T \text{ (with } T^{1/2} \ll X \ll T) \tag{5.1}$$

(cf. (3.9)), approximately, with corresponding expansions for the other flow variables in (2.3), (2.5), and appropriately scaled co-ordinates, then substitution into the latter nonlinear equations leads eventually to an Airy equation for $\tilde{A}(\eta)$, exactly consistent with the form (3.9) as expected, provided that the displacement in (5.1) is sufficiently small, specifically $N < N_0$ for some critical value N_0. If $N = N_0$, however, nonlinear terms enter to alter the governing equation. A similar estimate applies to the compressible regime of Section 4. Research on these matters is in progress, by F.T.S., who suggests that N_0 can equal $\frac{1}{3}$ or zero depending on the flow conditions. It can be shown further that the asymptote (3.9) and its counterpart (5.1) in the nonlinear case hold also for the full Euler system (2.1), such that at large times (t) there are in effect two main areas of activity in the spot, namely at distances of orders $t^{1/2}$ and t downstream, just as in Fig. 2.

Thanks are due to AFOSR (grant no. 89-0475) from F.T.S. and SERC from D.J.D. for support of this and related work.

References

Brown, S.N. and Smith, F.T., The inviscid instability of a Blasius boundary layer at large values of the Mach number. *J. Fluid Mech.* 219 (1990) 499–518.

Cheng, H.K. and Johnson, E.R., Inertial waves above an obstacle in an unbounded, rapidly rotating fluid. *Proc. Roy. Soc.* A 383 (1982) 71–87.

Duck, P.W., private communication (1987).

Gaster, M. and Grant, I., An experimental investigation of the formation and development of a wave packet in a laminar boundary layer. *Proc. Roy. Soc.* A 347 (1975) 253–269.

Gaster, M., A theoretical model of a wave packet in the boundary layer on a flat plate. *Proc. Roy. Soc.* A 347 (1975) 271–289.

Gaster, M., The nonlinear phase of wave growth leading to chaos and breakdown to turbulence in a boundary layer as an example of an open system. *Proc. Roy. Soc.* A 430 (1990) 3–24.

Glezer, A., Katz, Y. and Wygnanski, I., On the breakdown of the wave packet trailing a turbulent spot in a laminar boundary layer. *J. Fluid Mech.* 198 (1989) 1–26.

Hoyle, J.M., Smith, F.T. and Walker, J.D.A., On sublayer eruption and vortex formation. *Comp. Phys. Commns.* 65 (1991) 151–157.

Katz, Y., Seifert, A. and Wygnanski, I., On the evolution of the turbulent spot in a laminar boundary layer with a favourable pressure gradient. *J. Fluid Mech.* 221 (1990) 1–22.

Mack, L.M., Boundary-layer linear stability theory. AGARD, Report 709 (1984).

Malik, M.R., Prediction and control of transition in hypersonic boundary layers. A.I.A.A. paper no. 87-1414 (1987).

Peridier, V.J., Smith F.T. and Walker, J.D.A., Vortex-induced boundary-layer separation. Part 1: the unsteady limit problem Re→∞. Part 2: unsteady interacting boundary-layer theory. *J. Fluid Mech.* 232 (1991) 99–131 & 133–165.

Riley, J.J. and Gad-el-Hak, M., In S.H. Davis and J.L. Lumley (eds), *Frontiers in Fluid Mechanics*. Springer-Verlag (1985).

Rothmayer, A.P. and Smith, F.T., Strongly nonlinear wave-packets in boundary layers. *Trans. A.S.M.E., Forum on Unsteady Flow Separation, Cincinnati, Ohio, June 1987*, 52 (1987) 67–79.

Ryzhov, O.S., private communication (1987).

Smith, F.T., Theoretical aspects of steady and unsteady laminar separation. *A.I.A.A. 17th Fluid Dyn., Plasma Dyn. & Lasers Conf.*, 1984 (1984).

Smith, F.T., Theory and computations of 3D steady and unsteady interactive boundary layers. *R.T. Davis Mem. Symp., Cincinnati, Ohio, June 1987* (1987a) in press, *Computers & Fluids* (1991).

Smith, F.T., Break-up in unsteady separation. *Trans. A.S.M.E., Forum on Unsteady Flow Separation, Cincinnati, Ohio, June 1987*, 52 (1987b) 55–64.

Smith, F.T., Finite-time break-up can occur in any unsteady interacting boundary layer. *Mathematika* 35 (1988) 256–273.

Smith, F.T., On the first-mode instability in subsonic, supersonic or hypersonic boundary layers. *J. Fluid Mech.* 198 (1989) 127–153.

Smith, F.T. and Burggraf, O.R. (SB), On the development of large-sized short-scaled disturbances in boundary layers. *Proc. Roy. Soc.* A 399 (1985) 25–35.

Smith, F.T., Doorly, D.J. and Rothmayer, A.P. (SDR), On displacement-thickness, wall-layer and mid-flow scales in turbulent boundary layers, and slugs of vorticity in pipes and channels. *Proc. Roy. Soc.* A 428 (1990) 255–281.

Smith, F.T. and Stewart, P.A., The resonant-triad nonlinear interaction in boundary-layer transition. *J. Fluid Mech.* 179 (1987a) 227–252.

Smith, F.T. and Stewart, P.A., Three-dimensional instabilities in steady and unsteady nonparallel boundary layers, including effects of Tollmien–Schlichting disturbances and cross-flow. *Proc. Roy. Soc.* A 409 (1987b) 229–248.

Walker, J.D.A. and Smith, C.R., Theme Issue on Turbulent Flow. *Trans. Roy. Soc.* A, in press (1991).

Whitham, G.B., *Linear and Nonlinear Waves*. Wiley-Interscience (1974).

Computational results

Using the transformation

$$(X, Z) = T^{1/2}(\bar{X}, \bar{Z}),$$
$$(\alpha, \beta) = T^{-1/2}(\bar{\alpha}, \bar{\beta}),$$

the computational results show $T \cdot \hat{A}(\bar{X}, \bar{Z}, T)$, with $Q(\bar{\alpha}, \bar{\beta}, T) = \exp[-(\bar{\alpha}^2 + \bar{\beta}^2)/16 \cdot T]$, for various values of time, T, and Mach number M, in Figs 1, 3, 4.

Journal of Engineering Mathematics **26**: 107–118, 1992.
H.K. Kuiken and S.W. Rienstra (eds), Problems in Applied, Industrial and Engineering Mathematics.

Plasticity theory for fibre-reinforced composites

A.J.M. SPENCER

Department of Theoretical Mechanics, University of Nottingham, Nottingham NG7 2RD, UK

Abstract. Recent experiments reported in [1] show that for a boron-aluminium fibre-reinforced composite plastic yielding is effectively independent of tension in the fibre direction over a wide range of values of this tension. This confirms a long-standing conjecture by the author and colleagues. The results have major implications for the formulation of plasticity theories for fibre-reinforced materials. This paper reviews the theory of anisotropic plasticity based on the usual assumptions of plasticity theory, together with the property that yielding is not affected by a superposed tension in the fibre direction. Yield conditions, flow rules and hardening rules are formulated for uniaxial reinforcement; brief consideration is given to a material reinforced by two families of fibres.

1. Introduction

Dvorak, Bahei-el-din, Macheret and Liu [1] have recently reported some important experiments on the elastic-plastic behaviour of a fibre-reinforced boron-aluminium composite. The experiments were performed on axially reinforced tubular specimens loaded by axial tension, torsion and internal pressure. Among other results, they show that over a wide range of values of the direct stress in the fibre-direction (essentially over the range in which the yield or failure stress of the fibre is not exceeded) the yield behaviour of the composite is independent of this direct stress in the fibre direction. The effect is shown in Fig. 1, which is based on Fig. 7 of [1]. The yield surface in the relevant stress space is 'sausage-shaped', being a fairly long cylinder with closed ends. It is also shown by the extensive experiments described in [1] that subsequent to plastic deformation the yield surface retains this sausage-shape but translates in stress-space, demonstrating a large degree of kinematic hardening.

These results provide striking confirmation of a conjecture made by the author and colleagues over twenty years ago [2–5]. The results have important implications for the formulation of theories of plasticity for fibre-reinforced composites, with particular reference to metal-matrix composites. These implications have been explored in previous publications [2–15] but prior to the availability of experimental information, the theory had to be regarded in a rather tentative manner. As the experimental data now exist, it seems worthwhile to draw attention to this work. This paper is a summary of the theory proposed in the papers cited above.

A theory of the *rigid-plastic* plastic behaviour of materials reinforced by a single family of fibres was formulated by Mulhern, Rogers and Spencer [2]. This theory assumed the material to be inextensible in the fibre direction, and it was shown that, as a consequence of this assumption and the associated flow rule, the yield function is independent of the fibre tension (that is, the direct stress in the fibre direction). Subsequently the same authors [3] formulated an *elastic-plastic* theory of fibre-reinforced materials, in which it was assumed that the *plastic* part of the extensional strain in the fibre direction is zero, and in this case also it follows from the associated flow rule that plastic yielding is independent of fibre

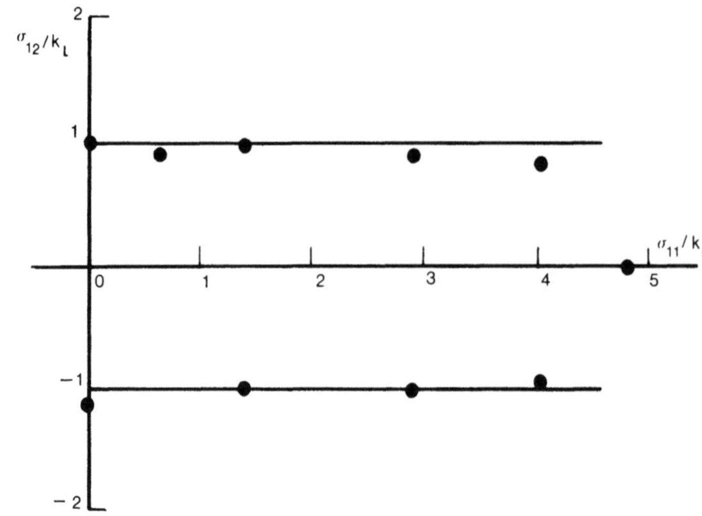

Fig. 1. Experimental results by Dvorak et al. [1] for initial yield of a boron-aluminium fibre-reinforced composite, with fibres parallel to the x_1-axis.

tension. However, this theory does *not* require inextensible fibres, because it admits an elastic strain in the fibre direction.

Smith and Spencer [4] formulated a *rigid-plastic* theory for material reinforced by two families of inextensible fibres, but adopted a slightly different point of view; they postulated that yielding is independent of fibre tensions. It then follows, if the associated flow rule is adopted, that the plastic extensional strain in a fibre direction is zero. The same viewpoint was taken in several subsequent publications, for example [5–15]. The assumptions of (a) inextensibility in the fibre direction, and (b) yield independent of fibre tension, are, in conjunction with the associated flow rule, virtually equivalent in rigid-plastic theory. However, in an elastic-plastic theory, (b) is less restrictive than (a), because it permits an axial elastic strain. Since (b) is susceptible to direct experimental observation, it seems to be the more natural choice of postulate for materials which exhibit this kind of behaviour.

2. Initial yield conditions – one family of fibres

We refer all vector and tensor quantities to a system of rectangular cartesian coordinates x_i. The Cauchy stress is denoted by $\boldsymbol{\sigma}$, with cartesian components σ_{ij}.

In most theories of metal plasticity, we postulate a yield function $f(\sigma_{ij})$ such that in admissible stress states $f \leqslant 0$, with $f = 0$ when plastic deformation is taking place. If the plastic material is isotropic then f can be expressed as a function of the stress invariants tr $\boldsymbol{\sigma}$, tr $\boldsymbol{\sigma}^2$ and tr $\boldsymbol{\sigma}^3$. In *isotropic* metal plasticity it is observed experimentally that for many materials yielding is effectively independent of a superposed hydrostatic pressure, or equivalently of tr $\boldsymbol{\sigma}$. This observation greatly simplifies the formulation and application of the theory. It is incorporated into the theory by restricting f to depend on $\boldsymbol{\sigma}$ only through the deviatoric stress $\mathbf{s}$, where

$$\mathbf{s} = \boldsymbol{\sigma} - \tfrac{1}{3}\mathbf{I} \operatorname{tr} \boldsymbol{\sigma} . \tag{2.1}$$

Then $\operatorname{tr} \mathbf{s} = 0$ and f can be expressed as a function of $\operatorname{tr} \mathbf{s}^2$ and $\operatorname{tr} \mathbf{s}^3$.

For anisotropic materials f is a function of $\boldsymbol{\sigma}$ (or $\mathbf{s}$) which is invariant under the appropriate transformation group which describes the anisotropy. A material reinforced by a single family of fibres, randomly distributed in their cross-sectional planes, is locally transversely isotropic with respect to the local fibre direction, which can be characterised by a unit vector $\mathbf{a}$. In this case f is a function of the five invariants [2, 5, 12]

$$\operatorname{tr}\boldsymbol{\sigma}\,, \qquad \operatorname{tr}\boldsymbol{\sigma}^2\,, \qquad \operatorname{tr}\boldsymbol{\sigma}^3\,, \qquad \mathbf{a}\cdot\boldsymbol{\sigma}\cdot\mathbf{a}\,, \qquad \mathbf{a},\boldsymbol{\sigma}^2\cdot\mathbf{a}\,. \tag{2.2}$$

For fibre-reinforced metals we expect initial yielding to remain independent of $\operatorname{tr}\boldsymbol{\sigma}$. We now have in [1] experimental confirmation that f is independent of the stress $\mathbf{a}\cdot\boldsymbol{\sigma}\cdot\mathbf{a}$ in the fibre direction over a wide range of values of this stress. The extra-stress $\mathbf{s}'$, defined as

$$\mathbf{s}' = \boldsymbol{\sigma} - \tfrac{1}{2}(\operatorname{tr}\boldsymbol{\sigma} - \mathbf{a}\cdot\boldsymbol{\sigma}\cdot\mathbf{a})\mathbf{I} + \tfrac{1}{2}(\operatorname{tr}\boldsymbol{\sigma} - 3\mathbf{a}\cdot\boldsymbol{\sigma}\cdot\mathbf{a})\mathbf{a}\otimes\mathbf{a}\,, \tag{2.3}$$

(where $\mathbf{a}\otimes\mathbf{a}$ denotes the tensor product) has the property that $\mathbf{s}'$ is unchanged if $\boldsymbol{\sigma}$ is replaced by $\boldsymbol{\sigma} - p\mathbf{I} + T\mathbf{a}\otimes\mathbf{a}$ for arbitrary p (representing a hydrostatic pressure) and T (representing a tension in the fibre direction). Thus $\mathbf{s}'$ is independent of hydrostatic pressure and fibre tension. Consequently, if $f(\sigma_{ij})$ is expressed as a function of $\mathbf{s}'$, then f is independent of hydrostatic pressure and fibre tension. Thus $\mathbf{s}'$ has the same role in relation to independence of hydrostatic pressure and fibre tension as $\mathbf{s}$ has in relation to independence of hydrostatic pressure only. Furthermore, it follows from (2.3) that

$$\operatorname{tr}\mathbf{s}' = 0\,, \qquad \mathbf{a}\cdot\mathbf{s}'\cdot\mathbf{a} = 0\,. \tag{2.4}$$

Now the set (2.2) is equivalent to

$$\operatorname{tr}\boldsymbol{\sigma}\,, \qquad \operatorname{tr}\mathbf{s}'^2\,, \qquad \operatorname{tr}\mathbf{s}'^3\,, \qquad \mathbf{a}\cdot\boldsymbol{\sigma}\cdot\mathbf{a}\,, \qquad \mathbf{a}\cdot\mathbf{s}'^2\cdot\mathbf{a}\,,$$

so, if f is independent of $\operatorname{tr}\boldsymbol{\sigma}$ and $\mathbf{a}\cdot\boldsymbol{\sigma}\cdot\mathbf{a}$, and the material is transversely isotropic with respect to the direction $\mathbf{a}$, then f may be expressed as a function of $\operatorname{tr}\mathbf{s}'^2$, $\mathbf{a}\cdot\mathbf{s}'^2\cdot\mathbf{a}$, $\operatorname{tr}\mathbf{s}'^3$. For our purposes it is more convenient to use the equivalent set

$$J_1 = \tfrac{1}{2}\operatorname{tr}\mathbf{s}'^2 - \mathbf{a}\cdot\mathbf{s}'^2\cdot\mathbf{a}\,, \qquad J_2 = \mathbf{a}\cdot\mathbf{s}'^2\cdot\mathbf{a}\,, \qquad J_3 = \tfrac{1}{3}\operatorname{tr}\mathbf{s}'^3\,. \tag{2.5}$$

The restriction to dependence on these three invariants, together with the relations (2.4), represents a considerable simplification compared to a (2.2).

In applications it is usually necessary to specify a particular form for f. In isotropic plasticity the commonly used yield functions are those which correspond to von Mises' and Tresca's yield conditions. For transversely isotropic materials, two forms which have been found useful in applications, and to agree well with experiment, are

$$\text{(a)}\quad f = \frac{1}{k_T^2}J_1 + \frac{1}{k_L^2}J_2 - 1\,, \tag{2.6}$$

and

$$\text{(b)}\quad f = \begin{cases} \dfrac{J_1^{1/2}}{k_T} - 1\,, & \text{for } J_2 \leqslant k_L^2\,,\\[2ex] \dfrac{J_2^{1/2}}{k_L} - 1\,, & \text{for } J_1 \leqslant k_T^2\,. \end{cases} \tag{2.7}$$

Here k_T and k_L are shear yield stresses for shear on planes containing the fibres, in directions transverse and parallel to the fibres respectively. The form (2.6) is the most general yield function of the required form that is quadratic in the stress components, and so may be regarded as an analogue of von Mises' condition. The form (2.7) is a maximum shear stress condition, and so is analogous to Tresca's condition for isotropic plasticity.

If the coordinate system is chosen so that **a** lies parallel to the x_1-axis (in conformity with the notation used in [1]), then $\mathbf{a} = (1, 0, 0)$ and (2.6) becomes

$$f = \frac{1}{k_T^2}\left\{\frac{1}{4}(\sigma_{22} - \sigma_{33})^2 + \sigma_{23}^2\right\} + \frac{1}{k_L^2}(\sigma_{12}^2 + \sigma_{13}^2) - 1\,, \tag{2.8}$$

whilst (2.7) takes the form

$$f = \begin{cases} \dfrac{1}{k_T}\left\{\dfrac{1}{4}(\sigma_{22} - \sigma_{33})^2 + \sigma_{23}^2\right\}^{1/2} - 1\,, & \text{for } (\sigma_{12}^2 + \sigma_{13}^2)^2 \leqslant k_L^2\,, \\ \dfrac{1}{k_L}(\sigma_{12}^2 + \sigma_{13}^2)^{1/2} - 1\,, & \text{for } \dfrac{1}{4}(\sigma_{22} - \sigma_{33})^2 + \sigma_{23}^2 \leqslant k_T^2\,. \end{cases} \tag{2.9}$$

The parameters k_T and k_L are identified as shear yield stresses for shear on planes containing the fibres, in directions transverse and parallel to the fibres respectively.

3. Associated flow rule

We denote the velocity vector by **v** and its cartesian components by v_i. The rate of deformation tensor **d** has components d_{ij} defined by

$$d_{ij} = \frac{1}{2}\left(\frac{\partial v_i}{\partial x_j} + \frac{\partial v_j}{\partial x_i}\right). \tag{3.1}$$

As is usual in plasticity, 'time' here need not be real time, but may be any parameter that orders the sequence of events. In the formulation of isotropic elastic-plastic theory it is usual to decompose **d** into an elastic part $\mathbf{d}^e$ and a plastic part $\mathbf{d}^p$. The decomposition may be done in various ways. When the elastic part of the strain is small there is little difference between the various formulations so we adopt a simple additive decomposition and set

$$\mathbf{d} = \mathbf{d}^e + \mathbf{d}^p\,. \tag{3.2}$$

The elastic strain-rate $\mathbf{d}^e$ is assumed to be related to the stress-rate by the elastic stress-strain law. For large deformations the stress-rate must be defined in a properly objective manner. Anisotropic behaviour does not raise any new issues in this regard so we do not elaborate on this subject. We observe that linear elastic stress-strain relations for transversely isotropic materials are well-known [e.g. 5, 9, 11–15].

It is usual to assume that the plastic strain-rate is related to the stress by the associated flow rule

$$d_{ij}^p = \dot{\lambda}\frac{\partial f}{\partial \sigma_{ij}}\,, \tag{3.3}$$

where for a perfectly plastic material $\dot{\lambda}$ is a positive scalar factor of proportionality, and for a strain-hardening plastic material $\dot{\lambda}$ depends on strain history. We consider the hardening case further below. The arguments used in support of the adoption of (3.3) are not affected by anisotropy or kinematic constraints. We follow common practice and assume (3.3). Then when f is expressed as a function of J_1, J_2 and J_3 we have

$$d_{ij}^p = \dot{\lambda} \sum_{\alpha=1}^{3} \frac{\partial f}{\partial J_\alpha} \frac{\partial J_\alpha}{\partial \sigma_{ij}} .$$

It is straightforward to verify that

$$\frac{\partial J_\alpha}{\partial \sigma_{ii}} = 0 , \qquad a_i a_j \frac{\partial J_\alpha}{\partial \sigma_{ij}} = 0 , \quad (\alpha = 1, 2, 3) .$$

Hence, in this model

$$\operatorname{tr} \mathbf{d}^p = 0 , \qquad \mathbf{a} \cdot \mathbf{d}^p \cdot \mathbf{a} = 0 . \tag{3.4}$$

Therefore, plastic incompressibility and plastic inextensibility in the fibre direction are *consequences* of the assumption that yield is independent of hydrostatic pressure and fibre tension, together with the associated flow rule.

For the particular forms of f given by (2.6) and (2.7), (3.3) gives, for the yield function (2.6)

$$\mathbf{d}^p = \dot{\lambda} \left\{ \frac{1}{k_T^2} (\mathbf{s}' - \mathbf{a} \otimes \mathbf{a} \cdot \mathbf{s}' - \mathbf{s}' \cdot \mathbf{a} \otimes \mathbf{a}) + \frac{1}{k_L^2} (\mathbf{a} \otimes \mathbf{a} \cdot \mathbf{s}' + \mathbf{s}' \mathbf{a} \otimes \mathbf{a}) \right\} , \tag{3.5}$$

and for the yield function (2.7)

$$\mathbf{d}^p = \begin{cases} \dot{\mu}(\mathbf{s}' - \mathbf{a} \otimes \mathbf{a} \cdot \mathbf{s}' - \mathbf{s}' \cdot \mathbf{a} \otimes \mathbf{a}) , & J_1 = k_T^2 , \quad J_2 < k_L^2 , \\ \dot{\nu}(\mathbf{a} \otimes \mathbf{a} \cdot \mathbf{s}' + \mathbf{s}' \cdot \mathbf{a} \otimes \mathbf{a}) , & J_1 < k_T^2 , \quad J_2 = k_L^2 , \\ \dot{\mu}\mathbf{s}' + (\dot{\nu} - \dot{\mu})(\mathbf{a} \otimes \mathbf{a} \cdot \mathbf{s}' + \mathbf{s}' \cdot \mathbf{a} \otimes \mathbf{a}) , & J_1 = k_T^2 , \quad J_2 = k_L^2 , \end{cases} \tag{3.6}$$

where $\dot{\mu}$ and $\dot{\nu}$ are positive multipliers.

If the coordinate system is chosen so that $\mathbf{a} = (1, 0, 0)$, then from (2.3)

$$\mathbf{s}' - \mathbf{a} \otimes \mathbf{a} \cdot \mathbf{s}' - \mathbf{s}' \cdot \mathbf{a} \otimes \mathbf{a} = \begin{bmatrix} 0 & 0 & 0 \\ 0 & \frac{1}{2}(\sigma_{22} - \sigma_{33}) & \sigma_{23} \\ 0 & \sigma_{23} & \frac{1}{2}(\sigma_{33} - \sigma_{22}) \end{bmatrix} ,$$

$$\mathbf{a} \otimes \mathbf{a} \cdot \mathbf{s}' + \mathbf{s}' \cdot \mathbf{a} \otimes \mathbf{a} = \begin{bmatrix} 0 & \sigma_{12} & \sigma_{13} \\ \sigma_{12} & 0 & 0 \\ \sigma_{13} & 0 & 0 \end{bmatrix} ,$$

and the resulting expressions for $\mathbf{d}^p$ are given by inserting these in (3.5) and (3.6).

4. Proportional hardening

The post-yield behaviour even of initially isotropic plastic materials is complicated. An idealisation often introduced in isotropic plasticity theory is that the current state of hardening can be described by a single parameter, usually taken to be the shear yield stress, k. It is further assumed that the current value of k depends on the deformation history through an 'equivalent strain', ε, such that the convected derivative $\dot{\varepsilon}$ is a function of the current plastic strain rate d_{ij}^P and homogeneous of degree -1 in time. It is usually not explicitly stated, but is clearly necessary, that for isotropic response $\dot{\varepsilon}$ must be an invariant of d_{ij}^P. Thus at most $\dot{\varepsilon}$ is a function of

$$d_{ii}^P, \quad d_{ij}^P d_{ij}^P, \quad d_{ij}^P d_{jk}^P d_{ki}^P. \tag{4.1}$$

However $d_{ii}^P = 0$, and so is discarded. The invariable practice is to identify (apart from a numerical factor which is unimportant)

$$\dot{\varepsilon} = (d_{ij}^P d_{ij}^P)^{1/2}, \tag{4.2}$$

although there seems no essential reason not to include also the third of (4.1). From (4.2) we then have

$$\varepsilon = \int_0^t (d_{ij}^P d_{ij}^P)^{1/2} \, \mathrm{d}t,$$

and $k = k(\varepsilon)$. This constitutive assumption is termed isotropic hardening.

We seek to extend these ideas to fibre-reinforced plastic materials. It is clear that in this case the description of the current state of hardening will require more than a single parameter; as a minimum the two shear yield stresses k_T and k_L are needed. Nor can it be assumed that k_T and k_L will depend on the deformation history through a single parameter, because, for example, shearing across and along the fibres may well affect the hardening differently. Thus we have to define several 'equivalent strain' parameters whose time derivatives are invariant under the symmetry transformation appropriate to transverse isotropy. These are the invariants of $\mathbf{d}^P$ analogous to the invariants (2.2) of $\boldsymbol{\sigma}$, namely

$$\operatorname{tr} \mathbf{d}^P, \quad \operatorname{tr}(\mathbf{d}^P)^2, \quad \operatorname{tr}(\mathbf{d}^P)^3, \quad \mathbf{a} \cdot \mathbf{d}^P \cdot \mathbf{a}, \quad \mathbf{a} \cdot (\mathbf{d}^P)^2 \cdot \mathbf{a}. \tag{4.3}$$

However $\operatorname{tr} \mathbf{d}^P$ and $\mathbf{a} \cdot \mathbf{d}^P \cdot \mathbf{a}$ are both zero. We follow the practice of isotropic plasticity and discard the third of (4.3), leaving two equivalent strain parameters which are derived from $\operatorname{tr}(\mathbf{d}^P)^2$ and $\mathbf{a} \cdot (\mathbf{d}^P)^2 \cdot \mathbf{a}$. A convenient choice is ε_T and ε_L, where

$$\dot{\varepsilon}_T^2 = \tfrac{1}{2} \operatorname{tr}(\mathbf{d}^P)^2 - \mathbf{a} \cdot (\mathbf{d}^P)^2 \cdot \mathbf{a}, \qquad \dot{\varepsilon}_L^2 = \mathbf{a} \cdot (\mathbf{d}^P)^2 \cdot \mathbf{a}. \tag{4.4}$$

In particular, if the coordinate system is chosen so that $\mathbf{a} = (1, 0, 0)$, then

$$\dot{\varepsilon}_T^2 = \tfrac{1}{4}(d_{22} - d_{33})^2 + d_{23}^2, \qquad \dot{\varepsilon}_L^2 = d_{12}^2 + d_{13}^2,$$

and it is clear that ε_T and ε_L are associated with transverse and longitudinal shearing respectively. Our constitutive assumption is then

$$k_T = k_T(\varepsilon_T, \varepsilon_L), \qquad k_L = k_L(\varepsilon_T, \varepsilon_L). \tag{4.5}$$

This appears to be the simplest analogue for transverse isotropy of the isotropic hardening theory of isotropic plasticity. Since it seems inconsistent to talk of isotropic hardening of an anisotropic material, we use the term 'proportional hardening' for this theory. Under proportional hardening the yield surface has a fixed centre at the origin of stress space and expands (for strain-hardening material) as plastic deformation proceeds. The expansion is uniform in any given direction in stress space, but differs with the direction. Thus for example a yield surface that was initially a hyper-ellipse would evolve to another hyper-ellipse, but in general the ratios of the lengths of the principal axes would not remain constant.

It remains to relate $\dot{\lambda}$ to the loading parameters. From (4.5)

$$\dot{k}_T = \frac{\partial k_T}{\partial \varepsilon_T}\,\dot{\varepsilon}_T + \frac{\partial k_T}{\partial \varepsilon_L}\,\dot{\varepsilon}_L\,, \qquad \dot{k}_L = \frac{\partial k_L}{\partial \varepsilon_T}\,\dot{e}_T + \frac{\partial k_L}{\partial \varepsilon_L}\,\dot{\varepsilon}_L\,. \tag{4.6}$$

We confine discussion to the yield functions (2.6) and (2.7) and their associated flow rules (3.5) and (3.6). For (2.6) we have, using (3.5) and (4.4)

$$\dot{\varepsilon}_T^2 = \dot{\lambda}^2 k_T^{-4} J_1\,, \qquad \dot{\varepsilon}_L^2 = \dot{\lambda}^2 k_L^{-4} J_2\,. \tag{4.7}$$

Hence from (2.6), when $f = 0$

$$\dot{\lambda}^2 = k_T^2 \dot{\varepsilon}_T^2 + k_L^2 \dot{\varepsilon}_L^2\,. \tag{4.8}$$

Also from (2.6), during plastic loading, when $\dot{f} = 0$ we have

$$k_T^{-2}\dot{J}_1 + k_L^{-2}\dot{J}_2 - 2(k_T^{-3}\dot{k}_T J_1 + k_L^{-3}\dot{k}_L J_2) = 0\,. \tag{4.9}$$

Hence, from (4.6), (4.8) and (4.9)

$$\begin{aligned}\dot{\lambda}\Big\{J_1^{1/2}k_T^{-2}\Big(k_T^{-3}J_1\,\frac{\partial k_T}{\partial \varepsilon_T} + k_L^{-3}J_2\,\frac{\partial k_L}{\partial \varepsilon_T}\Big) + J_2^{1/2}k_L^{-2}\Big(k_T^{-3}J_1\,\frac{\partial k_T}{\partial \varepsilon_L} + k_L^{-3}J_2\,\frac{\partial k_L}{\partial \varepsilon_L}\Big)\Big\}\\ = \frac{1}{2}\,(k_T^{-2}\dot{J}_1 + k_L^{-2}\dot{J}_2)\,.\end{aligned} \tag{4.10}$$

This determines $\dot{\lambda}$ in terms of the current stress, the current values of k_T and k_L, the hardening parameters $\partial k_T/\partial\varepsilon_T$, $\partial k_T/\partial\varepsilon_L$, $\partial k_L/\partial\varepsilon_T$, $\partial k_L/\partial\varepsilon_L$, and the rate of change of the stress invariants J_1 and J_2. The corresponding deformation-rate, in plastic loading, is then given by (3.5).

The plastic work-rate $\dot{W}_p$ is given by

$$\dot{W}_p = d_{ij}^p \sigma_{ij} = d_{ij}^p s'_{ij}\,, \tag{4.11}$$

and hence, from (2.5) and (3.5), for the yield function (2.6)

$$\dot{W}_p = 2\dot{\lambda}\Big(\frac{J_1}{k_T^2} + \frac{J_2}{k_L^2}\Big) = 2\dot{\lambda}\,, \tag{4.12}$$

and, from (4.8)

$$\dot{W}_p = 2(k_T^2 \dot{\varepsilon}_T^2 + k_L^2 \dot{\varepsilon}_L^2)^{1/2} . \tag{4.13}$$

Some possibilities for simplification suggest themselves. One is that k_T and k_L are functions only of the plastic work W_p. This is a plausible assumption in the case of a metal matrix reinforced by elastic fibres, if it is considered that the hardening of the composite originates in isotropic hardening of the matrix. In this case, from (4.12)

$$\dot{k}_T = 2 \frac{\mathrm{d}k_T}{\mathrm{d}W_p} \dot{\lambda} , \qquad \dot{k}_L = 2 \frac{\mathrm{d}k_L}{\mathrm{d}W_p} \dot{\lambda} , \tag{4.14}$$

and it follows from (4.9) and (4.14) that

$$\dot{\lambda} \left\{ k_T^{-3} J_1 \frac{\mathrm{d}k_T}{\mathrm{d}W_p} + k_L^{-3} J_2 \frac{\mathrm{d}k_L}{\mathrm{d}W_p} \right\} = \frac{1}{4} (k_T^{-2} \dot{J}_1 + k_L^{-2} \dot{J}_2) , \tag{4.15}$$

which determines $\dot{\lambda}$ in terms of the current stress and stress rate. A further possible simplification is to suppose that the ratio k_T/k_L remains constant, which is also plausible if the composite hardening results solely from isotropic hardening of the matrix.

Another possible simplification arises if it is assumed that the mechanisms for strain-hardening in shear in the directions parallel and normal to the fibres are independent. In this case

$$k_T = k_T(\varepsilon_T) , \qquad k_L = k_L(\varepsilon_L) ,$$

and (4.10) simplifies accordingly.

Similar considerations apply if the yield function (2.7) is adopted, with its associated flow rule (3.6). In this case, from (4.4) and (3.6)

$$\dot{\varepsilon}_T^2 = \begin{cases} \dot{\mu}^2 J_1 \\ 0 \\ \dot{\mu}^2 J_1 \end{cases} , \qquad \dot{\varepsilon}_L^2 = \begin{cases} 0 & J_1 = k_T^2 , \quad J_2 < k_L^2 \\ \dot{\nu}^2 J_2 , & J_1 < k_T^2 , \quad J_2 = k_L^2 \\ \dot{\nu}^2 J_2 & J_1 = k_T^2 , \quad J_2 = k_L^2 \end{cases} . \tag{4.16}$$

During plastic loading, when $\dot{f} = 0$, either

$$J_1 = k_T^2 , \qquad \dot{J}_1 = 2k_T \dot{k}_T , \qquad J_2 < k_L^2 ,$$

or

$$J_1 < k_T^2 , \qquad J_2 = k_L^2 , \qquad \dot{J}_2 = 2k_L \dot{k}_L , \tag{4.17}$$

or

$$J_1 = k_T^2 , \qquad J_2 = k_L^2 , \qquad \dot{J}_1 = 2k_T \dot{k}_T , \qquad \dot{J}_2 = 2k_L \dot{k}_L .$$

Hence from (4.6), (4.16) and (4.17), during plastic loading

$$\begin{aligned} \dot{\mu} \frac{\partial k_T}{\partial \varepsilon_T} &= \frac{1}{2} k_T^{-2} \dot{J}_1 \quad \text{if } J_1 = k_T^2 , \quad J_2 < k_L^2 , \\ \dot{\nu} \frac{\partial k_L}{\partial \varepsilon_L} &= \frac{1}{2} k_L^{-2} \dot{J}_2 \quad \text{if } J_1 < k_T^2 , \quad J_2 = k_L^2 , \end{aligned} \tag{4.18}$$

and at a vertex $J_1 = k_T^2$, $J_2 = k_L^2$ of the yield surface, $\dot{\mu}$ and $\dot{\nu}$ are given by

$$\begin{aligned} \dot{\mu} k_T \frac{\partial k_T}{\partial \varepsilon_T} + \dot{\nu} k_L \frac{\partial k_T}{\partial \varepsilon_L} &= \frac{1}{2} k_T^{-1} \dot{J}_1 , \\ \dot{\mu} k_T \frac{\partial k_L}{\partial \varepsilon_T} + \dot{\nu} k_L \frac{\partial k_L}{\partial \varepsilon_L} &= \frac{1}{2} k_L^{-1} \dot{J}_2 . \end{aligned} \tag{4.19}$$

The plastic work-rate associated with (3.6) is, using (4.16)

$$\dot{W}_p = \begin{cases} 2\dot{\mu} J_1 = 2\dot{\mu} k_T^2 = 2\dot{\varepsilon}_T k_T & \text{when } J_1 = k_T^2 , \quad J_2 < k_L^2 \\ 2\dot{\nu} J_2 = 2\dot{\nu} k_L^2 = 2\dot{\varepsilon}_L k_L & \text{when } J_1 < k_T^2 , \quad J_2 = k_L^2 \\ 2(\dot{\mu} J_1 + \dot{\nu} J_2) = 2(\dot{\mu} k_T^2 + \dot{\nu} k_L^2) = 2(\dot{\varepsilon}_T k_T + \dot{\varepsilon}_L k_L) & \text{when } J_1 = k_T^2 , \quad J_2 = k_L^2 \end{cases} . \tag{4.20}$$

In the special case in which k_T and k_L depend only on W_p, it follows from (4.20) that

$$\dot{k}_T = 2 \frac{\mathrm{d}k_T}{\mathrm{d}W_p} \begin{cases} \dot{\mu} k_T^2 , \\ \dot{\nu} k_L^2 , \\ \dot{\mu} k_T^2 + \dot{\nu} k_L^2 , \end{cases} \qquad \dot{k}_L = 2 \frac{\mathrm{d}k_L}{\mathrm{d}W_p} \begin{cases} \dot{\mu} k_T^2 , & J_1 = k_T^2 , \quad J_2 < k_L^2 \\ \dot{\nu} k_L^2 , & J_1 < k_T^2 , \quad J_2 = k_L^2 \\ \dot{\mu} k_T^2 + \dot{\nu} k_L^2 , & J_1 = k_T^2 , \quad J_2 = k_L^2 \end{cases} . \tag{4.21}$$

Then, from (4.17) and (4.21), $\dot{\mu}$ and $\dot{\nu}$ are given by

$$\dot{\mu} \frac{\mathrm{d}k_T}{\mathrm{d}W_p} = \frac{1}{4} k_T^{-3} \dot{J}_1 , \quad \text{when } J_1 = k_T^2 , \quad J_2 < k_L^2 ,$$

$$\dot{\nu} \frac{\mathrm{d}k_L}{\mathrm{d}W_p} = \frac{1}{4} k_L^{-3} \dot{J}_2 , \quad \text{when } J_1 < k_T^2 , \quad J_2 = k_L^2 ,$$

$$\left. \begin{aligned} (\dot{\mu} k_T^2 + \dot{\nu} k_L^2) \frac{\mathrm{d}k_T}{\mathrm{d}W_p} &= \frac{1}{4} \dot{J}_1 k_T^{-1} \\ (\dot{\mu} k_T^2 + \dot{\nu} k_L^2) \frac{\mathrm{d}k_L}{\mathrm{d}W_p} &= \frac{1}{4} \dot{J}_2 k_L^{-1} \end{aligned} \right\} \quad \text{when } J_1 = k_T^2 , \quad J_2 = k_L^2 .$$

Except perhaps when a vertex of the yield surface is involved, the 'Tresca-type' yield function (2.7) usually leads to simpler formulae than the 'von Mises-type' yield function (2.6).

5. Kinematic hardening

Kinematic hardening is characterised by a tensor $\boldsymbol{\alpha}$ termed the 'back stress' or 'shift tensor' which represents a translation of the yield surface in stress space that depends on the strain history. For materials of the class under consideration, in the yield condition $\mathbf{s}'$ is replaced by $\mathbf{s}' - \boldsymbol{\alpha}$. The tensor $\boldsymbol{\alpha}$ must satisfy the same constraints as $\mathbf{s}'$, so we require

$$\operatorname{tr} \boldsymbol{\alpha} = 0 , \qquad \mathbf{a} \cdot \boldsymbol{\alpha} \cdot \mathbf{a} = 0 . \tag{5.1}$$

It is necessary to specify evolution equations for $\boldsymbol{\alpha}$. For an isotropic material, it is common to adopt the rule proposed by Prager

$$\dot{\boldsymbol{\alpha}} = c(\varepsilon)\mathbf{d}^p . \tag{5.2}$$

For finite deformations, $\dot{\boldsymbol{\alpha}}$ must be an objective tensor-rate; the appropriate choice of $\dot{\boldsymbol{\alpha}}$ is a matter of discussion which we do not enter into here.

We seek a suitable generalisation of (5.2) for transversely isotropic materials. It seems clear that $\dot{\boldsymbol{\alpha}}$ must depend on both ε_L and ε_T. Also the natural generalisation of (5.2) is to propose that $\dot{\boldsymbol{\alpha}}$ is a linear tensor function of $\mathbf{d}^p$ with the appropriate invariance properties. This leads us to the form

$$\dot{\boldsymbol{\alpha}} = c_1(\varepsilon_T, \varepsilon_L)(\mathbf{d}^p - \mathbf{a}\otimes\mathbf{a}\cdot\mathbf{d}^p - \mathbf{d}^p\cdot\mathbf{a}\otimes\mathbf{a}) + c_2(\varepsilon_T, \varepsilon_L)(\mathbf{a}\otimes\mathbf{a}\cdot\mathbf{d}^p + \mathbf{d}^p\cdot\mathbf{a}\otimes\mathbf{a}) , \tag{5.3}$$

which is proposed as the appropriate form for fibre-reinforced plastic materials.

If the coordinate system is such that $\mathbf{a} = (1, 0, 0)$ then

$$\alpha_{11} = 0 , \qquad \alpha_{22} + \alpha_{33} = 0$$

and (5.3) becomes

$$\begin{bmatrix} 0 & \dot{\alpha}_{12} & \dot{\alpha}_{13} \\ \dot{\alpha}_{12} & \tfrac{1}{2}(\dot{\alpha}_{22} - \dot{\alpha}_{33}) & \dot{\alpha}_{23} \\ \dot{\alpha}_{13} & \dot{\alpha}_{23} & \tfrac{1}{2}(\dot{\alpha}_{33} - \dot{\alpha}_{22}) \end{bmatrix} = c_1(\varepsilon_T, \varepsilon_L) \begin{bmatrix} 0 & 0 & 0 \\ 0 & \tfrac{1}{2}(d^p_{22} - d^p_{33}) & d^p_{23} \\ 0 & d^p_{23} & \tfrac{1}{2}(d^p_{33} - d^p_{22}) \end{bmatrix}$$
$$+ c_2(\varepsilon_T, \varepsilon_L) \begin{bmatrix} 0 & d^p_{12} & d^p_{13} \\ d^p_{12} & 0 & 0 \\ d^p_{13} & 0 & 0 \end{bmatrix}$$

6. General transverse isotropy

In the theory of Sections 2–5 it was assumed that yielding is independent of the fibre tension, and this led to the result $\mathbf{a}\cdot\mathbf{d}^p\cdot\mathbf{a} = 0$. This assumption seems appropriate, and is supported by the results of [1], within the range of axial stress for which the yield or failure strain in the fibre is not exceeded. For brittle-elastic fibres, the theory applies up to failure by fibre breakage. For elastic-plastic fibres, an additional plastic flow regime involving plastic flow in the fibre direction must be included; for practical fibre composites, this will only become operative under fibre stress of large magnitude. Such an effect is incorporated in the bi-modal theory proposed by Dvorak and Bahei-el-din [16]. Extensions of the present theory to allow yield in the fibre direction have been outlined by Spencer [11, 12] and developed in more detail by Rogers [13, 14].

It is also demonstrated in [1] that fibre-reinforced materials show strong kinematic hardening, especially in relation to the stress in the fibre direction. This is due to residual stress in the fibre and matrix following plastic flow in the fibre direction. Brief discussions are given in [11], [13] and [14].

7. Reinforcement by two families of fibres

Similar considerations to those of Sections 2–5 can be applied in the case of an elastic-plastic material reinforced by two families of fibres (as for example, in a laminate comprised of many alternate layers of uniaxially reinforced material with alternating orientations). Such a material is, macroscopically, monoclinic in symmetry, or orthotropic if either (i) the two fibre directions are orthogonal or (ii) the two fibre directions are mechanically equivalent.

The two fibre directions are defined by unit vectors **a** and **b**, which are regarded as continuous vector fields. We denote by 2φ the angle between the two fibre directions, so that

$$\mathbf{a}\cdot\mathbf{b} = \cos 2\varphi \; . \tag{7.1}$$

The tensor

$$\begin{aligned} \mathbf{s}'' = \boldsymbol{\sigma} + (1 + 3\cos^2 2\varphi)^{-1}[&\{\mathbf{a}\cdot\boldsymbol{\sigma}\cdot\mathbf{a} + \mathbf{b}\cdot\boldsymbol{\sigma}\cdot\mathbf{b} - (1+\cos^2 2\varphi)\operatorname{tr}\boldsymbol{\sigma}\}\mathbf{I} \\ &+ \{\operatorname{tr}\boldsymbol{\sigma} - (2\operatorname{cosec}^2 2\varphi)\mathbf{a}\cdot\boldsymbol{\sigma}\cdot\mathbf{a} - (\operatorname{cosec}^2 2\varphi - 3\cot^2 2\varphi)\mathbf{b}\cdot\boldsymbol{\sigma}\cdot\mathbf{b}\}\mathbf{a}\otimes\mathbf{a} \\ &+ \{\operatorname{tr}\boldsymbol{\sigma} - (2\operatorname{cosec}^2 2\varphi)\mathbf{b}\cdot\boldsymbol{\sigma}\cdot\mathbf{b} - (\operatorname{cosec}^2 2\varphi - 3\cot^2 2\varphi)\mathbf{a}\cdot\boldsymbol{\sigma}\cdot\mathbf{a}\}\mathbf{b}\otimes\mathbf{b}] \end{aligned} \tag{7.2}$$

has the property that $\mathbf{s}''$ is independent of superposed hydrostatic pressure, of superposed fibre tension in the **a** direction, and of superposed fibre tension in the **b** direction, with

$$\operatorname{tr}\mathbf{s}'' = 0 \,, \qquad \mathbf{a}\cdot\mathbf{s}''\cdot\mathbf{a} = 0 \,, \qquad \mathbf{b}\cdot\mathbf{s}''\cdot\mathbf{b} = 0 \,. \tag{7.3}$$

If yielding is independent of hydrostatic pressure and the two fibre tensions, the yield function f is an isotropic invariant of $\mathbf{s}''$, $\mathbf{a}\otimes\mathbf{a}$ and $\mathbf{b}\otimes\mathbf{b}$. Taking into account (7.3) it follows [4] that f can be expressed as a function of

$$\begin{aligned} &J_1 = \tfrac{1}{2}\operatorname{tr}\mathbf{s}''^2 - \mathbf{a}\cdot\mathbf{s}''^2\cdot\mathbf{a} \,, \qquad J_2 = \mathbf{a}\cdot\mathbf{s}''^2\cdot\mathbf{a} \,, \qquad J_3 = \tfrac{1}{3}\operatorname{tr}\mathbf{s}''^3 \,, \\ &J_4 = \tfrac{1}{2}\operatorname{tr}\mathbf{s}''^2 - \mathbf{b}\cdot\mathbf{s}''^2\cdot\mathbf{b} \,, \qquad J_5 = \mathbf{a}\cdot\mathbf{s}''\cdot\mathbf{b}\cos 2\varphi \,, \qquad J_6 = \mathbf{a}\cdot\mathbf{s}''^2\cdot\mathbf{b}\cos 2\varphi \end{aligned} \tag{7.4}$$

and $\cos^2 2\varphi$. In the case when the two families of fibres are mechanically equivalent, then f has to be a symmetric function of **a** and **b**.

The most general *quadratic* yield function that satisfies these conditions is

$$f = \frac{J_1 + J_4}{c_1^2} + \frac{J_1 + 2J_2 - J_4}{c_2^2} + \frac{J_6}{c_3^2} - 1 \,, \tag{7.5}$$

where c_1, c_2 and c_3 have dimensions of stress and are functions of $\cos 2\varphi$. If we choose the coordinate system so that the fibres lie in the planes $x_3 = \text{constant}$ and make angles $\pm\varphi$ with the x_1 axis, so that

$$\mathbf{a} = (\cos\varphi, \sin\varphi, 0) \,, \qquad \mathbf{b} = (\cos\varphi, -\sin\varphi, 0) \tag{7.6}$$

then (7.5) can be written in the form

$$f = \frac{1}{Y^2}\{(\sigma_{11} - \sigma_{33})\sin^2\varphi - (\sigma_{22} - \sigma_{33})\cos^2\varphi\}^2 + \frac{1}{k_1^2}\sigma_{13}^2 + \frac{1}{k_2^2}\sigma_{23}^2 - 1 \,, \tag{7.7}$$

where Y, k_1 and k_2 can be related to c_1, c_2 and c_3. If we further denote

$$Y = Y_1 \sin^2\varphi = Y_2 \cos^2\varphi = Y_3|\cos^2\varphi - \sin^2\varphi| , \tag{7.8}$$

then k_1 and k_2 can be interpreted as shear yield stresses for shear on surfaces x_3 = constant in the x_1 and x_2 directions respectively, and Y_1, Y_2 and Y_3 as tensile yield stresses in the x_1, x_2 and x_3 directions respectively. The flow rule associated with (7.5) and (7.7) is given in [4] and [5]. Proportional and kinematic hardening rules can be formulated in a manner similar to that used in Sections 4–5, but we omit details.

References

1. G.J. Dvorak, Y.A. Bahei-el-din, Y. Macheret and C.H. Liu, An experimental study of elastic-plastic behaviour of a fibrous boron-aluminium composite. *J. Mech. Phys. Solids* 36 (1988) 655–687.
2. J.F. Mulhern, T.G. Rogers and A.J.M. Spencer, A continuum model for a fibre-reinforced plastic material. *Proc. Roy. Soc.* A301 (1967) 473–492.
3. J.F. Mulhern, T.G. Rogers and A.J.M. Spencer, A continuum theory of an elastic-plastic fibre-reinforced material. *Int. J. Engng. Sci.* 7 (1969) 129–152.
4. G.E. Smith and A.J.M. Spencer, A continuum theory of a plastic-rigid solid reinforced by two families of inextensible fibres. *Q.J. Mech. Appl. Math.* 23 (1969) 489–490.
5. A.J.M. Spencer, *Deformations of Fibre-reinforced Materials*. Clarendon Press, Oxford (1972).
6. A.J.M. Spencer, The formulation of constitutive equations for anisotropic solids. In: J.P. Boehler (ed.), *Mechanical Behaviour of Anisotropic Solids*. Martinus Nijhoff Publishers (1982) pp. 3–26.
7. A.J.M. Spencer, The formulation of constitutive equations in continuum models of fibre-reinforced composites. In: E. Kröner and K-H. Anthony (eds), *Proc. 3rd Int. Symposium on Continuum Modelling of Discrete Systems*. University of Waterloo Press (1980) pp. 479–489.
8. T.G. Rogers, A.J.M. Spencer and R.L. Moss, Elastic-plastic deformations of cross-ply fibre-reinforced cylinders. In: F.P.J. Rimrott and B. Tabarrok (eds), *Theoretical and Applied Mechanics*. North-Holland (1980) pp. 387–395.
9. A.J.M. Spencer, Constitutive theory for strongly anisotropic solids. In: A.J.M. Spencer (ed.), *Continuum Theory of the Mechanics of Fibre-reinforced Composites*. CISM Courses and Lecture Notes No. 282, Springer (1984) pp. 1–32.
10. A.J.M. Spencer, Modelling of finite deformations of anisotropic materials. In: J. Gittus, J. Zarka and S. Nemat-Nasser (eds), *Large Deformations of Solids: Physical Basis and Mathematical Modelling*. Elsevier (1986) pp. 41–51.
11. A.J.M. Spencer, Yield conditions and hardening rules for fibre-reinforced materials with plastic response. In: J.P. Boehler (ed.), *Failure Criteria of Structured Media* (in press).
12. A.J.M. Spencer, Kinematic constraints, constitutive equations and failure rules for anisotropic materials. In: J.P. Boehler (ed.), *Applications of Tensor Functions in Solid Mechanics*. CISM Courses and Lecture Notes No. 292, Springer (1987) pp. 187–201.
13. T.G. Rogers, Yield criteria flow rules and hardening in anisotropic plasticity. In: J.P. Boehler (ed.), *Yielding, Damage and Failure of Anisotropic Solids*. EFG Publications (1990) pp. 53–79.
14. T.G. Rogers, Plasticity of fibre-reinforced materials. In: G.A. Maugin (ed.), *Continuum Models of Discrete Systems*, Vol. 1. Longman (1990) pp. 87–102.
15. T.G. Rogers, Shear characterisation and inelastic torsion of fibre-reinforced materials. In: G.J. Dvorak (ed.), *Inelastic Deformation of Composite Materials*. Springer (1991) pp. 653–674.
16. G.J. Dvorak and Y.A. Bahei-el-din. A bimodal plasticity theory of fibrous composite materials. *Acta Mechanics* 69 (1987) 219–241.

Journal of Engineering Mathematics **26**: 119–130, 1992.
H.K. Kuiken and S.W. Rienstra (eds), Problems in Applied, Industrial and Engineering Mathematics.

Playing with nonuniform grids

A.E.P. VELDMAN[1] and K. RINZEMA[2]

[1]*Department of Mathematics,* [2]*Laboratory for Materia Technica, University of Groningen, P.O. Box 800, 9700 AV Groningen, The Netherlands*

Abstract. Numerical experiments with discretization methods on nonuniform grids are presented for the convection–diffusion equation. These show that the accuracy of the discrete solution is not very well predicted by the local truncation error. The diagonal entries in the discrete coefficient matrix give a better clue: the convective term should not reduce the diagonal. Also, iterative solution of the discrete set of equations is discussed. The same criterion appears to be favourable.

1. Introduction

Computational fluid dynamics has reached the level that simulation of flow around complex configurations is beginning to become routine. Grid generation is an important ingredient in these simulation methods. Often, boundary-conforming grids are generated, which necessarily will be nonuniform. Additionally, it is generally expected that a high density of grid points is only necessary in regions of large solution activity (steep gradients, large curvature, etc.), whereas in the smoother regions of the solution larger grid cells can be used. It is believed that the reduction of grid points also leads to a reduction of computational effort. Based on these expectations, adaptive grid generation methods are being developed, which further enhance the nonuniformity of the computational grid.

The next step in the simulation method is the discretization of the equations of motion. Discretization of nonuniform grids is not straightforward: e.g. there exist several ways to generalize the 'standard' central-difference formulas. In a number of papers such generalizations have been discussed; we mention [1–4]. These papers show that not every generalization conforms to the above expectations. Thus selections have been made, mainly based on the local truncation error of the discretization method.

This paper will reconsider some of these generalizations of the central-difference method. It appears that their performance can be completely different. 'Unlucky' generalizations can give rise to a dramatic increase of discretization error and computational effort when the number of grid points is reduced and the nonuniformity is increased, but much more benign behaviour can also be achieved. An explanation of this behaviour will be given in terms of the spectra of the discrete coefficient matrices. Additionally, the relation with finite-volume and finite-element discretizations will be discussed.

Subject to our numerical experiments is a one-dimensional convection–diffusion equation, in a convection-dominated case:

$$\frac{\mathrm{d}y}{\mathrm{d}x} - k\,\frac{\mathrm{d}^2y}{\mathrm{d}x^2} = 0\,, \quad 0 \leqslant x \leqslant 1\,, \tag{1.1}$$

with Dirichlet boundary conditions $y(0) = 0$ and $y(1) = 1$. Its solution reads

$$y(x) = \frac{e^{x/k} - 1}{e^{1/k} - 1} . \tag{1.2}$$

Solutions of singularity perturbed problems like (1.1) consist of a boundary-layer part and a smooth (inviscid) part outside the boundary layer. In order to resolve the former part a small mesh size is required; outside the boundary layer larger grid cells can be used.

Two discretization methods will be investigated – a simple one and a more sophisticated one – which for uniform grids both are equal to the central-difference discretization. We begin with an extreme example with only one interior gridpoint; thereafter more gridpoints are used.

2. Finite-difference formulations

Equation (1.1) will be discretized on a grid with grid points x_i ($i = 0, \ldots, N$), where $x_0 = 0$ and $x_N = 1$. The following abbreviations are introduced

$$h_- = x_i - x_{i-1} , \qquad h_+ = x_{i+1} - x_i , \qquad y_- = y_{i-1} , \qquad y_0 = y_i , \qquad y_+ = y_{i+1} .$$

Two finite-difference methods will be investigated, which differ in the discrete treatment of the first-order derivative:

$$\text{Method A:} \quad \frac{dy}{dx} = \frac{y_+ - y_-}{h_+ + h_-} - \frac{1}{2}(h_+ - h_-)y_{xx} - \frac{1}{6}\frac{h_+^3 + h_-^3}{h_+ + h_-} y_{xxx} + O(h^3)$$
$$= \frac{h_+}{h_+ + h_-}\frac{y_+ - y_0}{h_+} + \frac{h_-}{h_+ + h_-}\frac{y_0 - y_-}{h_-} + O(h_+ - h_-) ; \tag{2.1}$$

$$\text{Method B:} \quad \frac{dy}{dx} = \frac{h_-^2 y_+ + (h_+^2 - h_-^2)y_0 - h_+^2 y_-}{h_+ h_-(h_+ + h_-)} - \frac{1}{6} h_+ h_- y_{xxx} + O(h^3)$$
$$= \frac{h_-}{h_+ + h_-}\frac{y_+ - y_0}{h_+} + \frac{h_+}{h_+ + h_-}\frac{y_0 - y_-}{h_-} + O(h^2) . \tag{2.2}$$

Method A simply estimates the local slope from the values in the adjacent grid points. Method B estimates the slope by passing a parabola through the three points y_-, y_0 and y_+ (Fig. 1). The local truncation error of Method A looks larger than that of Method B: it contains an additional $O(h_+ - h_-)$ term proportional to y_{xx}, and its coefficient of y_{xxx} is

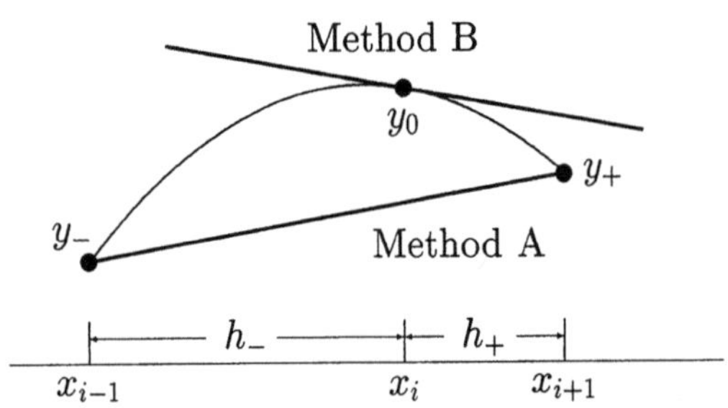

Fig. 1. Discrete approximations of a first-order derivative on a nonuniform grid.

never smaller than the corresponding coefficient for Method B. For a uniform grid both methods equal the second-order central-difference approximation. For a nonuniform grid the formal order of the truncation error depends on the smoothness of the grid during grid refinement. On an algebraic grid (i.e. a grid obtained from a coordinate transformation) both methods have a second-order local truncation error, but on an exponential grid with a fixed stretching rate $h_+/h_- \neq 1$ the local truncation error of Method A is only of first order.

In both cases, the discrete derivative can be written as a linear combination of the slopes on the two adjacent intervals. The difference between both methods becomes clearly visible when h_+ and h_- are significantly different: in Method A the discrete derivative approaches the slope on the coarsest interval, whereas for Method B it approaches the slope on the finest interval.

For both methods, the second derivative is discretized as

$$\frac{d^2y}{dx^2} = \frac{h_-y_+ - (h_+ + h_-)y_0 + h_+y_-}{\frac{1}{2}h_+h_-(h_+ + h_-)} - \frac{1}{3}(h_+ - h_-)y_{xxx} + O(h^2)\,. \tag{2.3}$$

Next to the *local* truncation error, the *global* discretization error should be addressed. Thereto, let us consider quasi-uniform grids, i.e. grids for which the ratio between the largest grid cell and the smallest grid cell is bounded during refinement; these include algebraic grids, but not exponential grids. Manteuffel and White [5] have proved the global discretization error of both methods to be of second order on quasi-uniform grids. Thus, asymptotically, for both methods the difference between the exact solution and its discrete approximation decays quadratically in the mesh size. However, as we will see, this gives only limited indication on the behaviour for finite, non-zero, mesh size.

3. Numerical experiments

3.1. *One internal grid point*

By just looking at the shape of the solution of (1.1), it should be possible to approximate it, at least qualitatively, by a piecewise linear polynomial with only one internal point ($N = 2$). The location of this point should be somewhere near the edge of the boundary layer, e.g. at a position where the exponential $\exp(x/k)$ in (1.2) is 10–20% of its value at $x = 1$. This yields a grid point somewhere between $x = 1 - 2.3k$ and $x = 1 - 1.6k$.

The discrete solution in this single grid point $x = 1 - h_+$ can be computed analytically:

$$\text{Method A:} \quad y = (1 - h_+)(1 - \tfrac{1}{2}h_+/k) = 1 - \tfrac{1}{2}h_+/k + O(k)\,; \tag{3.1}$$

$$\text{Method B:} \quad y = (1 - h_+)(1 - h_+ - 2k)/(1 - 2h_+ - 2k) = 1 + O(k)\,. \tag{3.2}$$

The asymptotic behaviour is derived under the assumption that $h_+ = O(k)$. It easily follows from (3.2) that, for small values of k, Method B will not be able at all to approximate the exact solution (1.2). Method A can do a better job, as can be seen from Table 1. Here the discrete solution (3.1) is compared with the exact solution (1.2) for two small values of k: $k = 10^{-2}$ and $k = 10^{-5}$. Taking into account that only one internal grid point is used, we cannot expect any method to be better than Method A: by choosing $h_+ \approx 1.6k$ the exact solution in this grid point can even be reproduced.

Table 1. Discrete versus exact solution of the convection-diffusion equation (1.1) using Method A with only one internal grid point

	$k = 10^{-2}$		$k = 10^{-5}$	
h_+/k	Discrete (3.1)	Exact (1.2)	Discrete (3.1)	Exact (1.2)
1	0.495	0.368	0.500	0.368
1.6	0.197	0.202	0.200	0.202
2	0.0	0.135	0.0	0.135

3.2. More grid points

Next, the nonuniform grid is refined to 10 grid cells to make the situation less extreme. Several grid point distributions will be investigated; some of them with abrupt changes in the size of the grid cells, some of them with more gradual changes.

Abrupt grids

The numerical experiments will begin with a grid consisting of two uniform parts. The 10 grid cells are divided into 5 equal cells ($h = k$) inside the boundary layer, and 5 equal cells ($h = 0.2 - k$) outside the boundary layer. Thus the grid becomes

$$\text{Grid 1:}\quad x_i = i(0.2 - k)\,,\quad (i = 0, \ldots, 5)\,;\qquad x_i = 1 - (10 - i)k\,,\quad (i = 6, \ldots, 10)\,.$$

Point $i = 5$ is the only point where the sizes of the adjacent grid cells are unequal: $h_- = 0.2 - k$ and $h_+ = k$. If we think this grid refined by halving the grid cells, the grid is quasi-uniform, but it is not algebraic.

Figure 2 shows the discrete solution for both methods on Grid 1. For comparison the exact solution is also indicated. The difference between the two methods is obvious; Method A produces better results.

In Grid 1, the point where the abrupt change in mesh size takes place is an odd-numbered point ($i = 5$). It could make a difference when this point would be an even-numbered point, since even-numbered grid points are better coupled to the boundary condition at $x = 0$ than odd-numbered grid points (due to the odd/even decoupling). Therefore we will also investigate a different grid (Grid 1′), with 4 grid cells of size $0.25 - 1.5k$ and 6 grid cells of size k. Figure 3 shows that the solution of Method A is not very much influenced, but for Method B this change in the grid indeed makes a difference: the discrete solution strongly deteriorates!

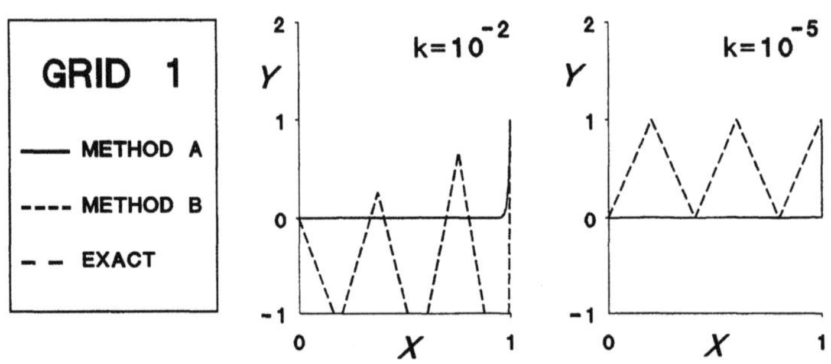

Fig. 2. Discrete solutions on an abrupt grid with 5 coarse grid cells, and 5 fine grid cells.

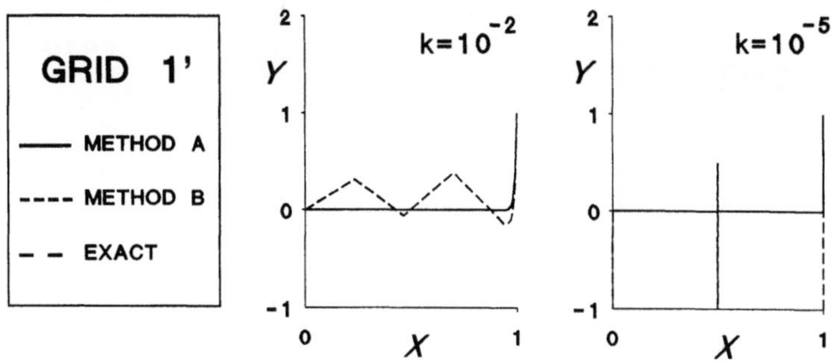

Fig. 3. Discrete solutions on an abrupt grid with 4 coarse grid cells, and 6 fine grid cells. A comparison with Fig. 2 shows a large sensitivity of Method B.

Remark. It is stressed that on both grids there is only one grid point ($i = 5$) where the discrete formulas differ from central discretization. This is the only grid point where the two methods are different. It is surprising to see that one single grid point can have such a large influence.

In a first attempt to explain the observed behaviour we refer to the limit form of both methods in case h_+ and h_- are significantly different. Equation (1.1) possesses a boundary layer at the right-hand side of the interval, so a situation where $h_+ < h_-$ is the natural one. Evaluating the coefficients in (2.2) it follows that, in the extreme case where $h_+ \ll h_-$, Method B approaches a downwind (!) discretization. This can explain its bad behaviour. In contrast, Method A yields a discretization in which the upwind direction has the largest weight.

In a grid point where $h_+ \gg h_-$ Method A uses mainly downwind information (although it does not become a downwind discretization), whereas Method B approaches an upwind discretization. To see whether this can bring Method A into difficulties, also a grid with fine grid cells near both ends of the interval has been tried.

Grid 2: $x_1, \ldots, x_9 = k,\ 2k,\ 3k,\ 0.25,\ 0.5,\ 0.75,\ 1-3k,\ 1-2k,\ 1-k\,.$

Results for this grid are shown in Fig. 4. We see that Method A is hardly affected, but Method B is extremely bad (we must remember that there is still one grid point where it approaches a downwind discretization).

In summary: On these quasi-uniform but non-algebraic grids, the local truncation error of

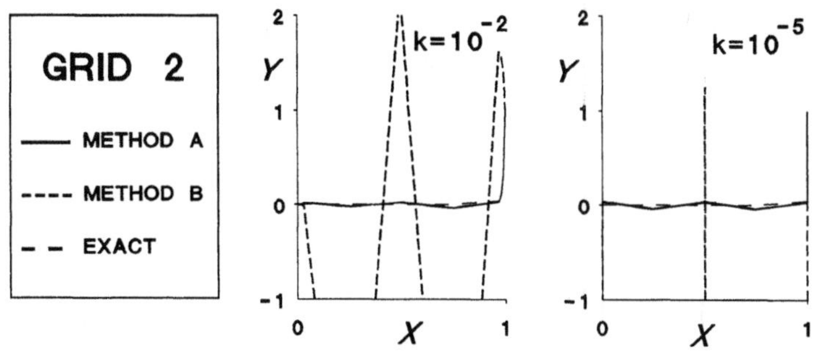

Fig. 4. Discrete solutions on an abrupt grid with small grid cells near both endpoints of the interval.

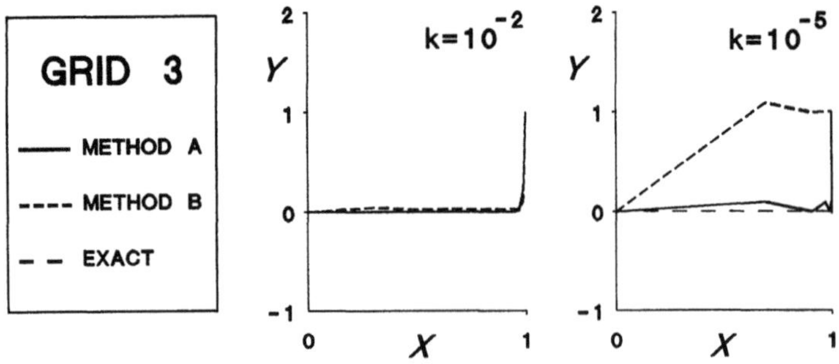

Fig. 5. Discrete solutions on an exponential grid.

Method A is only of first order, whereas that of Method B is of second order. Nevertheless, Method A gives good results for all three grids, whereas Method B is not able to produce an acceptable solution at all. Thus the *local* truncation error does not give a reliable indication about the behaviour of the *global* discretization error.

Exponential grids

In the above examples (Grids 1 and 1′) there is only one grid point with non-equal adjacent cells, but in that point the stretching rate h_+/h_- is extremely small. We will next present an example where all cells are different, but where the stretching rate is closer to unity. An exponentially stretched grid is chosen with a constant factor between the size of the successive grid cells. The grid points x_i are given by

$$\text{Grid 3:}\quad x_{i+1} = x_i + S(x_i - x_{i-1})\,,\quad (i = 1, \ldots, 9)\,,$$

where the stretching rate is $S = h_+/h_-$. Further, as before $x_0 = 0$ and $x_{10} = 1$. When S is kept constant during grid refinement, such a grid is not quasi-uniform.

Methods A and B are applied to (1.1) for $k = 10^{-2}$ on a grid for which $S = 0.7$, and for $k = 10^{-5}$ with $S = 0.3$. The coarsest grid cells have a size about 0.3 and 0.7, respectively. The finest grid cells are about 0.012 and 0.000014, respectively, and lie well inside the boundary layer. The results are shown in Fig. 5. For $k = 10^{-2}$ both methods do a good job. For $k = 10^{-5}$ Method B is having difficulties.

3.3. Conclusion

The results shown in Figs 2–5 have been summarized in Table 2, which shows the discretization error $\|y_{ex} - y\|_2$, computed with the trapezoidal rule. For all cases presented, Method A produces reasonable to fine results. Those of Method B are in most cases unacceptable, and can even be extremely inaccurate (e.g. on Grids 1′ and 2). From the

Table 2. Discretization error $\|y_{ex} - y\|_2$ for Methods A and B on various grids

k	10^{-2}				10^{-5}			
Grid	1	1′	2	3	1	1′	2	3
Method A	0.005	0.005	0.025	0.009	0.005	0.002	0.035	0.067
Method B	1.124	0.235	3.530	0.038	0.706	$>10^3$	$>10^3$	0.856

examples presented, it may be concluded that Method A is more accurate than Method B, although its local truncation error is larger. In the next section we will try to explain the observed behaviour.

4. Analysis of discretization error

To get a feeling about what is going on, it is good to have a closer look at the discretization error. Let the discrete system be given by

$$Ay = r .$$

The exact solution y_{ex} satisfies a related equation

$$Ay_{ex} = r + \tau_{loc} ,$$

where τ_{loc} is the local truncation error. The difference between the exact solution and its discrete approximation reads

$$y_{ex} - y = \mathbf{A}^{-1}\tau_{loc} . \tag{4.1}$$

Thus the discretization error is built from the product of the local truncation error and the inverse of the coefficient matrix. The above experiments give an impression of the behaviour of this product. Although available analytical techniques are only of modest power, we will first try to explain the observed behaviour theoretically.

Thereto, let us first consider the coefficient matrices of the above methods, denoted by $\mathbf{A}_A$ and $\mathbf{A}_B$ respectively. They possess a tri-diagonal structure

$$a_-y_- + a_0y_0 + a_+y_+ = 0 . \tag{4.2}$$

The coefficients are:

$$\text{Method A:} \quad a_- = \frac{-h_- - 2k}{h_-(h_+ + h_-)} ; \qquad a_0 = \frac{2k}{h_+h_-} ; \qquad a_+ = \frac{h_+ - 2k}{h_+(h_+ + h_-)} ;$$

$$\text{Method B:} \quad a_- = \frac{-h_+ - 2k}{h_-(h_+ + h_-)} ; \qquad a_0 = \frac{h_+ - h_- + 2k}{h_+h_-} ; \qquad a_+ = \frac{h_- - 2k}{h_+(h_+ + h_-)} .$$

In the sequel, we will in particular consider the spectra of the coefficient matrix **A** and of the shifted Jacobi matrix $(\text{diag}\,\mathbf{A})^{-1}\mathbf{A}$. These spectra can give information about the regularity of the coefficient matrix; also they play an important role in the convergence of iterative solution methods. In the discussion use will be made of the following Lemma:

LEMMA 1. *Let* **A** *be a positive real matrix (i.e.* $\mathbf{A} + \mathbf{A}^T$ *is positive definite). Then for any positive definite matrix* **Q**, *the matrix* **QA** *is N-stable (i.e. all eigenvalues have a positive real part).*

Proof. See Veldman [6]. □

We begin the analysis with Method A. Here the convective term does not contribute to the diagonal. This enables us to prove that its coefficient matrix $\mathbf{A}_A$ is N-stable. In the next section we will prove that the matrix $(\mathrm{diag}\,\mathbf{A}_A)^{-1}\mathbf{A}_A$ also is N-stable.

THEOREM 1. *The coefficient matrix* $\mathbf{A}_A$ *of* Method A *is N-stable.*

Proof. The proof starts by scaling $\mathbf{A}_A$ with a diagonal matrix $\mathbf{H} = \mathrm{diag}(h_+ + h_-)$. The matrix $\mathbf{HA}_A$ possesses an anti-symmetric part $(\mathbf{HA}_A)_a$ which stems from the convective term, and a symmetric part $(\mathbf{HA}_A)_s$ which stems from the diffusive term. The latter part is diagonally dominant and hence positive definite, so by definition $\mathbf{HA}_A$ is positive real. Since $\mathbf{H}^{-1}$ is positive definite, it follows from Lemma 1 that $\mathbf{A}_A = \mathbf{H}^{-1}(\mathbf{HA}_A)$ is N-stable. □

In Method B the convective term does contribute to the diagonal. Its contribution is negative when $h_+ < h_-$; it may even cause the diagonal to become negative. As we shall see, this is the reason that for Method B a similar theorem does not hold. This will be shown by determining the spectrum of $\mathbf{A}_B$ numerically.

Table 3 shows the eigenvalues of the coefficient matrices for Methods A and B on Grid 1. It is possible to associate the above eigenvalues with part of the grid. When the entries in a coefficient matrix like (4.2) are constant, under Dirichlet boundary conditions its eigenvalues are given by

$$a_0 + 2(a_- a_+)^{1/2}\cos(n\pi/N)\,, \quad (n = 1, \ldots, N-1)\,. \tag{4.3}$$

Grid 1, as used in Table 3, consists of two uniform parts with five equal grid cells. Therefore set $N = 5$, and substitute the corresponding values for the coefficients a_-, a_0 and a_+. Then for both parts of the grid, (4.3) yields four eigenvalues. Assuming k to be small, these are given by:

$$\text{eigenvalues coarse grid cells:} \quad 50k \pm 4.045\mathrm{i}\,; \qquad 50k \pm 1.545\mathrm{i}\,; \tag{4.4a}$$

$$\text{eigenvalues fine grid cells:} \quad (2 \pm 1.40126)/k\,; \quad (2 \pm 0.53523)/k\,. \tag{4.4b}$$

Comparing these values with those of Table 3 we can conclude:
- the eigenvalues 1–4 of Method B approach the coarse-grid values (4.4a) as $k \to 0$;
- the eigenvalues 6–9 of Method A approach the fine-grid values (4.4b) as $k \to 0$.

Thus we are tempted to associate eigenvalues 1–4 with the four points in the coarse part of

Table 3. Eigenvalues of the coefficient matrix for Methods A and B on Grid 1. The fifth eigenvalue is 'irregular'; for Method B it can become negative

	Method A		Method B	
#	$k = 10^{-2}$	$k = 10^{-5}$	$k = 10^{-2}$	$k = 10^{-5}$
1, 2	0.898 ± 4.386i	0.354 ± 4.163i	0.414 ± 4.633i	0.0005 ± 4.045i
3, 4	1.736 ± 2.314i	1.250 ± 2.141i	0.426 ± 2.819i	0.0005 ± 1.545i
5	2.258	1.836	0.52	−6194.4
6	61.71	59875.8	15.84	10441.8
7	148.42	146478.5	122.39	119929.1
8	254.64	253524.4	240.92	239565.7
9	340.44	340126.2	336.65	336272.8

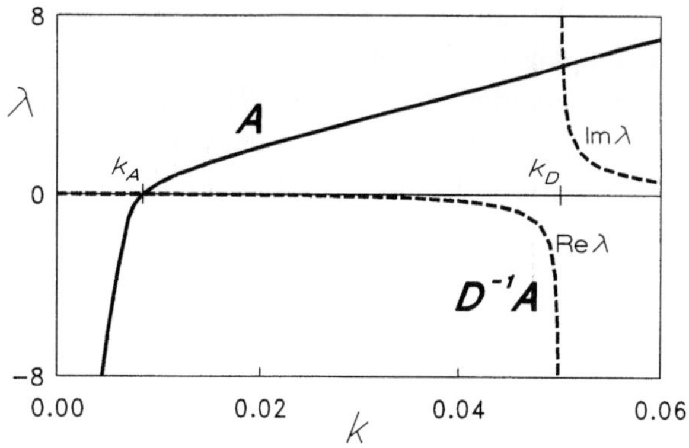

Fig. 6. 'Irregular' eigenvalue of coefficient matrix **A** (——) and shifted Jacobi matrix $\mathbf{D}^{-1}\mathbf{A}$ (- - -) for Method B on nonuniform Grid 1. Observe that both **A** and **D** can become singular.

the grid, and eigenvalues 6–9 with the four grid points in the fine part. Eigenvalue 5 is left over; we will call it the 'irregular' eigenvalue. Let us now concentrate on this eigenvalue.

For Method A the 'irregular' eigenvalue always has a positive real part, as this holds for each eigenvalue (Theorem 1), but Table 3 shows that for Method B the 'irregular' eigenvalue can become negative. To learn more about the behaviour of this eigenvalue we have computed it for a range of k-values. The result is presented in Fig. 6. We see that the 'irregular' eigenvalue vanishes when $k = k_A \approx 0.0084$, making $\mathbf{A}_B$ singular and the global error (4.1) increases without limit. For smaller values of k it becomes negative; $\mathbf{A}_B$ is no longer N-stable then.

5. Computational effort and iterative performance

Another aspect which is influenced by the discretization method is the way in which the set of equations can be inverted. As direct inversion is not always a feasible strategy, iterative techniques are often used. The structure of the matrix determines to a large extent which iterative techniques are applicable.

The problem to be solved may make it desirable to base the iteration method on time integration of the unsteady, semi-discretized version

$$\frac{\mathrm{d}y}{\mathrm{d}t} + \mathbf{A}y = r . \tag{5.1}$$

This is especially the case for highly non-linear problems, where existence or uniqueness of a steady-state solution cannot be guaranteed. Then time-integration methods are the best means to pursue the solution since they follow the physics more closely. Equation (5.1) does only possess a steady-state limit if and only if the matrix **A** is N-stable; hence this property is a necessary condition for time-integration methods to converge. However, as we will show below, for one of the above discretization methods this condition cannot always be satisfied.

As an example, consider first a grid with one internal grid point ($N = 2$, see Section 3.1). Method B yields a central coefficient a_0 which is negative. Hence, in this situation with one internal point, time integration in combination with Method B will *never* converge (unless one chooses $\Delta t < 0$, or selects an integration method with a large amount of numerical diffusion). Of course, for sufficiently fine (uniform) grids Method B can successfully be combined with time integration. Thus here a *decrease* of the number of grid points leads to an *unlimited increase* of computational effort!

There are more iterative methods for which N-stability of the coefficient matrix is necessary (and sufficient) for convergence, e.g. Chebyshev iteration [6]. Other iterative methods, like JOR and SOR, converge under a different criterion, as formulated in the next Lemma (Young [7], Ch. 6).

LEMMA 2. *For consistently ordered matrices* $\mathbf{A}$, *SOR and JOR converge for a sufficiently small relaxation parameter if and only if* $(\operatorname{diag}\mathbf{A})^{-1}\mathbf{A}$ *is N-stable.*

Lemma 2 implies that the last example ($N = 2$ and Method B) can be treated by SOR or JOR. In general, however, convergence of SOR or JOR applied to the coefficient matrix from Method B cannot be guaranteed, as we will see below. In contrast, again Method A gives no problems.

THEOREM 2. *The shifted Jacobi matrix* $(\operatorname{diag}\mathbf{A}_A)^{-1}\mathbf{A}_A$ *is N-stable.*

Proof. Let $\mathbf{H}$ be the scaling matrix defined in the proof of Theorem 1. As $\mathbf{H}$ and $\mathbf{D}_A = \operatorname{diag}\mathbf{A}$ are positive definite, also $(\mathbf{HD}_A)^{-1}$ is positive definite. Then it follows from Lemma 1 that $\mathbf{D}_A^{-1}\mathbf{A}_A = (\mathbf{HD}_A)^{-1}\mathbf{HA}_A$ is N-stable. □

Combining Lemma 2 and Theorem 2, it follows that the discrete equations created by Method A can always be solved iteratively by methods like JOR and SOR. As we saw earlier in Theorem 1, for these equations also time-integration methods are applicable.

Since Method B cannot be treated analytically, we have determined the eigenvalues of the shifted Jacobi matrix numerically. As an example, in Table 4 the eigenvalues corresponding with Methods A and B on Grid 1 are shown. They come in pairs of which the sum equals 2. Again, it is possible to associate the eigenvalues with part of the grid. In the same way as above, the eigenvalues corresponding with the coarse and fine parts of the grid can be computed. When $k \to 0$ we obtain:

eigenvalues coarse grid cells: $1 \pm 0.08090\mathrm{i}/k$; $1 \pm 0.03090\mathrm{i}/k$;

eigenvalues fine grid cells: 1 ± 0.7006 ; 1 ± 0.2676 .

The first four eigenvalues of both methods can clearly be associated with the coarse part of the grid. We have arranged the other five according to their distance from 1. The fifth and sixth eigenvalue form the interesting 'irregular' pair; for Method B one of these eigenvalues can become negative (call this one #5). Its adjoint eigenvalue (#6) then becomes larger than 2.

Table 4. Eigenvalues of $\mathbf{D}^{-1}\mathbf{A}$ for Methods A and B on Grid 1. Note that for Method B the fifth eigenvalue can become negative

	Method A		Method B	
#	$k = 10^{-2}$	$k = 10^{-5}$	$k = 10^{-2}$	$k = 10^{-5}$
1, 2	$1.0 \pm 7.68\mathrm{i}$	$1.0 \pm 8089.8\mathrm{i}$	$1.0 \pm 7.64\mathrm{i}$	$1.0 \pm 8092.0\mathrm{i}$
3, 4	$1.0 \pm 3.17\mathrm{i}$	$1.0 \pm 3090.3\mathrm{i}$	$1.0 \pm 2.91\mathrm{i}$	$1.0 \pm 3090.3\mathrm{i}$
5, 6	1.0 ± 0.792	1.0 ± 0.824	1.0 ± 1.001	1.0 ± 0.997
7, 8	1.0 ± 0.485	1.0 ± 0.509	1.0 ± 0.565	1.0 ± 0.564
9	1.0	1.0	1.0	1.0

To obtain more insight into its behaviour we have computed the 'irregular' fifth eigenvalue of $\mathbf{D}^{-1}\mathbf{A}$ for various values of k (Fig. 6). At $k = k_D = 0.05$ the diagonal element a_0 vanishes, making $\mathbf{D}$ singular. As a result, for $k \downarrow k_D$ the imaginary part of the 'irregular' eigenvalue (and its adjoint) grows without limit ($\lambda \rightarrow 1 \pm \infty i$). When k decreases below k_D this eigenvalue becomes negative real. For $k \uparrow k_D$ it approaches minus infinity (and its adjoint approaches positive infinity). Lowering k further, the 'irregular' eigenvalue vanishes at $k = k_A$ where $\mathbf{A}$ becomes singular, and is slightly positive for $k < k_A$.

In summary: The coefficient matrix of Method A is never singular; it even is N-stable, hence time integration can be applied. Also, $\mathbf{D}_A^{-1}\mathbf{A}_A$ is N-stable, therefore methods like SOR and JOR are applicable. For Method B it cannot be guaranteed that the coefficient matrix is N-stable, making time integration impossible. In fact, when the diagonal coefficient is negative, it turns out that either $\mathbf{A}_B$ or $\mathbf{D}_B^{-1}\mathbf{A}_B$ is no longer N-stable. Already for $k = 1/20$, where the stretching rate in the irregular grid point is $1/3$, the shifted Jacobi matrix becomes singular. Lowering k further, for $k \approx 0.0084$ (where the stretching rate is 0.04) the coefficient matrix $\mathbf{A}_B$ itself becomes singular.

6. Relation with other methods

Methods A and B can be considered as two special members of a family of discretization methods which approximate the first-order derivative in a grid point as a combination of the derivatives on the adjacent intervals:

$$\frac{\mathrm{d}y}{\mathrm{d}x} = w\,\frac{y_+ - y_0}{h_+} + (1-w)\,\frac{y_0 - y_-}{h_-}\,. \tag{6.1}$$

Method A is the only member of the family (6.1) for which the diagonal contribution vanishes. For all other members the diagonal is affected, and it is likely that they suffer from the same difficulties as Method B. This expectation has already been confirmed in a 2D finite-volume context. Rossow [8] has studied two cell-vertex methods: the method of Hall [9] which in one dimension equals Method A, and the method of Ni [10] which in one dimension fits in (6.1) for $w = 1/2$. His calculations show that the method of Hall is better than the method of Ni, which is in agreement with our findings.

In our experiments, the discretization has been performed in physical space. An alternative would have been to transform the equations to computational space, where the grid is uniform. However, now similar difficulties arise when the derivatives of the coordinate transformation have to be approximated; see e.g. Mynett et al. [11].

Also a link with finite-element methods can be made. These methods have become very popular, not in the least because of their good performance on irregular domains. Thus it is interesting to find out which finite-difference analogue corresponds with the 'standard' finite-element discretization: it turns out to be Method A.

7. Discussion

We have presented exploratory, one-dimensional calculations on nonuniform grids for two discretization methods. On uniform grids the methods are identical, but on nonuniform grids they appear to behave quite differently. One of them is very sensitive to the choice of the grid, the other one is more benign. Also they react totally different on iterative solution techniques. This behaviour can be explained by looking at the spectrum of the coefficient matrices.

Method A yields a matrix which is always *N*-stable and hence never singular. Even on extremely stretched grids this method produces acceptable discretization errors. Also the shifted Jacobi matrix $(\text{diag}\, \mathbf{A}_A)^{-1}\mathbf{A}_A$ is *N*-stable. As a result most iterative solution methods are applicable.

In Method B the convective term can reduce the diagonal of the coefficient matrix $\mathbf{A}_B$. As a consequence $\mathbf{A}_B$ can become singular, hence the discretization error can grow without limit. Also $\mathbf{A}_B$ and its shifted Jacobi matrix are not always *N*-stable, which restricts the number of applicable iterative solution methods.

Thusfar, discussions about discretization methods on nonuniform grids have concentrated on the local truncation error. In 1971, Crowder and Dalton [1] already observed that Method B behaves poorly on abrupt grids. The improvement has been sought in constructing smoother grids, e.g. grids obtained from a coordinate transformation (i.e. algebraic grids) as described in [2]. For more recent discussions on this subject, see [3] and [4]. Method A has often been mentioned, but each time it was rejected because of its local truncation error. The present experiments show that this rejection has been premature; Method A is much more powerful than generally assumed.

In conclusion: Nonuniform grids can be efficient, provided one chooses a suitable discretization method. The local truncation error, which is smallest for Method B, does not give a useful indication about the suitability of a method. Instead, the above results strongly suggest to use Method A. The underlying principle seems to be that a convective term should never reduce the diagonal of a coefficient matrix. We have demonstrated this with some one-dimensional examples. It is likely that the above elementary findings carry over directly to more-dimensional problems. Because of the somewhat surprising nature of the results, further detailed research is recommended.

References

1. H.J. Crowder and C. Dalton, Errors in the use of nonuniform mesh systems. *J. Comput. Phys.* 7 (1971) 32–45.
2. E. Kálnay de Rivas, On the use of nonuniform grids in finite difference equations. *J. Comp. Phys.* 10 (1972) 202–210.
3. E. Turkel, Accuracy of schemes with nonuniform meshes for compressible fluid flows. *ICASE Report* 85-43 (1985).
4. J. Pike, Grid adaptive algorithms for the solution of the Euler equations on irregular grids. *J. Comp. Phys.* 71 (1987) 194–223.
5. T.A. Manteuffel and A.B. White Jr., The numerical solution of second-order boundary value problems on nonuniform meshes. *Math. of Comp.* 47 (1986) 511–535.
6. A.E.P. Veldman, Who's afraid of non-symmetric matrices? A discussion of elementary iterative methods. *Report TWI* 88-49, Delft University of Technology (1988).
7. D.M. Young, *Iterative solution of large linear systems*. New York: Academic Press (1971).
8. C. Rossow, Comparison of cell centered and cell vertex finite volume schemes. In: M. Deville (ed.), *Proc. 7th GAMM Conf. on Numerical Fluid Dynamics*. Braunschweig: Vieweg Verlag (1987) pp. 327–333.
9. M.G. Hall, Cell-vertex multigrid scheme for solution of the Euler equations. In: K.W. Morton and M.J. Baines (eds), *Numerical Methods for Fluid Dynamics*. Oxford: University Press (1986) pp. 303–345.
10. R.-H. Ni, A multiple grid scheme for solving the Euler equations. *AIAA J.* 20 (1982) 1565–1571.
11. A.E. Mynett, P. Wesseling, A. Segal and C.G.M. Kassels, The ISNaS incompressible Navier–Stokes solver: invariant discretization. *Applied Scientific Research* 48 (1991) 175–191.

Journal of Engineering Mathematics **26**: 131–152, 1992.
H.K. Kuiken and S.W. Rienstra (eds), Problems in Applied, Industrial and Engineering Mathematics.

The Stewartson layer of a rotating disk of finite radius

A.I. van de VOOREN

Department of Mathematics, University of Groningen, P.O. Box 800, 9700 AV Groningen, The Netherlands

Abstract. It is shown that if a disk of finite radius and the surrounding medium rotate coaxially with slightly different angular velocities, an axial layer in the form of a cylindrical shell exists at the edge of the disk. This shell of thickness $O(E^{1/3})$ has length $O(E^{-1})$ in axial direction, where E is the Ekman number. Its most characteristic element is the axial velocity of $O(E^{1/6})$ which is larger than everywhere else in the field. We calculate the velocity components and the pressure in this layer.

1. Introduction

In 1957 Stewartson published a paper [1] in which he considered the shear layers existing between two coaxial rotating planes of which the center disks rotate with a slightly different angular velocity. He found that if the deviations of the angular velocities of the disks from that of the planes are equal but opposite, a shear layer of thickness $E^{1/3}$ exists, while if the deviations are equal in the same sense an additional layer of thickness $E^{1/4}$ appears. This last layer is necessary in order to fit the azimuthal velocity of the inner region to that of the outer region. In the following such shear layers will be denoted as Stewartson layers. E is the Ekman number.

Greenspan gave in his monograph [2] a clear account of these Stewartson layers while Moore and Saffman [3] presented an analysis of different possibilities for a variety of situations. Hide and Titman [4] performed an experimental investigation on a rotating disk of finite radius placed in a cylindrical tank which itself is rotating with a slightly different angular velocity. They showed the existence of the Stewartson layer.

In the present investigation a disk of radius a rotates with an angular velocity Ω in an unbounded medium which itself rotates coaxially with angular velocity $(1-\varepsilon)\Omega$. Our problem will be linearized in the small Rossby number ε. The configuration is clarified in Fig. 1, where the various regions of the flow field are indicated.

The rotation of the medium can physically be realized by thinking of a cylindrical tank rotating with an angular velocity $(1-\varepsilon)\Omega$. Top and bottom of the tank must have a distance to the disk larger than $O(E^{-1})$ in order that the Stewartson layer of the disk is not influenced by top and bottom of the tank (see also [3]).

It will be shown that for our configuration there exists a Stewartson layer of thickness $E^{1/3}$. There is no layer of thickness $E^{1/4}$ since at both sides of the Stewartson layer (regions III and IV) the angular velocities are equal, viz. $(1-\varepsilon)\Omega$. Velocity and pressure distributions are dependent upon a similarity parameter $\tau = z/r_1^3$, where r_1 is the stretched radial coordinate in the Stewartson layer. In this way expressions for velocity and pressure distributions are derived in the form of integrals, which have been evaluated by Romberg integration.

At the point $z=0$, $r_1=0$, which is the point, where the Stewartson layer is joined to the Ekman layer [3], there exists a logarithmic singularity in the pressure. This singularity is responsible for the deflection of the boundary layer flow to the axial flow in the Stewartson layer.

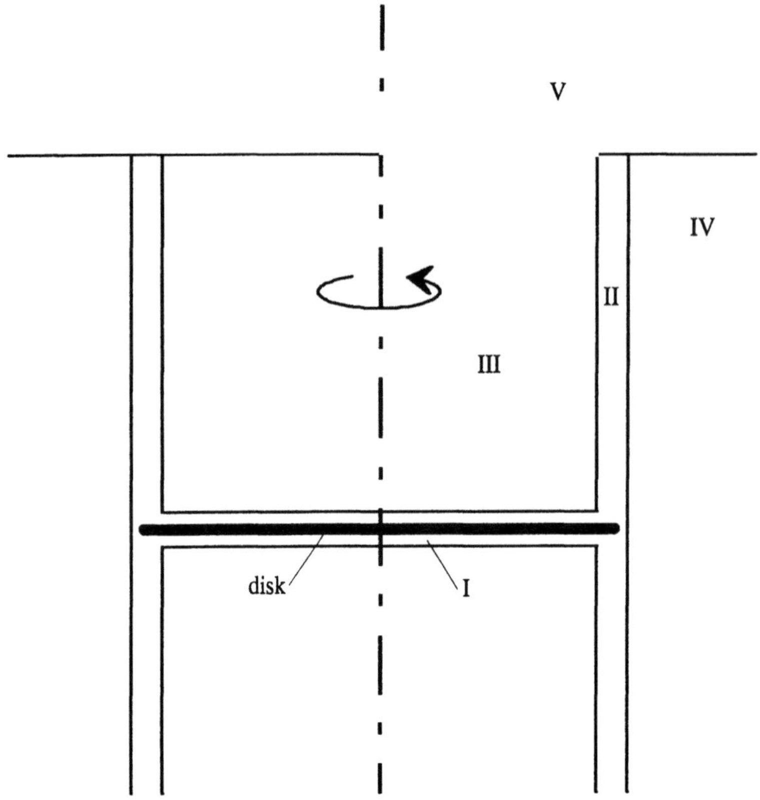

Fig. 1. The configuration. I = Ekman layer, II = Stewartson layer, III = inner region, IV = outer region, V = upper region.

The Stewartson layer II has small effects upon the region III and IV in the same way as the Ekman layer I induces an axial velocity in region III.

2. General equations

For an axially symmetric configuration the dimensionless equations of motion are in an inertial system of reference:

$$
\begin{aligned}
&u\,\frac{\partial u}{\partial r}+w\,\frac{\partial u}{\partial z}-\frac{v^2}{r}=-\frac{\partial p}{\partial r}+E\left\{\frac{\partial^2 u}{\partial r^2}+\frac{\partial}{\partial r}\left(\frac{u}{r}\right)+\frac{\partial^2 u}{\partial z^2}\right\},\\
&u\,\frac{\partial v}{\partial r}+w\,\frac{\partial v}{\partial z}+\frac{uv}{r}=E\left\{\frac{\partial^2 v}{\partial r^2}+\frac{\partial}{\partial r}\left(\frac{v}{r}\right)+\frac{\partial^2 v}{\partial z^2}\right\},\\
&u\,\frac{\partial w}{\partial r}+w\,\frac{\partial w}{\partial z}=-\frac{\partial p}{\partial z}+E\left\{\frac{\partial^2 w}{\partial r^2}+\frac{1}{r}\frac{\partial w}{\partial r}+\frac{\partial^2 w}{\partial z^2}\right\},
\end{aligned}
\tag{2.1}
$$

while the equation of continuity is

$$\frac{1}{r}\frac{\partial}{\partial r}(ur)+\frac{\partial w}{\partial z}=0\,. \tag{2.2}$$

u, v and w are the radial, azimuthal and axial velocities, respectively. p is the pressure, $E=\nu/\Omega a^2$ the Ekman number with ν the kinematical viscosity coefficient. Lengths have been made dimensionless with a, velocities with Ωa and the pressure with $\rho\Omega^2a^2$ where ρ is the fluid density. In order to satisfy the equation of continuity a stream function ψ is introduced by

$$u = \frac{1}{r}\frac{\partial\psi}{\partial z}, \qquad w = -\frac{1}{r}\frac{\partial\psi}{\partial r}. \tag{2.3}$$

The boundary conditions are
at the disk

$$\left.\begin{aligned} &z = 0, \quad r < 1: \quad u = 0, \quad v = r, \quad w = 0, \\ &\text{at infinity} \\ &\sqrt{z^2 + r^2} \to \infty: \quad u \to 0, \quad v \to (1-\varepsilon)r, \quad w \to 0, \quad p \to \frac{1}{2}(1-\varepsilon)^2 r^2. \end{aligned}\right\} \tag{2.4}$$

3. The Ekman layer

Since the Rossby number is infinitely small, all deviations from the original flow will be proportional to ε. Introducing the boundary layer coordinate $\tilde{z} = E^{-1/2}z$, we may write

$$\begin{aligned} &\psi = \frac{1}{2}\varepsilon E^{1/2} r^2 h(\tilde{z}), \qquad u = \frac{1}{2}\varepsilon r h'(\tilde{z}), \qquad v = r - \varepsilon r g(\tilde{z}), \\ &w = -\varepsilon E^{1/2} h(\tilde{z}), \qquad p = \frac{1}{2}(1 - 2\varepsilon)r^2. \end{aligned} \tag{3.1}$$

Substitution in the equation (2.1) leads to

$$-h''' + 4g = 4, \qquad h' + g'' = 0, \tag{3.2}$$

while the third equation (2.1) shows that $\partial p/\partial\tilde{z} = O(E)$. The linearization in (3.2) is valid for $|\varepsilon| < O(1)$. Boundary conditions for (3.2) are

$$\begin{aligned} &\tilde{z} = 0: \quad h = 0, \quad h' = 0, \quad g = 0, \\ &\tilde{z} \to \infty: \quad h' \to 0, \quad g \to 1. \end{aligned} \tag{3.3}$$

By elementary methods the solution of this system is obtained as

$$g = 1 - e^{-\tilde{z}}\cos\tilde{z}, \qquad h = 1 - e^{-\tilde{z}}(\sin\tilde{z} + \cos\tilde{z}). \tag{3.4}$$

We now investigate whether the boundary layer exists also for $r > 1$ as is the case for a rotating disk in a fluid at rest, see van de Vooren and Botta [5]. The boundary conditions (3.3) are replaced outside the disk by

$$\begin{aligned} &\tilde{z} = 0: \quad h = 0, \quad h'' = 0, \quad g' = 0, \\ &\tilde{z} \to \infty: \quad h' \to 0, \quad g \to 1. \end{aligned}$$

By the same elementary methods as used earlier, it is found that the only solution is $g = 1$, $h = 0$. This means that outside the disk we can only have the original flow with $v = (1 - \varepsilon)r$.

Hence, at $r = 1$ the Ekman layer suddenly ends, which means that there occur large changes in radial direction and this gives rise to a Stewartson layer.

Finally, we calculate the torque acting on the disk. The tangential shear stress at the disk is

$$\tau_0 = \rho\Omega^2 a^2 E^{1/2} \frac{\partial v}{\partial \tilde{z}}$$

and the torque is $M = 2\pi a^3 \int_0^1 \tau_0 r^2 \, dr$.

With $\partial v/\partial z = -\varepsilon r g'(0)$ and $g'(0) = 1$, we find

$$M = -\frac{1}{2} \varepsilon\pi\rho\Omega^2 a^5 E^{1/2} . \tag{3.5}$$

A negative value of M has a decelerating effect on the disk.

4. Equations in the Stewartson layer

The scaling of the various quantities in the Stewartson layer can be taken most easily from Greenspan [2], pp. 98 and 99. To comply with the rapid changes in radial direction, a stretched coordinate r_1 is introduced by

$$r = 1 + E^{1/3} r_1 . \tag{4.1}$$

r_1 and z are the independent variables in the Stewartson layer. The dependent variables are expanded as follows

$$\begin{aligned}
\psi &= \varepsilon E^{1/2}\psi_1 + \varepsilon E^{5/6}\psi_2 + \cdots \\
u &= \varepsilon E^{1/2}u_1 + \varepsilon E^{5/6}u_2 + \cdots \\
v &= (1 - \varepsilon)r + \varepsilon E^{1/6}v_1 + \varepsilon E^{1/2}v_2 + \cdots \\
w &= \varepsilon E^{1/6}w_1 + \varepsilon E^{1/2}w_2 + \cdots \\
p &= \frac{1}{2}(1 - \varepsilon)^2 r^2 + \varepsilon E^{1/2}p_1 + \varepsilon E^{5/6}p_2 + \cdots .
\end{aligned} \tag{4.2}$$

This gives an axial flux $O(E^{1/2})$ which is the deflected radial flux of $O(E^{1/2})$ existing in the Ekman layer. The second terms in the expansion to E are a factor $E^{1/3}$ smaller than the first terms. This is in agreement with (4.1) and (3.1) and, moreover, the second term in the expansion of w is required to match the term $w = -\varepsilon E^{1/2}$, present in the inner region as follows from (3.1) and (3.4).

Substitution of (4.2) into the equations (2.1) and (2.2) leads to the following set of equations for the first approximation

$$2v_1 = \frac{\partial p_1}{\partial r_1} , \qquad 2u_1 = \frac{\partial^2 v_1}{\partial r_1^2} , \qquad \frac{\partial p_1}{\partial z} = \frac{\partial^2 w_1}{\partial r_1^2} , \qquad \frac{\partial u_1}{\partial r_1} + \frac{\partial w_1}{\partial z} = 0 . \tag{4.3}$$

The terms of the third equation come from terms in (2.1) which are $O(\varepsilon E^{1/2})$, while the neglected non-linear terms in this equation are $O(\varepsilon^2 E^{1/3})$. Hence, it is required that $|\varepsilon| < O(E^{1/6})$.

For the second approximation the following set is obtained

$$2v_2 = \frac{\partial p_2}{\partial r_1}, \qquad 2u_2 = \frac{\partial^2 v_2}{\partial r_1^2} + \frac{\partial v_1}{\partial r_1}, \qquad \frac{\partial p_2}{\partial z} = \frac{\partial^2 w_2}{\partial r_1^2} + \frac{\partial w_1}{\partial r_1}, \qquad \frac{\partial u_2}{\partial r_1} + \frac{\partial w_2}{\partial z} + u_1 = 0. \tag{4.4}$$

This approximation is only valid if $|\varepsilon| < O(E^{1/2})$.

From equations (2.3) and (4.1) we find

$$u_1 = \frac{\partial \psi_1}{\partial z}, \qquad u_2 = \frac{\partial \psi_2}{\partial z} - r_1 \frac{\partial \psi_1}{\partial z}, \qquad w_1 = -\frac{\partial \psi_1}{\partial r_1}, \qquad w_2 = -\frac{\partial \psi_2}{\partial r_1} + r_1 \frac{\partial \psi_1}{\partial r_1}. \tag{4.5}$$

Elimination of all variables except ψ_1 and ψ_2 yields as fundamental equations

$$\frac{\partial^6 \psi_1}{\partial r_1^6} + 4\frac{\partial^2 \psi_1}{\partial z^2} = 0, \qquad \frac{\partial^6 \psi_2}{\partial r_1^6} + 4\frac{\partial^2 \psi_2}{\partial z^2} = 3\frac{\partial^5 \psi_1}{\partial r_1^5}. \tag{4.6}$$

The Ekman layer which has thickness $O(E^{1/2})$ is reduced in the z-coordinate to $z = 0$. Hence $z = 0$ must correspond to the outer edge of the Ekman layer. Moreover, the equations in the Ekman layer are only modified by a stretched coordinate

$$r = 1 + E^{1/2}\tilde{r},$$

since only then the second derivatives to r become of the same order as second derivatives to z. Thus, in the r_1-coordinate this modification occurs at $r_1 = 0$. For $r_1 < 0$ we have at the outer edge of the Ekman layer using (3.1), (3.4) and (4.1)

$$z = 0: \quad \psi = \frac{1}{2}\epsilon E^{1/2} + \varepsilon E^{5/6} r_1 + \frac{1}{2}\varepsilon E^{7/6} r_1^2.$$

Thus $\psi_1 = \frac{1}{2}$, $\psi_2 = r_1$, $w_1 = 0$, $w_2 = -1$.

For $r_1 > 0$ where there is no Ekman layer we have

$$\psi_1 = 0, \quad \psi_2 = 0, \quad w_1 = 0, \quad w_2 = 0.$$

Hence, the boundary condition for ψ_1 is

$$z = 0: \quad \psi_1 = \frac{1}{2}\{1 - U(r_1)\} = \frac{1}{2}U(-r_1), \tag{4.7}$$

where $U(x)$ is the unit step function

$$U(x) = 1 \quad \text{for } x > 0, \qquad U(x) = 0 \quad \text{for } x < 0.$$

The other boundary conditions are

$$z \to \infty: \qquad \psi_1 \text{ is bounded},$$

$r_1 \to -\infty$: ψ_1 is bounded ,

$r_1 \to \infty$: $\psi_1 \to 0$.

Boundary conditions for ψ_2 will be given later.

5. Main solution in the Stewartson layer

The stream function ψ_1 is solved by aid of Fourier transformation

$$F_1(\omega, z) = \frac{1}{\sqrt{2\pi}} \int_{-\infty}^{\infty} \psi_1(r_1, z)\, e^{i\omega r_1}\, dr_1 .$$

We take Im $\omega < 0$ since $\psi_1 \neq 0$ for $r_1 \to -\infty$.

The transformed equation becomes

$$4 \frac{d^2 F_1}{dz^2} - \omega^6 F_1 = 0 .$$

The boundary condition for $z = 0$ is

$$F_1(\omega, 0) = \frac{1}{\sqrt{2\pi}} \int_{-\infty}^{\infty} \frac{1}{2} \{1 - U(r_1)\}\, e^{i\omega r_1}\, dr_1 = \frac{1}{2i\omega\sqrt{2\pi}} .$$

Hence,

$$F_1(\omega, z) = \frac{1}{2i\omega\sqrt{2\pi}}\, e^{-|\omega|^3 z/2} \quad \text{for } z \geq 0 .$$

Transforming back we obtain

$$\psi_1(r_1, z) = \frac{1}{4\pi i} \int_{-\infty}^{\infty} \frac{e^{-i\omega r_1}}{\omega}\, e^{-|\omega|^3 z/2}\, d\omega , \tag{5.1}$$

where the path of integration has to pass below the pole $\omega = 0$. We write

$$\psi_1(r_1, z) = \frac{1}{4\pi i} \int_{-\infty}^{\infty} \frac{e^{-i\omega r_1}}{\omega} (e^{-|\omega|^3 z/2} - 1)\, d\omega + \frac{1}{4\pi i} \int_{-\infty}^{\infty} \frac{e^{-i\omega r_1}}{\omega}\, d\omega .$$

In the first integral $\omega = 0$ is no longer a singular point, so we can integrate straight along the ω-axis. The integration path of the second integral is closed by the infinitely large semi-circle in the half plane Im $\omega > 0$ if $r_1 < 0$ and by the semi-circle in the half plane Im $\omega < 0$ if $r_1 > 0$. By the residu theorem this gives $\frac{1}{2}U(-r_1)$ as a result for the second integral. Then

$$\psi_1(r_1, z) = \frac{1}{4\pi i} \int_{-\infty}^{\infty} \frac{\cos \omega r_1 - i \sin \omega r_1}{\omega} (e^{-|\omega|^3 z/2} - 1)\, d\omega + \frac{1}{2} U(-r_1) .$$

Since the integrand with $\cos \omega r_1$ is odd in ω it gives no contribution and we retain

$$\psi_1(r_1, z) = \frac{1}{2\pi} \int_0^{\infty} \frac{\sin \omega r_1}{\omega} (1 - e^{-\omega^3 z/2})\, d\omega + \frac{1}{2} U(-r_1) . \tag{5.2}$$

For $z \to \infty$ the value of the integral is $\pi/2$ if $r_1 > 0$ and $-\pi/2$ if $r_1 < 0$. This means that

$$\psi_1(r_1, \infty) = \frac{1}{4} \quad \text{for all finite values of } r_1 \text{ and}$$

$$\psi_1(0, z) = \frac{1}{4} \quad \text{for all finite values of } z.$$

The integral in (5.2) is an odd function of r_1 and hence needs only to be calculated for $r_1 > 0$. Putting $\omega r_1 = y$, we can write for $r_1 > 0$

$$\begin{aligned} \psi_1(r_1, z) &= \frac{1}{2\pi} \int_0^\infty \frac{\sin y}{y} (1 - \mathrm{e}^{-y^3\tau/2})\, \mathrm{d}y \\ &= \frac{1}{4} - \frac{1}{2\pi} \int_0^\infty \frac{\sin y}{y} \mathrm{e}^{-y^3\tau/2}\, \mathrm{d}y\,. \end{aligned} \tag{5.3}$$

This formula shows that ψ_1 only depends upon the similarity parameter

$$\tau = z/r_1^3\,. \tag{5.4}$$

Moreover, for $r_1 \to 0$, $\tau \to \infty$, (5.3) gives again $\psi_1 = 1/4$, so (5.3) is valid for $r_1 \geqslant 0$. For $r_1 \leqslant 0$ we have

$$\psi_1(r_1, z) = \frac{1}{4} + \frac{1}{2\pi} \int_0^\infty \frac{\sin y}{y} \mathrm{e}^{y^3\tau/2}\, \mathrm{d}y\,.$$

For any finite value of z and r_1 varying from $-\infty$ to $+\infty$, ψ_1 varies from $1/2$ to 0. Thus, the axial mass flow in the Stewartson layer is for any finite z equal to $2\pi\varepsilon E^{1/2}/2$. This is exactly equal but opposite to the axial mass flow in the inner region which is

$$-2\pi\varepsilon E^{1/2} \int_0^1 h(\infty) r\, \mathrm{d}r\,,$$

where $h(\infty) = 1$ is determined in the Ekman layer. For finite z there is no interchange between the two mass flows. That the axial velocity is constant in the inner region is due to the Taylor–Proudman theorem. However, there is a small viscosity $\nu = O(E)$ and this causes the axial velocity in the inner region to diminish at a distance $z = O(E^{-1})$, that is where the upper region V (Fig. 1) begins. The axial velocity in the Stewartson region diminishes with increasing z as a result of the widening of this layer proportional to $z^{1/3}$. In the upper region the two axial mass flows begin to annihilate each other.

Numerical values for ψ_1 are obtained by Romberg integration of the integral in (5.3). The infinite integration interval is ended when the integrand becomes smaller than 10^{-9}. Results for $\psi_1(\tau)$ are presented in Table 1 and Fig. 2.

For negative values of $\tau (r_1 < 0)$ we have

$$\psi_1(\tau) = \frac{1}{2} - \psi_1(-\tau)\,.$$

By expansion of the exponential in (5.3) we obtain in the limit $\tau \downarrow 0$

$$\psi_1(\tau) = -\frac{\tau}{2\pi} + O(\tau^3)\,.$$

Table 1. The function $\psi_1(\tau)$

τ	$\psi_1(\tau)$
0	0
0.001	−0.000 158 744
0.01	−0.002 505 524
0.03	−0.014 295 372
0.043 99	−0.016 061 169
0.05	−0.015 834 700
0.1	−0.006 200 428
0.125 785	0
0.25	0.025 184 891
0.438 175	0.05
1.339 016	0.10
5.205 917	0.15
44.867 564	0.20
∞	0.25

Negative values of $\psi_1(\tau)$ occur for $0<\tau<0.125\,785$. The streamlines $0<\psi_1(\tau)<0.5$ originate from the Ekman layer by way of the point $r_1=0$, $z=0$. The main flow takes some fluid with it (negative values of ψ_1) which originates from the outer region. The minimal value of ψ_1 occurs for $\tau=0.04399$ and this corresponds to a streamline with zero velocity. At the other side of the Stewartson layer some streamlines with $\psi_1>0.5$ exist, which means that some fluid is attracted from the inner region.

We can find the behaviour of ψ_1 for large values of τ by using the series expansion for sin y in (5.3). We obtain

$$I=\int_0^\infty \frac{\sin y}{y}\,e^{-y^3\tau/2}\,dy=\frac{2^{1/3}}{3}\int_0^\infty \sum_{n=0}^\infty (-1)^n\,\frac{2^{2n/3}\xi^{2(n-1)/3}}{(2n+1)!}\,e^{-\xi\tau}\,d\xi\,,$$

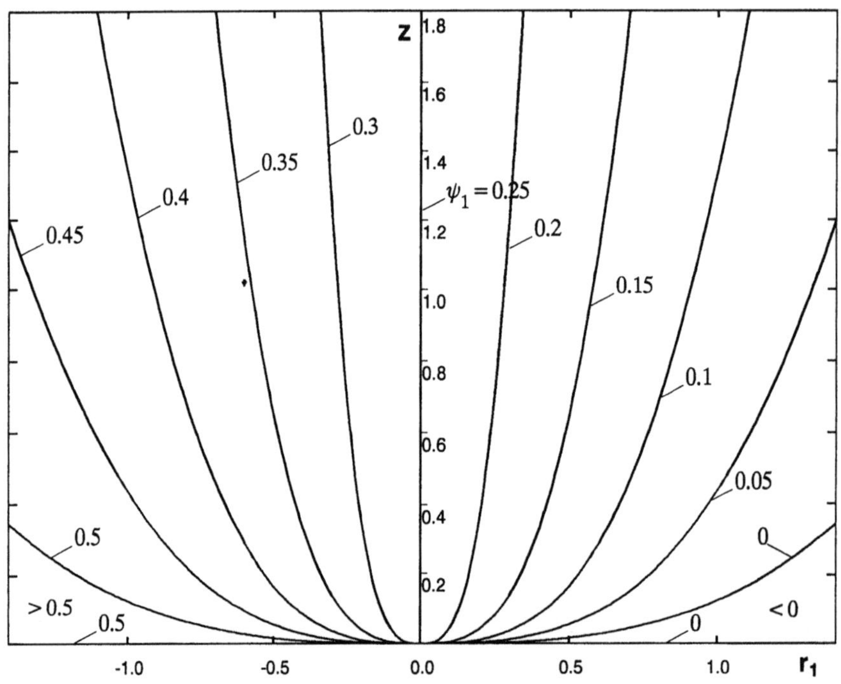

Fig. 2. Streamlines in the r_1-z plane.

where $\xi = y^3/2$. Introducing $\xi\tau = u$ as a new variable the integrals lead to Γ-functions and the result is

$$I = \frac{2^{1/3}\Gamma(\frac{1}{3})}{3\tau^{1/3}} - \frac{1}{9\tau} + \frac{2^{2/3}\Gamma(\frac{5}{3})}{180\tau^{5/3}} - \cdots$$

and

$$\psi_1 = \frac{1}{4} - \frac{1}{2\pi}\left\{\frac{2^{1/3}\Gamma(\frac{1}{3})}{3}\frac{r_1}{z^{1/3}} - \frac{r_1^3}{9z} + \frac{2^{2/3}\Gamma(\frac{5}{3})}{180}\frac{r_1^5}{z^{5/3}} - \cdots\right\}. \tag{5.5}$$

For $r_1 > 0$ the radial velocity u_1 is equal to

$$u_1 = \frac{\partial \psi_1}{\partial z} = \frac{1}{r_1^3}\frac{d\psi_1}{d\tau} = \frac{1}{4\pi r_1^3}\int_0^\infty y^2 \sin y\, e^{-y^3\tau/2}\, dy\,. \tag{5.6}$$

Furthermore, u_1 is an odd function of r_1, so $u(-r_1) = -u(r_1)$. For $\tau\downarrow 0$ which occurs if $z\downarrow 0$, $r \neq 0$ but also for z finite and $r_1 \to \infty$, we find

$$u_1 = -\frac{1}{2\pi r_1^3}\,. \tag{5.7}$$

For $\tau\to\infty$ we obtain an expression in the same way as we did for ψ_1. The result is

$$u_1 = \frac{1}{2\pi}\left\{\frac{2^{1/3}\Gamma(\frac{1}{3})}{9}\frac{r_1}{z^{4/3}} - \frac{1}{9}\frac{r_1^3}{z^2} + \frac{2^{2/3}\Gamma(\frac{5}{3})}{108}\frac{r_1^5}{z^{8/3}} - \cdots\right\}, \tag{5.8}$$

which is also obtained by direct differentiation of (5.5). Figure 3 shows u_1 as a function of r_1 for some values of z.

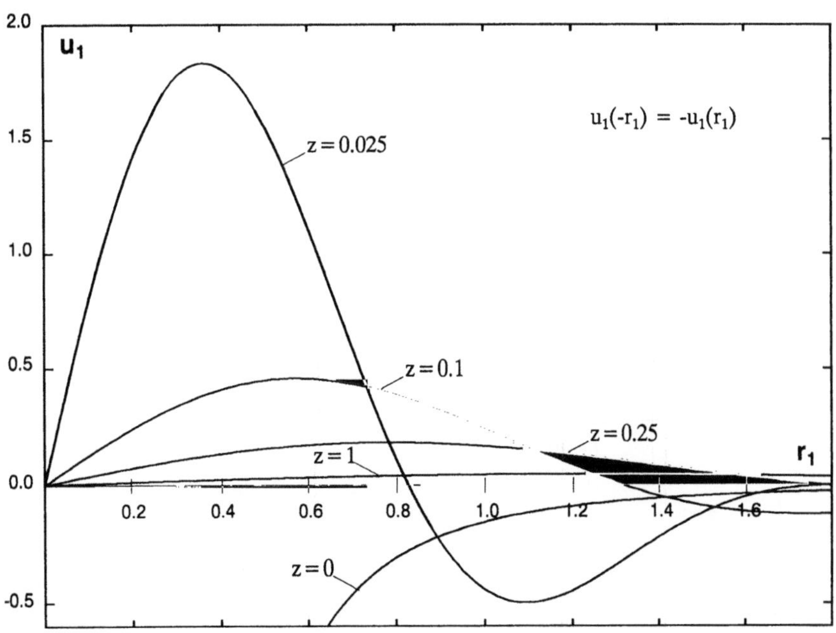

Fig. 3. The radial velocity u_1 as function of r_1 for some values of z.

For $r_1 > 0$ the axial velocity w_1 is equal to

$$w_1 = -\frac{\partial \psi_1}{\partial r_1} = \frac{3z}{r_1^4}\frac{d\psi_1}{d\tau} = \frac{3z}{4\pi r_1^4}\int_0^\infty y^2 \sin y \, e^{-y^3\tau/2}\, dy . \tag{5.9}$$

For $z = 0$ this gives $w_1 = 0$, while for z finite and $r_1 \to \infty$ $(\tau \to 0)$ we obtain

$$w_1 = -\frac{3z}{2\pi r_1^4} . \tag{5.10}$$

w_1 is an even function of r_1, so $w(-r_1) = w(r_1)$.

Finally, by differentiation of (4.7) we have for $z = 0$, $w_1 = \frac{1}{2}\delta(r_1)$, where $\delta(x)$ is the Dirac δ-function.

The expansion of w_1 for $\tau \to \infty$ becomes

$$w_1 = \frac{1}{2\pi}\left\{\frac{2^{1/3}\Gamma(\frac{1}{3})}{3z^{1/3}} - \frac{r_1^2}{3z} + \frac{2^{2/3}\Gamma(\frac{5}{3})}{36}\frac{r_1^4}{z^{5/3}} - \cdots\right\} . \tag{5.11}$$

Figure 4 shows w_1 as a function of r_1 for some values of z.

For calculating the remaining variables v_1 and p_1 we proceed as follows. In order to differentiate w_1 to r_1, we replace y by ωr_1 in (5.9)

$$w_1 = \frac{3z}{4\pi r_1}\int_0^\infty \omega^2 \sin \omega r_1 \, e^{-\omega^3 z/2}\, d\omega .$$

By using

$$\frac{d\, e^{-\omega^3 z/2}}{d\omega} = -\frac{3}{2}\omega^2 z \, e^{-\omega^3 z/2}$$

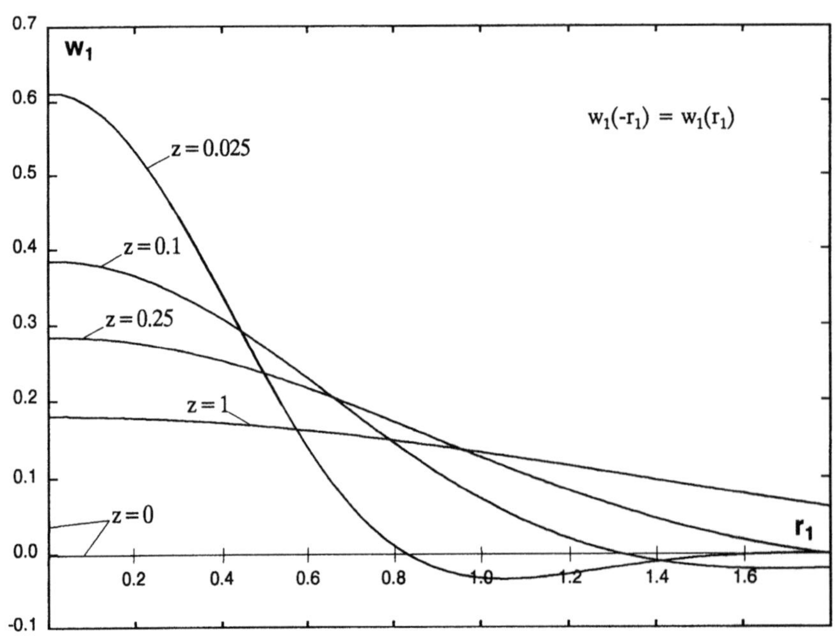

Fig. 4. The axial velocity w_1 as function of r_1 for some values of z.

and applying partial integration the result for w_1 becomes

$$w_1 = \frac{1}{2\pi}\int_0^\infty \cos \omega r_1 \, e^{-\omega^3 z/2} \, d\omega \, . \tag{5.12}$$

Then, from (4.3) we obtain

$$\frac{\partial p_1}{\partial z} = \frac{\partial^2 w_1}{\partial r_1^2} = -\frac{1}{2\pi}\int_0^\infty \omega^2 \cos \omega r_1 \, e^{-\omega^3 z/2} \, d\omega \, . \tag{5.13}$$

Since direct integration to z produces a divergent integral, we first calculate

$$\frac{\partial^2 p_1}{\partial r_1 \, \partial z} = \frac{1}{2\pi}\int_0^\infty \omega^3 \sin \omega r_1 \, e^{-\omega^3 z/2} \, d\omega$$

and then integrate to z. Hence

$$\frac{\partial p_1}{\partial r_1} = -\frac{1}{\pi}\int_0^\infty \sin \omega r_1 \{ e^{-\omega^3 z/2} + C(r_1)\} \, d\omega \, . \tag{5.14}$$

Again from (4.3)

$$v_1 = -\frac{1}{2\pi}\int_0^\infty \sin \omega r_1 \{ e^{-\omega^3 z/2} + C(r_1)\} \, d\omega \, .$$

Furthermore, we have $2u_1 = \partial^2 v_1 / \partial r_1^2$ and thus

$$u_1 = \frac{1}{4\pi}\int_0^\infty \omega^2 \sin \omega r_1 \, e^{-\omega^3 z/2} \, d\omega + \frac{1}{4\pi}\int_0^\infty \sin \omega r_1 \, e^{-\omega^3 z/2} \left\{ \omega^2 C(r_1) - \frac{d^2 C_1}{d r_1^2} \right\} d\omega \, .$$

Comparing this result with (5.6) we see that the second term must vanish. This leads to $C(r_1) = 0$ since the possibilities $C(r_1) = \sinh wr_1$ or $\cosh wr_1$ are excluded as $v_1 \to 0$ for $r_1 \to \infty$. Hence

$$v_1 = -\frac{1}{2\pi r_1}\int_0^\infty \sin y \, e^{-y^3 \tau/2} \, dy \, . \tag{5.15}$$

Writing (5.12) also in terms of y, it follows that w_1 and v_1 are the real and imaginary parts of

$$\frac{1}{2\pi r_1}\int_0^\infty e^{-iy} \, e^{-y^3 \tau/2} \, dy \, ,$$

which is in agreement with [3].

Finally, (5.15) shows that for $\tau \downarrow 0$ ($z \downarrow 0$, $r_1 > 0$ or z finite and $r_1 \to \infty$) we have

$$v_1 = -\frac{1}{2\pi r_1} \, . \tag{5.16}$$

The expansion of v_1 for $\tau \to \infty$ is

$$v_1 = -\frac{1}{3\pi}\left\{ \frac{\Gamma(\frac{2}{3})}{2^{1/3}} \frac{r_1}{z^{2/3}} - \frac{2^{1/3}\Gamma(\frac{4}{3})}{6} \frac{r_1^3}{z^{4/3}} + \frac{1}{60} \frac{r_1^5}{z^2} - \cdots \right\} . \tag{5.17}$$

Figure 5 shows v_1 as a function of r_1 for some values of z. v_1 is an odd function of r_1, $v_1(-r_1) = -v_1(r_1)$.

It follows also from (5.13) that

$$\frac{\partial p_1}{\partial z} = -\frac{1}{2\pi r_1^3} \int_0^\infty y^2 \cos y \, \mathrm{e}^{-y^3\tau/2} \, \mathrm{d}y \, .$$

Expanding this again for $\tau \to \infty$ the result is

$$\frac{\partial p_1}{\partial z} = -\frac{1}{3\pi} \left\{ \frac{1}{z} - \frac{\Gamma(\frac{5}{3})}{2^{1/3}} \frac{r_1^2}{z^{5/3}} + \frac{2^{1/3}\Gamma(\frac{7}{3})}{12} \frac{r_1^4}{z^{7/3}} - \cdots \right\},$$

which agrees with differentiation of (5.11).

Integrating, we obtain as expansion of p_1 near $r_1 = 0$

$$p_1 = -\frac{1}{3\pi} \left\{ \ln z + \frac{3\Gamma(\frac{5}{3})}{2^{4/3}} \frac{r_1^2}{z^{2/3}} - \frac{2^{1/3}\Gamma(\frac{7}{3})}{16} \frac{r_1^4}{z^{4/3}} + \cdots \right\} + C_1(r_1) \, .$$

For arbitrary values of r_1 and $z > 0$, we can write

$$p_1 = -\frac{1}{3\pi} \ln z + \int_0^{r_1} \frac{\partial p_1}{\partial r_1} \, \mathrm{d}r_1 + C_1(0) \, .$$

Since the pressure is only fixed up to an arbitrary constant, we may take $C_1(0) = 0$. Substituting the value for $\partial p_1 / \partial r_1$ from (5.14) with $C(r_1) = 0$ and performing the integration to r_1, we find for positive values of r_1

$$p_1 = -\frac{1}{3\pi} \ln z - \frac{1}{\pi} \int_0^\infty \frac{1 - \cos y}{y} \, \mathrm{e}^{-y^3\tau/2} \, \mathrm{d}y \, , \tag{5.18}$$

where again y has been written for ωr_1.

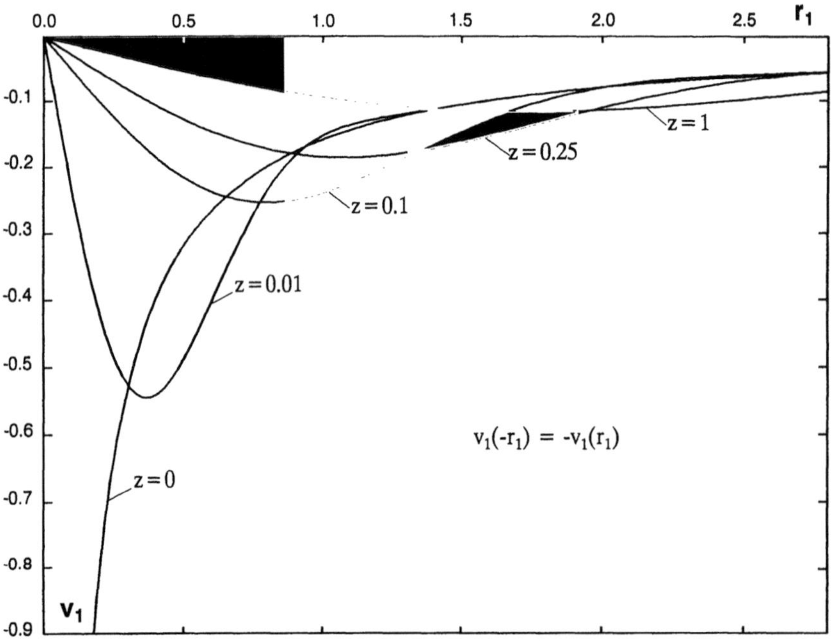

Fig. 5. The azimuthal velocity v_1 as function of r_1 for some values of z.

In order to investigate the behaviour of p_1 for finite z and $r_1 \to \infty$, we have to evaluate the integral in (5.18) for small values of τ. This integral is written as

$$\lim_{a \to 0} \left\{ \int_a^\infty \frac{1}{y} \mathrm{e}^{-y^3\tau/2} \, \mathrm{d}y - \int_a^\infty \frac{\cos y}{y} \mathrm{e}^{-y^3\tau/2} \, \mathrm{d}y \right\}, \quad a > 0 . \tag{5.19}$$

The first integral is

$$\int_a^\infty \frac{1}{y} \mathrm{e}^{-y^3\tau/2} \, \mathrm{d}y = \frac{1}{3} \int_{\frac{1}{2}a^3\tau}^\infty \frac{e^{-u}}{u} \, du = \frac{1}{3} E_1\left(\frac{1}{2} a^3\tau\right),$$

according to [6], formula (5.1.1). For small values of a we have, see [6], (5.1.11)

$$\frac{1}{3} E_1\left(\frac{1}{2} a^3\tau\right) = \frac{1}{3}\left\{-\gamma - \ln \frac{1}{2} a^3\tau + O(a^3\tau)\right\},$$

where γ is Euler's constant. Hence in the limit $a \to 0$

$$\int_a^\infty \frac{1}{y} \mathrm{e}^{-y^3\tau/2} \, \mathrm{d}y = -\frac{1}{3}\gamma + \frac{1}{3}\ln 2 - \ln a - \frac{1}{3}\ln\tau + O(a^3\tau) .$$

For small values of τ the second integral in (5.19) is reduced as follows

$$\lim_{a \to 0} \int_a^\infty \frac{\cos y}{y} \mathrm{e}^{-y^3\tau/2} \, \mathrm{d}y = \lim_{a \to 0} \int_a^\infty \frac{\cos y}{y} \, \mathrm{d}y - 15\tau^2 + O(\tau^4),$$

where the exponential has been expanded. From [6], formulae (5.2.27) and (5.2.16) we have for $a \to 0$

$$\int_a^\infty \frac{\cos y}{y} \, \mathrm{d}y = -\mathrm{Ci}(a) = -\gamma - \ln a .$$

The conclusion is that

$$\int_0^\infty \frac{1 - \cos y}{y} \mathrm{e}^{-y^3\tau/2} \, \mathrm{d}y = \frac{2}{3}\gamma + \frac{1}{3}\ln 2 - \frac{1}{3}\ln\tau + 15\tau^2$$

and, substituting in (5.18), for $\tau \to 0$ the pressure becomes equal to

$$p_1 = -\frac{1}{\pi}\ln r_1 - \frac{1}{3\pi}(2\gamma + \ln 2) - \frac{15z^2}{\pi r_1^6} .$$

Since the pressure is an even function of r_1 we can write for $|\tau| \to 0$

$$p_1 = -\frac{1}{\pi}\ln|r_1| - 0.1960341 - \frac{15z^2}{\pi r_1^6} . \tag{5.20}$$

For arbitrary values of r_1 and $z > 0$ we have (5.18). For some values of z, p_1 is given as function of r_1 in Fig. 6. It may be remarked that (5.10) and (5.20) satisfy the relation $\partial p_1/\partial z = \partial^2 w_1/\partial r_1^2$ in (4.3).

The infinitely large pressure at the singularity $r_1 = 0$, $z_1 = 0$ is the fundamental reason of the deviation of the flow from the Ekman boundary layer toward the Stewartson layer.

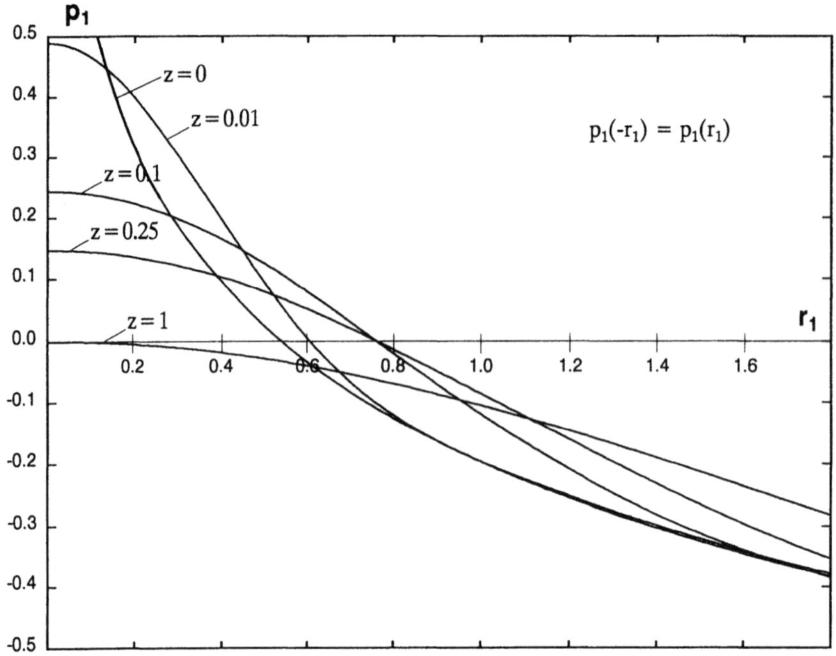

Fig. 6. The pressure p_1 as function of r_1 for some values of z.

6. The inner, outer and upper regions

The flow in the inner region for $r \uparrow 1$ should be matched to the flow in the Stewartson layer for $r_1 \to -\infty$. With $r_1 = -E^{-1/3}(1-r)$ and taking into account the even or odd character of the functions, we find from (4.2), (5.7), (5.10), (5.16) and (5.20), for $r \uparrow 1$

$$\left.\begin{aligned}
\psi &= \frac{1}{2}\,\varepsilon E^{1/2}\,, \\
u &= \varepsilon E^{3/2}\,\frac{1}{2\pi(1-r)^3}\,, \\
v &= (1-\varepsilon)r + \varepsilon E^{1/2}\,\frac{1}{2\pi(1-r)}\,, \\
w &= -\varepsilon E^{1/2} - \varepsilon E^{3/2}\,\frac{3z}{2\pi(1-r)^4}\,, \\
p &= \frac{1}{2}\,(1-\varepsilon)^2 r^2 + \frac{1}{3\pi}\,\varepsilon E^{1/2}\ln E - \varepsilon E^{1/2}\left\{\frac{1}{\pi}\ln(1-r) + 0.1960341\right\}.
\end{aligned}\right\} \qquad (6.1)$$

In general, we can write in the inner region

$$\left.\begin{aligned}
\psi &= \frac{1}{2}\,\varepsilon E^{1/2} r^2 + \varepsilon E^{3/2}\psi_i\,, \\
u &= \varepsilon E^{3/2} u_i\,, \\
v &= (1-\varepsilon)r + \varepsilon E^{1/2} v_i\,, \\
w &= -\varepsilon E^{1/2} + \varepsilon E^{3/2} w_i\,, \\
p &= \frac{1}{2}\,(1-\varepsilon)^2 r^2 + \frac{1}{3\pi}\,\varepsilon E^{1/2}\ln E + \varepsilon E^{1/2} p_i\,,
\end{aligned}\right\} \qquad (6.2)$$

where the limiting values of the variables u_i etc. for $r\uparrow 1$ are given by (6.1) while

$$\lim_{r\uparrow 1} \psi_i = \frac{z}{2\pi(1-r)^3} .$$

The equations to be satisfied by the variables are

$$\left.\begin{aligned}
&2v_i = \frac{\partial p_i}{\partial r} ,\\
&2u_i = \frac{\partial^2 v_i}{\partial r^2} + \frac{\partial}{\partial r}\left(\frac{v_i}{r}\right),\\
&\frac{\partial p_i}{\partial r} = 0 ,\\
&u_i = \frac{1}{r}\frac{\partial \psi_i}{\partial z} , \qquad w_i = -\frac{1}{r}\frac{\partial \psi_i}{\partial r} .
\end{aligned}\right\} \tag{6.3}$$

Since p_i is independent of z, it follows from the first equation, that also v_i is independent of z which is the reason that the term $\partial^2 v_i/\partial z^2$ could be omitted in the second equation. Only ψ_i and w_i are linear in z, the other variables being independent of z.

In the outer region we have

$$\begin{aligned}
&\psi = \varepsilon E^{3/2}\psi_0 ,\\
&u = \varepsilon E^{3/2}u_0 ,\\
&v = (1-\varepsilon)r + \varepsilon E^{1/2}v_0 ,\\
&w = \varepsilon E^{3/2}w_0 ,\\
&p = \frac{1}{2}(1-\varepsilon)^2 r^2 + \frac{1}{3\pi}\varepsilon E^{1/2}\ln E + \varepsilon E^{1/2}p_0 .
\end{aligned}$$

The limiting values for $r\downarrow 1$ are

$$\begin{aligned}
&\psi_0 \to -\frac{z}{2\pi(r-1)^3} , \qquad u_0 \to -\frac{1}{2\pi(r-1)^3} , \qquad v_0 \to -\frac{1}{2\pi(r-1)} ,\\
&w_0 \to -\frac{3z}{2\pi(r-1)^4} , \qquad p_0 \to -\left\{\frac{1}{\pi}\ln(r-1) + 0.1960341\right\} .
\end{aligned} \tag{6.4}$$

The variables in the outer region also satisfy (6.3). Again ψ_0 and z_0 are linear in z, the other variables are independent of z.

The equations (6.3) are only modified when z becomes $O(E^{-1})$. We then have for the upper region

$$\psi = \varepsilon E^{1/2}\psi_u , \qquad u = \varepsilon E^{3/2}u_u , \qquad v = (1-\varepsilon)r + \varepsilon E^{1/2}v_u ,$$

$$w = \varepsilon E^{1/2}w_u , \qquad p = \frac{1}{2}(1-\varepsilon)^2 r^2 + \frac{1}{3\pi}\varepsilon E^{1/2}\ln E + \varepsilon E^{1/2}p_u .$$

With $z = E^{-1}z_u$ the equations become

$$2v_u = \frac{\partial p_u}{\partial r} ,$$

$$2u_u = \frac{\partial^2 v_u}{\partial r^2} + \frac{\partial}{\partial r}\left(\frac{v_u}{r}\right),$$

$$\frac{\partial p_u}{\partial z_u} = \frac{\partial^2 w_u}{\partial r^2} + \frac{1}{r}\frac{\partial w_u}{\partial r},$$

$$u_u = \frac{1}{r}\frac{\partial \psi_u}{\partial z_u}, \qquad w_u = -\frac{1}{r}\frac{\partial \psi_u}{\partial r}.$$

Only the third equation is changed in comparison with (6.3). For $z_u \downarrow 0$ the limits of ψ_u and w_u are different in case r is smaller or larger than 1

$$r<1: \quad \psi_u \to \frac{1}{2}r^2, \quad w_u \to -1$$

$$r>1: \quad \psi_u \to 0, \quad w_u \to 0.$$

It follows that the solution in the upper region contains a singularity at the point $r=1$, $z_u = 0$. At the scale of the upper region, the Stewartson layer is reduced to the point $r=1$, $z_u = 0$. In the upper region it no longer exists as a layer but its influence in the whole region is apparent through the singularity. The variable $\tau = z/r_1^3$ of the Stewartson layer becomes $z_u/(r-1)^3$ in the upper region.

7. Second approximation in the Stewartson layer

It was shown in Section 6 that the Stewartson layer gives rise to axial velocities $O(E^{3/2})$ in the inner and outer regions. However, in the inner region there is a more important axial velocity $w = -\varepsilon E^{1/2}$, due to the Ekman layer, which is lacking in the outer region. The second approximation in the Stewartson layer will show how the transition from $w = -\varepsilon E^{1/2}$ to $w = o(E^{1/2})$ occurs. The fundamental equation is given by (4.6) as

$$\frac{\partial^6 \psi_2}{\partial r_1^6} + 4\frac{\partial^2 \psi_2}{\partial z^2} = 3\frac{\partial^5 \psi_1}{\partial r_1^5}. \tag{7.1}$$

Boundary conditions are

$z = 0$: $\quad \psi_2 = r_1 U(-r_1)$,

$z \to \infty$: $\quad \psi_2$ is bounded,

$r_1 \to -\infty$: $\quad \psi_2 \to r_1$,

$r_1 \to \infty$: $\quad \psi_2 \to 0$.

The solution of ψ_2 is again found by aid of Fourier transformation

$$F_2(\omega, z) = \frac{1}{\sqrt{2\pi}}\int_{-\infty}^{\infty} \psi_2(r_1, z)\, e^{i\omega r_1}\, dr_1, \quad \text{Im}\, \omega < 0$$

The transformed equation is

$$4\frac{d^2F_2}{dz_2} - \omega^6 F_2 = -3i\omega^5 F_1 .$$

Substituting F_1 from Section 5 we obtain

$$4\frac{d^2F_2}{dz^2} - \omega^6 F_2 = -\frac{3\omega^4}{2\sqrt{2\pi}}\, e^{-|\omega|^3 z/2} .$$

Due to the boundedness of F_2 for $z \to \infty$, the solution can be written in the form

$$F_2 = A\, e^{-|\omega|^3 z/2} + Bz\, e^{-|\omega|^3 z/2} .$$

Fourier transformation of the boundary condition for $z = 0$ yields

$$F_2(\omega, 0) = A = \frac{1}{\omega^2\sqrt{2\pi}} ,$$

while substitution of F_2 in the differential equation gives

$$B = \frac{3|\omega|}{8\sqrt{2\pi}} .$$

The solution for ψ_2 then becomes

$$\begin{aligned}\psi_2(r_1, z) &= \frac{1}{2\pi}\int_{-\infty}^{\infty}\left(\frac{1}{\omega^2} + \frac{3|\omega|z}{8}\right) e^{-i\omega r_1}\, e^{-|\omega|^3 z/2}\, d\omega \\ &= \frac{1}{2\pi}\left[\left[\int_{-\infty}^{\infty}\frac{1}{\omega^2}\, e^{-i\omega r_1}(e^{-|\omega|^3 z/2} - 1)\, d\omega + \int_{-\infty}^{\infty}\frac{3|\omega|z}{8}\, e^{-i\omega r_1}\, e^{-|\omega|^3 z/2}\, d\omega\right.\right. \\ &\quad \left. + \int_{-\infty}^{\infty}\frac{1}{\omega^2}\, e^{-i\omega r_1}\, d\omega\right] .\end{aligned}$$

The last integral has a double pole at $z = 0$ with residue $-ir_1$. For $r_1 < 0$ the integration path is closed by the infinitely large semi-circle in the half plane $\mathrm{Im}\ \omega > 0$ and by the semi-circle in the half plane $\mathrm{Im}\ \omega < 0$ if $r_1 > 0$. The result is $2\pi r_1 U(-r_1)$. In the other integrals we replace $e^{-i\omega r_1}$ by $\cos \omega r_1 - i \sin \omega r_1$. Remarking that the integrals with $\cos \omega r_1$ are even in ω but those with $\sin \omega r_1$ odd, the result is

$$\psi_2(r_1, z) = \frac{1}{\pi}\int_0^{\infty}\frac{\cos \omega r_1}{\omega^2}(e^{-\omega^3 z/2} - 1)\, d\omega + \frac{3z}{8\pi}\int_0^{\infty}\omega \cos \omega r_1\, e^{-\omega^3 z/2}\, d\omega + r_1 U(-r_1) . \tag{7.2}$$

Expanding the exponential we find for $z \downarrow 0$

$$\psi_2(r_1, z) = r_1 U(-r_1) + \frac{z}{8\pi r_1^2} + O\left(\frac{z^3}{r_1^8}\right) .$$

Except for the first term, $\psi_2(r_1, z)$ is even in r_1. The axial velocity will be calculated from the formula $w_2 = -\partial\psi_2/\partial r_1 + r_1\, \partial\psi_1/\partial r_1$, see (4.5).

From (7.2) and (5.2) we have

$$\frac{\partial \psi_2}{\partial r_1} = -\frac{1}{\pi}\int_0^\infty \frac{\sin \omega r_1}{\omega}(e^{-\omega^3 z/2}-1)\,d\omega - \frac{3z}{8\pi}\int_0^\infty \omega^2 \sin \omega r_1\, e^{-\omega^3 z/2}\,d\omega + U(-r_1)\,,$$

$$\frac{\partial \psi_1}{\partial r_1} = \frac{1}{2\pi}\int_0^\infty \cos \omega r_1 (1-e^{-\omega^3 z/2})\,d\omega - \frac{1}{2}\,\delta(r_1)\,.$$

Using

$$\int_0^\infty \frac{\sin \omega r_1}{\omega}\,d\omega = \frac{\pi}{2} - \pi U(-r_1)$$

and

$$\int_0^\infty \cos \omega r_1\,d\omega = \pi\delta(r_1)\,,$$

we obtain

$$w_2 = \frac{1}{2\pi}\int_0^\infty \left(2\frac{\sin \omega r_1}{\omega} - r_1 \cos \omega r_1\right) e^{-\omega^3 z/2}\,d\omega + \frac{3z}{8\pi}\int_0^\infty \omega^2 \sin \omega r_1\, e^{-\omega^3 z/2}\,d\omega - \frac{1}{2}\,.$$

Partial integration of the last integral gives

$$w_2 = \frac{1}{4\pi}\int_0^\infty \left(4\frac{\sin \omega r_1}{\omega} - r_1 \cos \omega r_1\right) e^{-\omega^3 z/2}\,d\omega - \frac{1}{2}$$

or

$$w_2 = \frac{1}{4\pi}\int_0^\infty \left(4\frac{\sin y}{y} - \cos y\right) e^{-y^3\tau/2}\,dy - \frac{1}{2} \quad \text{if } \tau = \frac{z}{r_1^3} > 0$$

and

$$w_2 = -\frac{1}{4\pi}\int_0^\infty \left(4\frac{\sin y}{y} - \cos y\right) e^{y^3\tau/2}\,dy - \frac{1}{2} \quad \text{if } \tau = \frac{z}{r_1^3} < 0\,.$$

It is seen that w_2 only depends on τ. For $\tau \downarrow 0$ we have

$$w_2 = \frac{7\tau}{4\pi} + O(\tau^3)\,,$$

while for $\tau < 0$

$$w_2(\tau) = -1 - w_2(-\tau) = -1 + \frac{7\tau}{4\pi} + O(\tau^3)$$

holds. Results for w_2 are presented in Table 2 and Fig. 7.

The contribution to w for $|r_1| \to \infty$ is

$$\varepsilon E^{1/2}\left\{-U(-r_1) + \frac{7z}{4\pi r_1^3}\right\}.$$

Table 2. The function $w_2(\tau)$, $\tau = 0.027171$ gives maximum of $w_2(\tau)$

τ	$w_2(\tau)$
0	0
0.000 01	0.000 005 570
0.027 171	0.041 872 659
0.075 277	0
0.1	−0.023 631 987
0.219 096	−0.1
0.617 158	−0.2
2.281 694	−0.3
19.107 978	−0.4
∞	−0.5

Matching to the inner region gives for $r \uparrow 1$

$$w = -\varepsilon E^{1/2} - \varepsilon E^{3/2} \frac{7z}{4\pi(1-r)^3}$$

and the outer region

$$w = \varepsilon E^{3/2} \frac{7z}{4\pi(r-1)^3} \,.$$

This yields further terms in the expansions for $r \to 1$ of the solutions in the inner and outer regions as given by (6.1) and (6.4).

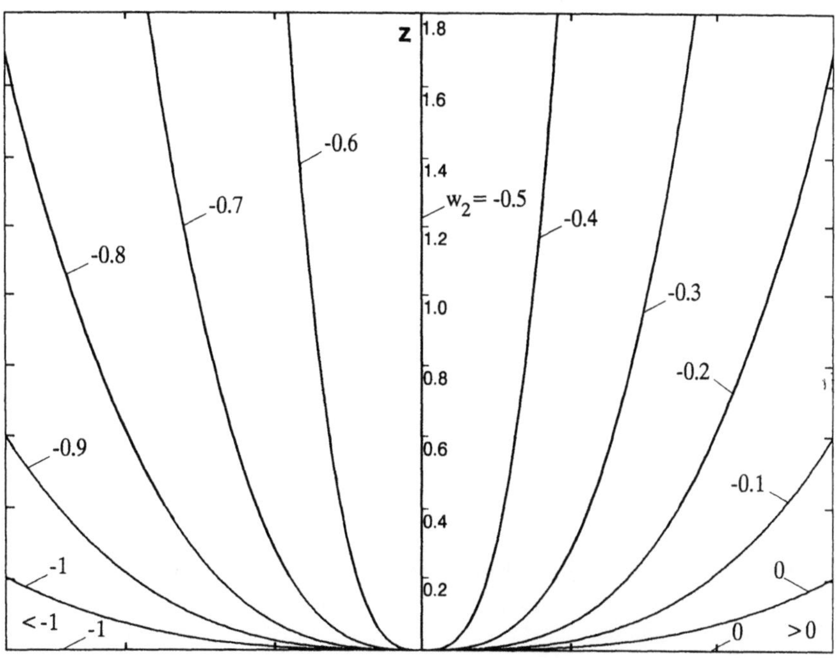

Fig. 7. The axial velocity w_2 in the $r_1 - z$ plane.

The radial velocity follows from $u_2 = \partial\psi_2/\partial z - r_1\,\partial\psi_1/\partial z$. Using (7.2) and (5.6) the result turns out to be

$$u_2 = -\frac{1}{8\pi}\int_0^\infty (3\omega\cos\omega r_1 + \omega^2 r_1 \sin\omega r_1)\,\mathrm{e}^{-\omega^3 z/2}\,\mathrm{d}\omega$$

or

$$u_2 = -\frac{1}{8\pi r_1^2}\int_0^\infty (3y\cos y + y^2\sin y)\,\mathrm{e}^{-y^3\tau/2}\,\mathrm{d}y \quad \text{if } \tau > 0\,.$$

u_2 is an even function of r_1. The contribution to u for $|r_1| \to \infty$ is

$$\varepsilon E^{5/6}\,\frac{5}{8\pi r_1^2}\,,$$

which gives by matching to the inner and outer regions a term

$$\varepsilon E^{3/2}\,\frac{5}{8\pi(r-1)^2}$$

in addition to the terms already obtained in (6.1) and (6.4).

After elaborate calculations along the same lines as in Section 5 we find the following results for p_2 and v_2

$$p_2 = -\frac{r_1}{2\pi}\left\{\ln z + \int_0^\infty \frac{y(3+\cos y) - 4\sin y}{y^2}\,\mathrm{e}^{-y^3\tau/2}\,\mathrm{d}y + C_3\right\} \quad \text{if } \tau > 0\,,$$

$$v_2 = -\frac{1}{4\pi}\left\{\ln z + \int_0^\infty \frac{3(1-\cos y) - y\sin y}{y}\,\mathrm{e}^{-y^3\tau/2}\,\mathrm{d}y + C_3\right\} \quad \text{if } \tau > 0\,.$$

p_2 is odd and v_2 is even in r_1. For small numbers of τ we obtain

$$p_2 = \frac{r_1}{2\pi}\,(-3\ln r_1 + 27\tau^2 + C_4)\,, \quad \tau\downarrow 0\,,$$

$$v_2 = \frac{1}{4\pi}\,(-3\ln r_1 - 135\tau^2 + C_4 - 3)\,, \quad \tau\downarrow 0\,,$$

where $C_4 = 4 - 2\gamma - \ln 2 - C_3$. The constant C_3 is determined by the flow outside the Stewartson layer.

Matching of p_2 and v_2 gives further terms in the asymptotic expansions of p_i, p_0, v_i and v_0 for $r \to 1$ in the inner and outer regions.

8. Conclusions

A rotating disk placed in a fluid rotating coaxially with a slightly different angular velocity shows at its edge a Stewartson layer of width $O(E^{1/3})$ and height $O(E^{-1})$ provided $|\varepsilon| < O(E^{1/6})$. This layer is due to the sudden deflection of the boundary layer flow into an axial flow if $\varepsilon > 0$ and reversely if $\varepsilon < 0$. The deflection is caused by a logarithmic pressure singularity at $r_1 = 0$, $z = 0$.

Velocity and pressure distributions in the Stewartson layer have been evaluated and calculated for some values of z as functions of r_1, see Figs 2 to 6. Due to the occurrence of the similarity parameter τ they have a simple analytic form.

The orders of magnitude of the various quantities in the Stewartson layer are

stream function	ψ	$O(E^{1/2})$,
radial velocity	u	$O(E^{1/2})$,
azimuthal velocity	v	$O(E^{1/6})$,
axial velocity	w	$O(E^{1/6})$,
pressure	p	$O(E^{1/2})$.

The everywhere present azimuthal velocity $v = (1 - \varepsilon)r$ and corresponding pressure $p = \frac{1}{2}(1 - \varepsilon)^2 r^2$ have been left out of account. The orders of magnitude in the inner and outer regions are respectively

	ψ	u	v	w	p
inner	$O(E^{1/2})$	$O(E^{3/2})$	$O(E^{1/2})$	$O(E^{1/2})$	$O(E^{1/2} \ln E) + O(E^{1/2})$,
outer	$O(E^{3/2})$	$O(E^{3/2})$	$O(E^{1/2})$	$O(E^{3/2})$	$O(E^{1/2} \ln E) + O(E^{1/2})$.

The reduction of the axial velocity of $O(E^{1/2})$ in the inner region to $O(E^{3/2})$ in the outer region has been investigated with the aid of the second approximation of the solution of the differential equations, see Section 5. This approximation, which is valid for $|\varepsilon| < O(E^{1/2})$ gives contributions in the Stewartson layer of the following orders of magnitude

ψ	u	v	w	p
$O(E^{5/6})$	$O(E^{5/6})$	$O(E^{1/2})$	$O(E^{1/2})$	$O(E^{5/6})$.

In the upper region the orders of magnitude are the same as in the inner region, which means that w is also $O(E^{1/2})$.

Note added in proof

It appears that the homogeneous differential equation

$$\frac{\partial^6 \psi}{\partial r_1^6} + 4 \frac{\partial^2 \psi}{\partial z^2} = 0$$

has additional solutions. These might be excited by the Ekman layer at the singular point $r_1 = 0$, $z = 0$. It means that the solution ψ_2 might be modified by an additional term containing an unknown factor. Whether this is the case should follow from an investigation of the region connecting the Stewartson and Ekman layers. Such investigation is being performed.

Acknowledgement

The author wishes to thank Kees Visser for programming the figures for this paper.

References

1. K. Stewartson, On almost rigid rotations. *J. Fluid Mech.* 3 (1957) 17–26.
2. H.P. Greenspan, *The Theory of Rotating Fluids*. Cambridge University Press, Cambridge (1968).
3. D.W. Moore and P.G. Saffman, The structure of free vertical shear layers in a rotating fluid and the motion produced by a slowly rising body. *Phil. Trans. Roy. Soc.* A264 (1969) 597–634.
4. R. Hide and C.W. Titman, Detached shear layers in a rotating fluid. *J. Fluid Mech.* 29 (1967) 39–60.
5. A.I. van de Vooren and E.F.F. Botta, Fluid flow induced by a rotating disk of finite radius. *J. Eng. Math.* 24 (1990) 55–71.
6. M. Abramowitz and J.A. Stegun, *Handbook of Mathematical Functions*. Dover (1965).

Journal of Engineering Mathematics **26**: 153–158, 1992.
H.K. Kuiken and S.W. Rienstra (eds), Problems in Applied, Industrial and Engineering Mathematics.

Causality and the radiation condition

John V. WEHAUSEN*

Department of Naval Architecture and Offshore Engineering, University of California, Berkeley, California, USA

Abstract. For a hydrostatically stable floating body making small oscillations about a fixed position as a result of external forces and moments, it is shown that the radiation condition implies that the motion at time t depends only upon the forces and moments at times $\leqslant t$, i.e. that the future does not determine the present.

We wish to consider here one aspect of the motion of a body floating in a heavy fluid. The motions will be assumed to be small enough so that the equations may be linearized. In order to describe the motion of both body and fluid, we shall adopt a right-handed coordinate system with Oz directed against gravity, Ox to the right, and Oy into the paper. The plane Oxy lies in the undisturbed free surface. The small excursions that the body makes about its fixed equilibrium position will be denoted by $\alpha_1, \ldots, \alpha_6$, where $\alpha_1, \alpha_2, \alpha_3$ represent translational displacements and $\alpha_4, \alpha_5, \alpha_6$ angular ones. The dynamical constants of the body will be denoted by m_{ik} where $m_{11} = m_{22} = m_{33} = m$, the mass of the body, and

$$m_{ik} = \int \rho[r^2\delta_{ik} - x_i x_k]\,\mathrm{d}V\,, \quad i, k = 4, 5, 6\,,$$

where ρ is the density distribution of the body, $r^2 = x_i x_i$, and the integral is taken over the body. All other m_{ik} are zero. In the equations to be given below c_{ik} are the hydrostatic coefficients, μ_{ik} are the added masses as defined by Cummins (1962) (i.e., the added masses at infinite frequency), and $L_{ik}(t)$ is a weighting function defined in terms of the velocity potential for the fluid motion. Its definition as well as those of μ_{ik} and c_{ik} may be found in Wehausen (1971 or 1967). An important property of L_{ik} is that it is zero for $t<0$. Let $X_i(t)$ be the force ($i = 1, 2, 3$) or moment ($i = 4, 5, 6$) to which the body is subjected. $X_i(t)$ may be a result of oncoming waves, of wind, or of some other exterior forcing mechanism. We suppose $X_i(t)$ to be absolutely integrable.

In order to accommodate the possibility that only certain modes of motion are allowed, we shall suppose that the subscripts i, j or k may be restricted to some subset A of the integers $1, 2, \ldots, 6$. Then the linearized equations of motion for the body are as follows (see Wehausen, loc. cit.):

$$(m_{ik} + \mu_{ik})\ddot{\alpha}_k(t) + c_{ik}\alpha_k + \int_0^t L_{ik}(t-\tau)\ddot{\alpha}_k(\tau)\,\mathrm{d}\tau = X_i(t)\,, \quad i, k \in A\,. \tag{1}$$

Repeated indices are summed over the integers belonging to A, non-repeated indices take on successively the integers in A.

The equations are a natural candidate for a Fourier or Laplace transform. We shall use the Fourier transform:

---

*Editorial member from 1967–1986.

$$X_k(t) = \int_{-\infty}^{\infty} \tilde{X}_k(\sigma)\, \mathrm{e}^{-\mathrm{i}\sigma t}\, \mathrm{d}\sigma\,. \tag{2}$$

After taking the Fourier transform of the equations of motion, we find the following:

$$\{-\sigma^2[m_{ik} + \mu_{ik}(\sigma)] + c_{ik} - \mathrm{i}\sigma\lambda_{ik}(\sigma)\}\tilde{\alpha}_k(\sigma) = \tilde{X}_i(\sigma)\,, \quad i, k \in A\,, \tag{3}$$

where

$$\mu_{ik}(\sigma) - \mu_{ik}(\infty) + \mathrm{i}\sigma^{-1}\lambda_{ik}(\sigma) = \int_0^{\infty} L_{ik}(\tau)\, \mathrm{e}^{\mathrm{i}\sigma\tau}\, \mathrm{d}\tau\,. \tag{4}$$

It follows from this that $\mu_{ik}(-\sigma) = \mu_{ik}(\sigma)$ and $\lambda_{ik}(-\sigma) = \lambda_{ik}(\sigma)$. We introduce the following notation:

$$\tilde{M}_{ik} = -\sigma^2[m_{ik} + \mu_{ik}(\sigma)] + c_{ik}\,, \qquad \tilde{N}_{ik} = \sigma\lambda_{ik}(\sigma)\,, \qquad \tilde{S}_{ik} = \tilde{M}_{ik} - \mathrm{i}\tilde{N}_{ik}\,. \tag{5}$$

The transformed equation (3) then reads:

$$\tilde{S}_{ik}\tilde{\alpha}_k = \tilde{X}_i\,, \quad i, k \in A\,, \tag{6}$$

and its solution is evidently

$$\tilde{\alpha}_i = \tilde{T}_{ik}\tilde{X}_k\,, \quad i, k \in A\,, \tag{7}$$

where $\tilde{T} = \tilde{S}^{-1}$, i.e. $\tilde{T}_{ij}\tilde{S}_{jk} = \delta_{ik}$. It now follows easily from the above and from known properties of L_{ik} that $\tilde{S}_{ik}(-\sigma) = \bar{\tilde{S}}_{ik}(\sigma)$, $\tilde{S}_{ki} = \tilde{S}_{ik}$ and similarly for $\tilde{T}_{ik}$. Because of this property of T_{ik} we find

$$\begin{aligned} T_{ik}(t) &= \int_{-\infty}^{\infty} \tilde{T}_{ik}(\sigma)\, \mathrm{e}^{-\mathrm{i}\sigma t}\, \mathrm{d}\sigma \\ &= \int_0^{\infty} [\bar{\tilde{T}}_{ik}(\sigma)\, \mathrm{e}^{\mathrm{i}\sigma t} + \tilde{T}_{ik}(\sigma)\, \mathrm{e}^{-\mathrm{i}\sigma t}]\, \mathrm{d}\sigma\,, \end{aligned}$$

i.e. $T_{ik}(t)$ is real. Even though $\tilde{S}_{ik}$ is independent of the choice of A, this is not true for $\tilde{T}_{ik}$ or T_{ik}, which will be different for different choices of A. Although it might be helpful to indicate this, we have not done so in order to avoid a cluttered notation.

Having found $\tilde{\alpha}_i(\sigma)$ above, we may now calculate $\alpha_i(t)$:

$$\begin{aligned} \alpha_i(t) &= \int_{-\infty}^{\infty} \tilde{\alpha}_i(\sigma)\, \mathrm{e}^{-\mathrm{i}\sigma t}\, \mathrm{d}\sigma = \int_{-\infty}^{\infty} \tilde{T}_{ik}\tilde{X}_k\, \mathrm{e}^{-\mathrm{i}\sigma t}\, \mathrm{d}\sigma \\ &= \frac{1}{2\pi}\int_{-\infty}^{\infty} \tilde{T}_{ik}(\sigma)\, \mathrm{e}^{-\mathrm{i}\sigma t}\left[\int_{-\infty}^{\infty} X_k(\tau)\, \mathrm{e}^{\mathrm{i}\sigma\tau}\, \mathrm{d}\tau\right] \mathrm{d}\sigma \\ &= \frac{1}{2\pi}\int_{-\infty}^{\infty} \mathrm{d}\tau\, X_k(\tau) \int_{-\infty}^{\infty} \mathrm{d}\sigma\, \tilde{T}_{ik}(\sigma)\, \mathrm{e}^{-\mathrm{i}\sigma(t-\tau)} \\ &= \frac{1}{2\pi}\int_{-\infty}^{\infty} X_k(\tau) T_{ik}(t-\tau)\, \mathrm{d}\tau \\ &= \frac{1}{2\pi}\int_{-\infty}^{\infty} T_{ik}(\tau) X_k(t-\tau)\, \mathrm{d}\tau\,, \quad i, k \in A\,. \end{aligned} \tag{8}$$

The preceding development is well known and has been reviewed in order to display the last formula for $\alpha_i(t)$, for this formula seems to indicate that α_i at time t depends upon the value of the exciting force X_k at all future as well as all past times unless we can show that $T_{ik}(t)=0$ for all $t<0$. It is this problem that we wish to address and to which we now turn.

First we shall show that the desired property of T_{ik} is equivalent to a certain property of $\tilde{T}_{ik}$. The reasoning is well known and can be found in books on control theory (e.g., Solodovnikov, 1960, pp. 24–28). Consider the transform

$$\tilde{T}_{ik}(\sigma)=\frac{1}{2\pi}\int_{-\infty}^{\infty} T_{ik}(t)\,\mathrm{e}^{\mathrm{i}\sigma t}\,\mathrm{d}t\,.$$

Although heretofore we have thought of σ as being real, we shall now take it to be complex. $\tilde{T}_{ik}(\sigma)$ is then defined in the whole σ-plane. Let us write $\sigma=\rho\,\mathrm{e}^{\mathrm{i}\theta}=\rho(\cos\theta+\mathrm{i}\sin\theta)$. Consider now

$$\int_C \tilde{T}_{ik}(\sigma)\,\mathrm{e}^{-\mathrm{i}\sigma t}\,\mathrm{d}\sigma=\int \tilde{T}_{ik}(R\,\mathrm{e}^{\mathrm{i}\theta})\,\mathrm{e}^{-\mathrm{i}Rt\cos\theta}\,\mathrm{e}^{Rt\sin\theta}Ri\,\mathrm{e}^{\mathrm{i}\theta}\,\mathrm{d}\theta\,,$$

where the path of integration is either along the semicircle C_+: $\rho=R$, $0<\theta<\pi$, or the semicircle C_-: $\rho=R$, $2\pi>\theta>\pi$. These paths are now completed by paths along the real axis from $-R$ to R. Evidently, as $R\to\infty$ the integral along C_+ converges to zero if $t<0$ and that along C_-converges to zero if $t>0$. It then follows that for $t<0$

$$T_{ik}(t)=\lim_{R\to\infty}\left[\int_{-R}^{R}\tilde{T}_{ik}(\sigma)\,\mathrm{e}^{-\mathrm{i}\sigma t}\,\mathrm{d}\sigma+\int_{C_+}\tilde{T}_{ik}(\sigma)\,\mathrm{e}^{-\mathrm{i}\sigma t}\,\mathrm{d}\sigma\right].$$

If $\tilde{T}_{ik}$ is analytic in the upper half-plane, then $T_{ik}(t)=0$ for $t<0$. The converse of this is also true, i.e. if $T_{ik}(t)=0$ for $t<0$, then $\tilde{T}_{ik}(\sigma)$ is analytic in the upper half-plane.

Our problem has now been transformed to that of showing that $\tilde{T}_{ik}$ is analytic in the upper half-plane. Since $\tilde{T}=\tilde{S}^{-1}$, we shall search for an equivalent property of $\tilde{S}$. Let P_{ik} be the cofactor of the element $\tilde{S}_{ik}$ in the determinant $\det\tilde{S}$ where $i,k\in A$. Then it is known that

$$\tilde{T}_{ik}=P_{ki}/\det\tilde{S}\,,\quad i,k\in A\,. \tag{9}$$

As we shall see, $\tilde{S}$ and hence P_{ik} are analytic in the upper half-plane. Thus what remains to be shown is that $\det\tilde{S}$ has no zeros in the upper half-plane. How do we know that $\tilde{S}$ is analytic in the upper half-plane? From the earlier formulas (4) and (5) defining μ_{ik}, λ_{ik} and $\tilde{S}_{ik}$ it follows that

$$\tilde{S}_{ik}=-\sigma^2\int_0^{\infty}L_{ik}(t)\,\mathrm{e}^{\mathrm{i}\sigma t}\,\mathrm{d}t+c_{ik}-\sigma^2[m_{ik}+\mu_{ik}(\infty)]\,,\quad i,k\in A\,. \tag{10}$$

It has already been mentioned that $L_{ik}(t)=0$ for $t<0$, so that its transform is analytic in the upper half-plane. Since the other terms in the equation above are obviously analytic, $\tilde{S}_{ik}$ is also. We recall that $\tilde{S}$ is a matrix of order between 1 and 6, depending upon A. It will be one of the main-diagonal matrices of the matrix that one would obtain when A consists of all the integers 1 to 6.

We turn now to $\det\tilde{S}$. The matrix $\tilde{S}=\tilde{M}-\mathrm{i}\tilde{N}$ is symmetric but is not hermitian, so that no easy conclusion can be drawn from the fact of symmetry alone. However, the matrix

$$\tilde{S}\bar{\tilde{S}}^* = [\tilde{M}(\sigma) - \mathrm{i}\tilde{N}(\sigma)][\bar{\tilde{M}}^T(\sigma) + \mathrm{i}\bar{\tilde{N}}^T(\sigma)] \tag{11}$$

is hermitian, and we shall be able to exploit this fact. Here $\tilde{M}^T$ is the transpose of $\tilde{M}$ and is, of course, equal to $\tilde{M}$ when $\tilde{M}$ is symmetric. We write $\tilde{S}^* = \bar{\tilde{S}}^T$, a usual notation. We note that the product above is hermitian even when $\tilde{S}$ is not symmetric.

Associated with any hermitian matrix $H_{ik} = \bar{H}_{ki}$, $i, k = 1, \ldots, n$, is a so-called hermitian form

$$Q = x_i H_{ik} \bar{x}_k$$

where repeated indices are to be summed from 1 to n. It is easy to show that $Q = \bar{Q}$, so that Q is real. Within this class of forms one distinguishes positive (negative) definite and non-negative (non-positive) definite forms. A non-negative definite form is one such that $Q \geqslant 0$ for any choice of $x_1, \ldots, x_n$; a positive definite form is one such that $Q > 0$ for any choice of $x_1, \ldots, x_n$ except $x_1 = x_2 = \cdots = x_n = 0$. Analogously for the terms in parentheses. A classical theorem about hermitian forms states that such a form is positive definite if and only if all the determinants formed with the first minors along the main diagonal are positive, i.e.,

$$H_{11} > 0\,, \qquad \begin{vmatrix} H_{11} & H_{12} \\ H_{21} & H_{22} \end{vmatrix} > 0, \ldots, \begin{vmatrix} H_{11} & \cdots & H_{1n} \\ H_{21} & \cdots & H_{2n} \\ \vdots & & \vdots \\ H_{n1} & \cdots & H_{nn} \end{vmatrix} > 0\,.$$

It then follows that *all* main-diagonal minors are positive. There is an analogous theorem for negative definite forms and a somewhat more complicated one for non-negative and non-positive definite forms.

Consider now the special hermitian form

$$\begin{aligned} Q &= x_i \tilde{S}_{ij} \bar{\tilde{S}}_{kj} \bar{x}_k = x_i[\tilde{M}_{ij}(\sigma) - \mathrm{i}\tilde{N}_{ij}(\sigma)][\bar{\tilde{M}}_{kj}(\sigma) + \mathrm{i}\bar{\tilde{N}}_{kj}(\sigma)]\bar{x}_k \\ &= \sum_j |x_i[\tilde{M}_{ij}(\sigma) - \mathrm{i}\tilde{N}_{ij}(\sigma)]|^2 \geqslant 0\,, \quad i, j, k \in A\,. \end{aligned} \tag{12}$$

Evidently Q is a sum of squares and hence $Q \geqslant 0$. Suppose that for some particular $\sigma = s + \mathrm{i}r$, $r > 0$, there exists a set of complex numbers x_i, $i \in A$, not all zero, such that $Q = \Sigma_j = 0$. Then

$$x_i[\tilde{M}_{ij}(\sigma) - \mathrm{i}\tilde{N}_{ij}(\sigma)] = x_i \tilde{S}_{ij} = 0\,, \quad i, j \in A\,,$$

and also

$$\bar{x}_i[\bar{\tilde{M}}_{ij}(\sigma) + \mathrm{i}\bar{\tilde{N}}_{ij}(\sigma)] = \bar{x}_i \bar{\tilde{S}}_{ij} = 0\,, \quad i, j \in A\,.$$

The following quadratic forms are then also zero:

$$x_i \tilde{S}_{ij}(\sigma) \bar{x}_j = 0\,, \qquad \bar{x}_i \bar{\tilde{S}}_{ij}(\sigma) x_j = 0\,. \tag{13}$$

Interchanging i and j in the second equation, we obtain

$$x_i \bar{\tilde{S}}_{ji}(\sigma)\bar{x}_j = 0\,. \tag{14}$$

After first adding and then subtracting (13) and (14) we obtain:

$$x_i\{\tilde{S}_{ij}(\sigma) + \bar{\tilde{S}}_{ji}(\sigma)\}\bar{x}_j = 0\,, \tag{15}$$

$$x_i\{\tilde{S}_{ij}(\sigma) - \bar{\tilde{S}}_{ji}(\sigma)\}\bar{x}_j = 0\,, \quad i, j \in A\,. \tag{16}$$

We now substitute the expression from (10) into (15) and (16):

$$x_i\Big\{-\sigma^2 \int_0^\infty L_{ij}(t)\, \mathrm{e}^{\mathrm{i}\sigma t}\, \mathrm{d}t \mp \sigma^2 \int_0^\infty L_{ji}(t)\, \mathrm{e}^{-\mathrm{i}\sigma t}\, \mathrm{d}t$$

$$+ c_{ij} \pm c_{ji} - \sigma^2[m_{ij} + \mu_{ij}(\infty)] \mp \sigma^2[m_{ji} + \mu_{ji}(\infty)]\Big\}\bar{x}_j = 0\,, \tag{17}$$

where the top signs go with (15) and the bottom ones with (16). If we now let $\sigma = s + \mathrm{i}r$, equation (17) becomes the following:

$$x_i\Big\{-(s^2 - r^2 + 2\,\mathrm{i}rs) \int_0^\infty L_{ij}(t)\, \mathrm{e}^{\mathrm{i}st}\, \mathrm{e}^{-rt}\, \mathrm{d}t \mp (s^2 - r^2 - 2\mathrm{i}rs) \int_0^\infty L_{ji}(t)\, \mathrm{e}^{-\mathrm{i}st}\, \mathrm{e}^{-rt}\, \mathrm{d}t$$

$$+ c_{ij} \pm c_{ji} - (s^2 - r^2 + 2\mathrm{i}rs)[m_{ij} + \mu_{ij}(\infty)] \mp (s^2 - r^2 - 2\mathrm{i}rs)[m_{ji} + \mu_{ji}(\infty)]\}\bar{x}_j = 0\,. \tag{18}$$

We now exploit the fact that when there is no mean forward motion all matrices are symmetric, i.e., $L_{ij} = L_{ji}$, $c_{ij} = c_{ji}$, etc. It is then easy to deduce the following two equations from (18):

$$x_i\Big\{-2(s^2 - r^2) \int_0^\infty L_{ij}(t)\, \mathrm{e}^{-rt} \cos(st)\, \mathrm{d}t + 4rs \int_0^\infty L_{ij}(t)\, \mathrm{e}^{-rt} \sin(st)\, \mathrm{d}t$$

$$+ 2c_{ij} - 2(s^2 - r^2)[m_{ij} + \mu_{ij}(\infty)]\Big\}\bar{x}_j = 0\,, \tag{19}$$

$$x_i\Big\{2(s^2 - r^2) \int_0^\infty L_{ij}(t)\, \mathrm{e}^{-rt} \sin(st)\, \mathrm{d}t + 4rs \int_0^\infty L_{ij}(t)\, \mathrm{e}^{-rt} \cos(st)\, \mathrm{d}t$$

$$+ 4rs[m_{ij} + \mu L_{ij}(\infty)]\Big\}\bar{x}_j = 0\,. \tag{20}$$

Between these two equations we may now eliminate first the integrals with cos(st) and next the integrals with sin(st) to obtain the following equations:

$$x_i\Big\{(s^2 + r^2)^2 \int_0^\infty L_{ij}(t)\, \mathrm{e}^{-rt} \sin(st)\, \mathrm{d}t + 2rsc_{ij}\Big\}\bar{x}_j = 0\,, \tag{21}$$

$$x_i\Big\{(s^2 + r^2)^2 \int_0^\infty L_{ij}(t)\, \mathrm{e}^{-rt} \cos(st)\, \mathrm{d}t + (s^2 - r^2)c_{ij} - (s^2 + r^2)^2[m_{ij} + \mu_{ij}(\infty)]\Big\}\bar{x}_j = 0\,. \tag{22}$$

From (4) we easily find

$$L_{ij}(t) = \frac{2}{\pi} \int_0^\infty s'^{-1} \lambda_{ij}(s') \sin(s't)\, \mathrm{d}s'$$

$$= \frac{2}{\pi} \int_0^\infty [\mu_{ij}(s') - \mu_{ij}(\infty)] \cos(s't)\, \mathrm{d}s' . \tag{23}$$

If we substitute the first of these into (21) and integrate with respect to t, we find the following equation:

$$x_i c_{ij} \bar{x}_j + \frac{2}{\pi} \int_0^\infty x_i \lambda_{ij}(s') \bar{x}_j \frac{(r^2 + s^2)}{[r^2 + (s + s')^2][r^2 + (s - s')^2]} \mathrm{d}s' = 0 \tag{24}$$

for the hypothesized values of r, s, and x_i, $i \in A$. Equation (22) does not seem to lead to anything useful for our purpose.

We now recall that for a hydrostatically stable floating body the quadratic form $x_i c_{ij} \bar{x}_j$ is non-negative. In fact, if the set A includes only those modes of motion for which there is a hydrostatic restoring force, then it is positive. Moreover, the radiation condition implies that the quadratic form $x_i \lambda_{ij}(s') \bar{x}_j > 0$. Hence the equation (24) cannot hold, that is, the assumption that the hermitian form Q can be zero for some σ in the upper half-plane has led to a contradiction. Thus Q is positive definite and all the main-diagonal determinants of $\tilde{S}\tilde{S}^*$ are >0. But then also $\det \tilde{S} \neq 0$, which is what we wanted to prove. A somewhat weaker version of the radiation condition would be sufficient, namely $x_i \lambda_{ij}(s) \bar{x}_j \geqslant 0$ but with the $>$ holding for at least some intervals of s, so that the integral on the right in (24) is >0.

We have shown that the radiation condition implies that $T_{ij}(t) = 0$ if $t < 0$, i.e. that the future does not determine the present, at least in this particular water-wave problem. One would also like to show the converse, that if $T_{ij} = 0$ for $t < 0$ and for $i, j \in A$ for any A, then the radiation condition holds in the weakened form. We have not been able to resolve this problem.

Note that the non-negativeness of the buoyancy quadratic form $x_i c_{ij} \bar{x}_j$ plays an essential role in the proof. On the other hand, the positive definiteness of $x_i m_{ij} \bar{x}_j$ does not seem to be called upon.

Note: The material in this paper was first presented at the First Workshop on Water Waves and Floating Bodies (1986). The author is much indebted to Mr Gyeong Joong Lee of the Department of Naval Architecture of Seoul National University for having pointed out to him not only the inadequacy of the proof presented at the First Workshop but also of a purportedly corrected proof. The present analysis was presented at the Fourth Workshop (1989).

References

Cummins, W.E., The impulse response function and ship motions. *Schiffstechnik* 9 (1962) 101–109.

Solodovnikov, V.V., *Introduction to the Statistical Dynamics of Automatic Control Systems*. Dover, New York (1960) ix + 307 pp.

Wehausen, J.V., Initial-value problem for the motion in an undulating sea of a body with fixed equilibrium position. *J. Engrg. Math.* 1 (1967) 1–17.

Wehausen, J.V., The motion of floating bodies. *Ann. Rev. Fluid Mech.* 3 (1971) 237–268.

Wehausen, J.V., Causality and the radiation condition. The First International Workshop on Water Waves and Floating Bodies, Mass. Inst. Tech., 1986, pp. 181–186, disc. 187–189.

Wehausen, J.V., Causality and the radiation condition. II. The Fourth International Workshop on Water Waves and Floating Bodies, Øystese, Norway, 1989, pp. 251–252, disc. 253–257.

Journal of Engineering Mathematics **26**: 159–194, 1992.
H.K. Kuiken and S.W. Rienstra (eds), Problems in Applied, Industrial and Engineering Mathematics.

Recent mathematical results in the nonlinear theory of flat and curved elastic membranes of revolution

H.J. WEINITSCHKE[†] and H. GRABMÜLLER

Institut für Angewandte Mathematik, Universität Erlangen-Nürnberg, D-8520 Erlangen, Germany
([†]Deceased)

Abstract. The present article reviews some recent developments in nonlinear elastic membrane theory with special emphasis on axisymmetric deformation of flat circular and annular membranes subjected to a vertical surface load and with prescribed radial stresses or radial displacements at the edges. The nonlinear Föppl membrane theory of small finite deflections as well as a simplified version of Reissner's finite-rotation theory is employed, assuming linear stress-strain relations. The main analytical techniques are reported which have been applied recently in order to determine the ranges of those boundary parameters for which solutions of the relevant nonlinear boundary value problems exist, and ranges of parameters for which the principal stresses are nonnegative everywhere. Concerning plane membranes, it is shown how the mathematical theory of existence and uniqueness was nearly completed in recent works in contrast to curved membranes where references can be given to rather few results.

1. Introduction

In the linear theory of elastic membranes and thin shells, the deformation at any given place of the body is proportional to the magnitude of the applied load. As in three-dimensional elasticity, the load must be sufficiently small for linear membrane or shell theory to be applicable; otherwise a nonlinear theory is required. If the strain-displacement relations are nonlinear, but linear stress-strain relations are adequate, we have a *geometrically* nonlinear theory. When the latter relations are also nonlinear, one deals with a *physically* nonlinear theory. Many conventional materials for engineering thin shell and membrane designs are linearly elastic for small strain so that linear stress-strain relations are appropriate. Thus for sufficiently thin structures there is good reason to formulate geometrically nonlinear theories assuming finite deflections (rotations) but small strains, and to investigate the solution structure of the basic boundary value problems. This is the purpose of the present review, where we restrict ourselves to the class of problems described by nonlinear membrane theories. For simplicity, we only consider membranes of revolution under axisymmetric loads (aximembranes). Recently, materially nonlinear deformation has been studied for rubber-like membranes, but this topic is beyond the scope of this paper.

In 1859, Kirchhoff first proposed a nonlinear plate theory [28] which was reduced by von Kármán to a set of two simultaneous equations for the normal displacement w and a stress function f [26]. Somewhat earlier, A. Föppl had derived a nonlinear theory for flat membranes [13]. These are all geometrically nonlinear theories for *small finite deflections*, that is, the displacement components u and v tangential to the plane of the membrane or plate are assumed small compared to the normal displacement w; furthermore, only quadratic terms in w and in the change of angles of rotation are retained.

The von Kármán equations have been studied extensively by both mathematicians[1] and engineers because their simple structure makes them amenable to both theoretical and

[1] For a recent survey see P.G. Ciarlet [7].

numerical analysis. In cartesian coordinates, these equations are

$$D\Delta^2 w - [f, w] = p(x, y), \qquad A\Delta^2 f + \frac{1}{2}[w, w] = 0, \tag{1.1}$$

where D and A are elasticity constants, f the Airy stress function, p is the surface load normal to the (undeformed) plate, Δ^2 is the biharmonic operator and $[f, g] = f_{xx}g_{yy} + f_{yy}g_{xx} - 2f_{xy}g_{xy}$. The Föppl membrane equations [13] are obtained by setting the flexural rigidity D equal to zero in (1.1).

In 1938, a nonlinear theory of shallow shells was given by Marguerre [31], a few years after Donnell [12] had derived a nonlinear theory for circular cylindrical shells to investigate certain buckling problems. Again these papers are restricted to a nonlinear theory with *small finite deflections*. The Marguerre equations are

$$\begin{aligned} &D\Delta^2 w - [z, f] - [w, f] = p(x, y), \\ &A\Delta^2 f + [z, w] + \frac{1}{2}[w, w] = 0, \end{aligned} \tag{1.2}$$

where $z(x, y)$ represents the (undeformed) shape of the middle surface. Due to the presence of the (linear) curvature terms $[z, f]$ and $[z, w]$, the solution structure of (1.2) changes drastically from that of (1.1), including, in particular, buckling under normal pressure.

In a related development, the need for a stress analysis of thin inflatable sheets led Bromberg and Stoker to develop a geometrically nonlinear theory for aximembranes [5].

The first geometrically nonlinear shell theory not restricted to *small finite deflections* was given by E. Reissner [37]. This work concerns axisymmetric bending and stretching of shells of revolution (axishells). The basic equilibrium and compatibility equations and stress-strain relations were reduced essentially to a system of two coupled second order ordinary differential equations for the meridional angle of rotation Φ and a stress function F. Various simplifications of the system derived in [37] have been proposed in recent years. One by Reissner [38], others by Koiter [29] and by Libai and Simmonds [30]. Some of these simplifications are parallel to the linear theory, as far as neglecting terms involving Poisson's ratio ν is concerned, others are obtained by neglecting $O(\varepsilon)$ terms compared to unity, where ε is a measure for the strain in the shell. The simplified Reissner equations [30] can be written in the form

$$\begin{aligned} &r(\Phi - \varphi)'' + (\Phi - \varphi)' \cos\varphi = \frac{1}{r}\cos\Phi + \frac{1}{D}(F\sin\Phi - rV\cos\Phi), \\ &rF'' + F'\cos\varphi - \frac{1}{r}F = \frac{1}{A}(\cos\Phi - \cos\varphi) + (r^2 p_H)' + \nu r p_s. \end{aligned} \tag{1.3}$$

In cylindrical coordinates (r, θ, z) the parametric representation of the surface of revolution is taken in the form $r = r(s)$, $z = z(s)$, where s is the arclength along a meridian of the surface. Primes denote differentiation with respect to s. Thus we have $r' = \cos\varphi$ and $z' = \sin\varphi$, where φ is the angle made by the meridional tangent with the base plane of the surface of revolution. The components of the surface load p are denoted by appropriate subscripts (H = horizontal, V = vertical; s, n meridional and normal to the deformed surface). We note the relations

$$p_H = p_s\cos\Phi + p_n\sin\Phi, \qquad p_V = p_s\sin\Phi - p_n\cos\Phi, \tag{1.4}$$

$F = rR$, R and V are the radial (horizontal) and axial stress resultants, respectively. Depending on the type of loading, the effect of the Poisson's ratio term in (1.3) may be significant for moderate to large values of the load. In (1.3) meridionally uniform shell properties have been assumed (constant thickness, isotropy and homogeneity of the material).

In order to solve the differential equations (1.3), boundary conditions consistent with the small strain assumption must be formulated. In many cases of practical interest, these boundary conditions are nonlinear, for example, if the horizontal displacement u is prescribed at the edge. In terms of Φ and F, one has

$$u = rA[F' + rp_H - \nu(r^{-1}F\cos\Phi + V\sin\Phi)]\,. \tag{1.5}$$

Very thin shells have negligible bending stiffness. Therefore, an important special case is *nonlinear membrane theory*, which is obtained from (1.3) by setting $D = 0$, which yields $rV\cos\Phi = F\sin\Phi$. Substituting this into the second equation of (1.3) we find the basic equation for *geometrically nonlinear aximembranes*

$$(rF')' - \frac{1}{r}F = \frac{1}{A}\left\{\frac{F}{[F^2 + (rV)^2]^{1/2}} - \cos\varphi\right\} + (r^2p_H)' + \nu rp_s\,. \tag{1.6}$$

The equations of the approximate *small finite deflection* theory can now be obtained as follows. Introducing $\beta = \varphi - \Phi$, one may write $\cos\Phi = \cos\varphi + \beta\sin\varphi - \frac{1}{2}\beta^2\cos\varphi + \cdots$ and a similar expansion for $\sin\Phi$. Retaining only terms up to the second degree in β and F in equations (1.3), we get the equations of small finite deflection theory. It is a simple exercise to transform equations (1.2) to cylindrical coordinates. Set $z = z(r)$, $w = w(r)$, $f = f(r)$ (axishells) and integrate the resulting equations twice with respect to r. The result will be the small finite deflection version of equations (1.3) for shallow shells of revolution. The *small finite deflection membrane equation* can be obtained directly from (1.6) by expanding the square root and retaining only quadratic terms. What comes out is a generalization of the Föppl membrane theory to curved membranes:

$$(rF')' - \frac{1}{r}F = \frac{1}{A}\left[1 - \cos\varphi - \frac{1}{2}\left(\frac{rV}{F}\right)^2\right] + (r^2p_H)' + \nu rp_s\,. \tag{1.7}$$

For the special case of a plane membrane $z = \varphi = 0$ under vertical load $p_H = p_s = 0$, we get

$$(rF')' - \frac{1}{r}F = -\frac{1}{2A}\left(\frac{rV}{F}\right)^2. \tag{1.8}$$

This review paper is organized as follows. The mathematical theory for the boundary value problems of plane membranes is nearly complete, while for curved membranes relatively few results are available. Accordingly, Sections 2, 3, and 4 are devoted to circular and annular membranes. In the first part of Section 2, results of the small finite deflection theory (1.7), called the Föppl theory, are presented for circular membranes. In the second part, results for a (simplified) geometrically nonlinear theory involving arbitrary finite rotations (1.6), called the Reissner theory, are discussed for circular membranes. Similarly, annular membranes for the Föppl and Reissner theories are discussed in Section 3. The tensile solutions of Sections 2 and 3 are characterized by $\sigma_r \geqslant 0$ where σ_r is the radial stress component. In Section 4, the additional condition that the circumferential stress σ_θ satisfies $\sigma_\theta \geq 0$ is imposed, leading to what we call wrinkle-free solutions. Again these solutions are discussed both within the framework of the Föppl and the Reissner theories. Finally, some results for curved membranes are presented in Section 5.

2. Circular membranes

Consider a circular membrane of radius a and thickness d subjected to a vertical pressure $p = p(r)$. At the edge $r = a$, either a constant radial stress σ_r or a constant radial displacement u is prescribed. Assuming *small finite deflections*, we have the governing equation (1.8). Here $s = r$ and $F = r\sigma_r d$. Dimensionless variables are introduced by $r = ax$, $\sigma_r/E = k^2y/4$, $p(r) = p_0\bar{p}(x)$, where E is Young's modulus, related to A in (1.8) by $AEd = 1$, $p_0 > 0$ is the maximum of $|p(r)|$, so that $|\bar{p}(x)| \leq 1$, and $k = (2p_0a/Ed)^{1/3}$. $\bar{p}(x)$ is assumed piecewise continuous for $0 \leq x \leq 1$. In terms of x, y and $\bar{p}$, equation (1.8) can be reduced to the form

$$Ly := -y'' - \frac{3}{x}\, y' = \frac{2}{y^2}\, Q^2(x)\,, \qquad 0 < x < 1\,, \qquad Q(x) := \frac{2}{x^2}\int_0^x t\bar{p}(t)\,\mathrm{d}t\,. \tag{2.1}$$

Henceforth primes denote differentiation with respect to x, unless stated otherwise. If the surface load is *uniform*, then $\bar{p} = 1$ and therefore $Q = 1$. The circumferential stress σ_θ, the radial and normal displacements u and w are related to the variables x and y through

$$\sigma_\theta/E = k^2(xy' + y)/4\,, \qquad u = ak^2x[xy' + (1-\nu)y]/4\,,$$
$$w = ak\int_x^1 tQ(t)[y(t)]^{-1}\,\mathrm{d}t\,, \qquad 0 \leq \nu \leq 1/2\,. \tag{2.2}$$

Solutions of (2.1) are sought satisfying the boundary conditions

$$y'(0) = 0\,, \quad y(1) = S > 0 \quad \text{or} \quad y'(0) = 0\,, \quad y'(1) + (1-\nu)y(1) = H \in \mathbb{R}\,, \tag{2.3}$$

depending on whether σ_r or u is prescribed at the edge $r = a$. Hence, (2.1) and (2.3) define two different boundary value problems for the circular membrane, called Problem S and Problem H in what follows. An integration of (2.1), making use of $x^3Ly = -(x^3y')'$, shows that

$$x^3y'(x) = -\int_0^x \frac{2t^3}{y^2(t)}\, Q^2(t)\,\mathrm{d}t \leq 0\,. \tag{2.4}$$

Accordingly, the solutions of both Problems S and H are monotone decreasing in the interval $[0, 1]$. In particular, we have $y(x) \geq S > 0$ in Problem S.

The first solution of (2.1, 2.3) for uniform pressure ($Q = 1$) and $H = 0$ was apparently given by Hencky [23], using *formal power series* in x. In a formulation slightly different from [23], we introduce $z = 2/y$ and seek a solution in the form

$$y(x) = \sum_{n=0}^{\infty} y_nx^{2n}\,, \qquad z(x) = \sum_{n=0}^{\infty} z_nx^{2n}\,, \qquad z_0 = 2/y_0\,. \tag{2.5}$$

The *convergence* of this series solution was first established in 1965 by Weinitschke [46], yielding first existence theorems for Problem S and Problem H for $H = 0$. A different and simpler convergence proof for both fixed edge ($H = 0$) and loaded edge (Problem S) was published later in [47], where uniqueness of solutions was also established.

The first step is to introduce the series (2.5) into $Ly = z^2/2$ and $yz = 2$, obtain recurrence relations for the coefficients y_n, z_n and show by induction that for suitable choices of A, B and $q > 1$

$$|y_n| \leqslant \frac{A}{(n+1)^{q+1}}, \qquad z_n \leqslant \frac{B}{(n+1)^q}, \quad n = 1, 2, \ldots . \tag{2.6}$$

This has the immediate consequence that the series (2.5) and the differentiated series $y'(x)$ are all uniformly convergent for $|x| \leqslant 1$. The next step is to show that there exist suitable choices of A, B and q such that the boundary conditions for Problems S and H, respectively, are satisfied. In Problem S it is easy to see that convergence will occur if S is sufficiently large. In Problem H, the boundary term $y' + (1 - \nu)y$ at $x = 1$ must be shown to change sign for two different choices of A, B and q. The details are quite technical, so we refer to [47] and [44]. The results can be summarized as follows:

THEOREM 2.1. *Problem S for $Q = 1$ and $S > S_0 = 4/5$ has a unique solution $y(x) > 0$, which admits a uniformly convergent power series expansion for $|x| \leqslant 1$, with $S < y_0 \leqslant S + (1/2S)^2$.*

THEOREM 2.2. *Problem H for $Q = 1$ and $H = 0$ has a unique solution $y(x)$. The solution is positive and can be represented by a power series, uniformly convergent for $|x| \leqslant 1$, with* $20/19 < y_0 < 2$ for $\nu = 1/3$.

The series solutions may be used for getting accurate numerical approximations to Problems S and H. A number of different numerical treatments of both circular and annular membrane problems have appeared in the engineering literature (see [25] and the references given there).

A different and more elegant existence proof for solutions of Problems S and H can be obtained by an *integral equation method*. This was first done by Dickey [10], again for the case of uniform pressure $Q = 1$. He showed existence of a unique $C^1[0, 1]$-solution for Problem S provided $S^3 > 4j^{-2}$, where j denotes the first zero of the Bessel function J_1, which amounts to $S > S_0 \doteq 0.648$, thus improving on Theorem 2.1. By an interpolation between values of S well above S_0, he also showed existence of a unique solution of Problem H for $H = 0$. The method will be discussed below.

In order to remove the above restrictions $Q = 1$ and $S > S_0$, Callegari and Reiss [6] employed a *shooting technique*. In this way, they first proved existence and uniqueness of solutions of Problem S for *all* $S > 0$, and for *variable load* $p(r)$. The proofs in [6] are quite technical and will not be described here. They are much more complicated than the ones given below by the integral equation method. Although the shooting method is constructive with respect to the initial value problem, no converging iteration scheme was given in [6] to solve the relevant boundary value problems.

A simple and standard way to find an *integral equation* for Problems S and H is to calculate the Green's functions $G(x, t)$ for the problems $x^3Ly = 0$ with homogeneous boundary conditions (2.3). It follows that the solutions of Problem S and Problem H, respectively, are equivalent to an integral equation [49]

$$y(x) = q + \int_0^1 \frac{2t^3}{y^2(t)} Q^2(t) G(x,t)\,dt =: Ty\,, \tag{2.7}$$

where $q = S$ for Problem S and $q = H/(1-\nu)$ for Problem H, and

$$G(x,t) = \begin{cases} (x^{-2} + m)/2\,, & 0 \leqslant t \leqslant x \leqslant 1\,, \\ (t^{-2} + m)/2\,, & 0 \leqslant x \leqslant t \leqslant 1\,, \end{cases} \tag{2.8}$$

where $m = -1$ or $m = (1+\nu)/(1-\nu)$ for Problem S or Problem H, respectively. In the case $Q = 1$ the integral equation (2.7) for Problem S reduces to that obtained by Dickey [10].

With a view towards getting numerical solutions for Problems S and H, we first consider a constructive proof of the existence of solutions of (2.7). Starting with $y_0 = q > 0$, we define $y_n(x)$ by $y_{n+1} = Ty_n$.

LEMMA 2.3. *The operator T is* antitone: *If $0 < y \leqslant z$, then $Ty \geqslant Tz$.*

A general theorem on antitone operators (e.g., see [9]) implies that the sequence y_n has the following basic property.

LEMMA 2.4. *If $0 < y_0 \leqslant y_2 \leqslant y_1$ then for any positive integer n*

$$y_0 \leqslant y_2 \leqslant y_4 \leqslant \cdots \leqslant y_{2n} \leqslant y_{2n+1} \leqslant \cdots \leqslant y_3 \leqslant y_1\,. \tag{2.9}$$

In view of Lemma 2.3 and the choice of y_0 the condition $y_0 \leqslant y_2 \leqslant y_1$ is satisfied for any $S > 0$ or any $H > 0$. In order to apply the Banach fixed point theorem we introduce a norm

$$\|y\| = \sup_{0 \leqslant x \leqslant 1} \{W(x)^{-1}|y(x)|\}\,, \quad W(x) > 0\,.$$

The objective then is to find an optimal $W(x)$ such that T is contractive for a maximal range of the parameters S and H. The result of this calculation is [49]:

THEOREM 2.5. *Assume $Q^2(x) \leqslant c$, then the integral equation* (2.7) *has a unique solution for all $S > S_0 = 0.648c^{1/3}$ in Problem S and for all $H > H_0 = H_0(\nu)c^{1/3}$ in Problem H, where $0.648 < H_0(\nu) < 1.057$ for $1/2 \geqslant \nu \geqslant 0$. In these ranges of S and H, $y_n(x)$ converges uniformly to the solution $y(x)$. The convergence is alternating and $y(x) \in M_n$ where*

$$M_n := [y_{2n}, y_{2n+1}] := \{y(x) \mid y_{2n} \leqslant y \leqslant y_{2n+1}\}\,. \tag{2.10}$$

In particular, for $n = 0$, we have for the solution of Problem S

$$\begin{aligned} &S \leqslant y(x) \leqslant S + \frac{1}{S^2} \int_0^1 2t^3 Q^2(t) G(x,t)\,dt = y_1\,, \\ &S \leqslant y(x) \leqslant S + \frac{1}{4S^2}(1 - x^2)\,, \quad \text{for } Q = 1\,. \end{aligned} \tag{2.11}$$

The restrictions on the size of S and H can now be removed by applying Schauder's fixed point theorem. In view of Lemma 2.4, the operators T map the convex set M_n into itself, that is, for any positive integer n we have $TM_n \subseteq M_n$, provided $S>0$ $(H>0)$. T is completely continuous on the set $M_0 \supseteq M_n$ (e.g., see [9]). Hence the Schauder fixed point theorem is applicable and yields [49]:

THEOREM 2.6. *Problem S (Problem H) has at least one solution* $y(x)>0$ *for* all $S>0$ $(H>0)$. *The solution is contained in* M_n.

The uniqueness is proved by a classical argument. Let y_1, y_2 be two positive solutions of (2.1), (2.3) then $w = y_1 - y_2$ satisfies

$$Lw = -M(x)w\,, \qquad M(x) = \frac{2Q^2(x)}{y_1^2 y_2^2}(y_1 + y_2) \geq 0\,, \tag{2.12}$$

$$w'(0) = 0 \quad \text{and} \quad w(1) = 0 \quad \text{or} \quad w'(1) + (1-\nu)w(1) = 0\,. \tag{2.13}$$

By the maximum principle (see Protter and Weinberger [36]) it follows that $w \equiv 0$.

THEOREM 2.7. *Any positive solution of Problem S (Problem H) is* unique, *if* $\bar{p}(x)$ *is piece-wise continuous.*

The unique solutions of Problems S and H guaranteed by Theorems 2.6 and 2.7 cannot be constructed by a convergent iteration, yet we obtain upper and lower bounds to the exact solution by iteration.

It should be mentioned that the existence part of Theorem 2.6 was obtained independently by Stuart [43], as an application of his general theory of integral equations with decreasing nonlinearities. He also obtained convergence of the sequence $y_n(x)$ from Problem S in the range $S > S_0 = 0.648c^{1/3}$ (see Theorem 2.5).

It is worth noting that $S>0$ covers the complete physically meaningful range of tensile solutions for Problem S, while $H>0$ does not. In the latter case, a physically significant situation is $H=0$. On the other hand, the condition $q>0$ in (2.7) is essential for starting the iteration. If $H=0$, the operator T is still antitone, but it does not seem possible to choose y_0 and y_1 such that $y_0 \leq y_2 \leq y_1$. T is not contractive, nor is the Schauder theorem applicable. Hence it appears that Problem H for $H=0$ cannot be solved directly via an integral equation method. However, there is a simple idea, by which this case can be covered. Any solution $y(x; S)$ of Problem S is also a solution of Problem H for a value of H given by

$$N(S) := y'(1; S) + (1-\nu)S = H\,. \tag{2.14}$$

A simple estimate yields $y'(1; S) \to -\infty$ as $S \to 0$ [49]. Since $N(S) \to +\infty$ as $S \to \infty$, there exists for any given real value of H a corresponding value $S>0$ such that $N(S) = H$. This result, together with Theorem (2.7), yields

THEOREM 2.8. *Problem H has a unique positive solution for all real H.*

A substantial improvement of the convergence properties of the iteration $y_{n+1} = Ty_n$ can be achieved by an interpolated (averaged) iteration of the type

$$y_{n+1} = \alpha T y_n + (1-\alpha) y_n \,. \tag{2.15}$$

This was first pointed out by Ostrowski [34]. It was rediscovered in connection with an elastic plate problem by Keller and Reiss [27], who found numerically, that the range of convergence was drastically increased by an appropriate choice of α. But the convergence has remained an open problem until Novak [33] recently proved the convergence of the iteration (2.15) for a class of operators T which includes the operator T defined in (2.7). More precisely, he proved the following

THEOREM 2.9. *Problem S has a unique solution $y(x)$ for all $S>0$ which can be obtained as the limit of the iteration* (2.15). *The interpolated iteration converges for all $\alpha>0$ which are sufficiently small.*

Hence we have arrived at a constructive method to solve Problem S in the full range $S>0$. In addition it turns out that the iteration (2.15) also converges for the operator T corresponding to Problem H *including* the important case $H=0$ [49]. Numerical results based on (2.15) are also given in [49]. The value of S in (2.14) corresponding to the fixed edge problem $H=0$ is approximately $S=0.8549$ (for $\nu=1/3$).

Next we turn to the boundary value problems posed by Reissner's *finite rotation theory*. The governing equation is (1.6), which can also be derived directly (without reference to shell theory), as shown by Clark and Narayanaswamy [8]. Furthermore, the equations in [8] can be shown to be equivalent to those derived earlier by Bromberg and Stoker [5]. We consider again a circular membrane under vertical pressure $p_V = p(r)$, with prescribed σ_r (Problem S) or prescribed u (Problem H) at the edge $r=a$. Hence $p_H=0$ and $p_s = p_V \sin\Phi$ in equation (1.6). Introducing the same dimensionless variables as before, we obtain the basic differential equation and the boundary conditions for Problems S and H,

$$\begin{aligned} &Ly = f(x,y) - 2\nu k^2 \bar{p}(x) Q(x)/D(x,y)\,, \quad 0<x<1\,,\\ &f(x,y) := (2/kx)^2[1 - y/D(x,y)]\,, \qquad D(x,y) := [y^2 + k^2x^2Q^2(x)]^{1/2}\,,\\ &y'(0)=0 \quad \text{and} \quad y(1)=S \quad \text{or} \quad y'(1)+y(1)-\nu D(1,y(1)) = H\,. \end{aligned} \tag{2.16}$$

The meridional and circumferential (Piola–Kirchhoff) stress resultants S_ϕ and S_θ, the angle of rotation Φ and the radial and axial displacements u and w are related to y as follows

$$\begin{aligned} &S_\phi/Ed = k^2 D(x,y)/4\,, \qquad S_\theta/Ed = k^2(xy'+y)/4\,,\\ &\cos\Phi = y/D(x,y)\,, \qquad u = ak^2x[xy' + y - \nu D(x,y)]/4\,,\\ &w = a\int_x^1 \left[1 + \frac{1}{4}k^2 D(t,y(t)) - \nu \frac{\mathrm{d}}{\mathrm{d}t}(ty)\right] \sin\Phi(t)\,\mathrm{d}t\,. \end{aligned} \tag{2.17}$$

In contrast to the Föppl theory, the boundary value problem (2.16) contains, besides $Q(x)$, the load parameter $k=(2p_0a/Ed)^{1/3}$. It is seen that in the limit case $k\to 0$ equations (2.16) reduce to Problems S and H for the Föppl theory (2.1), (2.3). For finite k, the solution structure of (2.16) turns out to be *markedly* different from that of (2.1), (2.3), in particular with respect to Problem H.

A mathematical analysis of Problems S and H defined by (2.16) was virtually nonexistent until 1980. Without surface load, the problem simplifies considerably. Two cases were treated by Clark and Narayanaswamy [8]: a membrane of revolution with a uniform radial edge load and a membrane with uniform edge load parallel to the axis of revolution. The first problem is reducible to a linear equation solvable in closed form. The second problem is nonlinear, it includes the special case of a flat annular membrane with transverse edge load treated by Schwerin [39] for the Föppl small deflection theory. In the presence of surface load, the integral equation technique used for the Föppl membrane equations was extended by Weinitschke [48] to obtain first existence results for $\nu = 0$ and a rather restricted range of boundary data.

A theory of existence and uniqueness of positive solutions for the boundary value problems (2.16) for the full range of physically meaningful data is still lacking. It has been argued that terms involving ν in the differential equations for finite rotations can often be neglected. Simmonds has given a rigorous justification of neglecting terms multiplied by ν in linear shell theory [40]. Although in the nonlinear theory his technique is not applicable, one might argue heuristically in the present case that the ν-term in the differential equation (2.16), which is $O(k^2)$, is small compared to the term $f(x, y)$, as the dominant load component acts in the direction normal to the membrane. For sufficiently small k this is certainly justified, but the influence of that term might increase for larger values of k. In the remaining part of this paper, we omit the term $2\nu k^2\bar{p}Q/D$ and refer to (2.16) as the *simplified* Reissner theory.

The nonlinear function $f(x, y)$ has the same monotonicity property as in the Föppl theory because of

$$\frac{\partial f}{\partial y} = -\frac{k^2x^2Q^2(x)}{[y^2 + k^2x^2Q^2(x)]^{3/2}} \leq 0 . \tag{2.18}$$

Hence the integral equation method employed for the Föppl theory is applicable so long as the boundary conditions are linear, that is for Problem S, $S > 0$ and for Problem H, $H > 0$, $\nu = 0$. The integral equation equivalent to $Ly = f$ and boundary conditions is

$$y(x) = q + \int_0^1 K(x, t)\tilde{f}(t, y(t))\,\mathrm{d}t + q_0[y^2(1) + \Theta^2]^{1/2} =: Ty , \tag{2.19}$$

where $0 \leq x \leq 1$, $\tilde{f}(t, y) := [1 - y/D(t, y)]/t$ and $\Theta := k|Q(1)|$. The Green's function $K(x, t)$ is defined by

$$K(x, t) = \begin{cases} 2k^{-2}(1 + nx^2)(t/x)^2 , & 0 \leq t \leq x \leq 1 , \\ 2k^{-2}(1 + nt^2) , & 0 \leq x \leq t \leq 1 . \end{cases} \tag{2.20}$$

For Problem S, we have

$$q = S , \qquad q_0 = 0 , \qquad n = -1 . \tag{2.21}$$

For Problem H, we have

$$q = H , \qquad q_0 = \nu , \qquad n = +1 . \tag{2.22}$$

$K(x, t)$ is essentially the same as $G(x, t)$ defined in (2.8), with $\nu = 0$ in Problem H.

Problem S and Problem H for $\nu = 0$ can now be solved as before. In view of $K \geqslant 0$ and (2.18) the operator T defined in (2.19) is antitone, so that Lemma 2.3 and Lemma 2.4 hold, provided $q > 0$. Application of the Banach fixed point theorem yields a theorem analogous to Theorem 2.5. The integral equation (2.19) has a unique positive solution for restricted ranges $S > S_0(k)c^{1/3}$, Problem S, and $H > H_0(k)c^{1/3}$, Problem H. In these ranges of S and H, the sequence $y_n(x)$ defined by $y_{n+1} = Ty_n$ converges uniformly to the solution $y(x)$, and $y \in M_n$ for $n = 0, 1, \ldots$ [49].

The restriction on S and H can again be removed by employing the Schauder fixed point theorem. The arguments are exactly as in the Föppl theory, the uniqueness follows as in (2.12), (2.13), replacing M by

$$M(x) = \frac{4(y_1 + y_2)Q^2(x)}{D(x, y_1)D(x, y_2)(y_1 D(x, y_2) + y_2 D(x, y_1))} \geqslant 0 . \tag{2.23}$$

THEOREM 2.10. *In the simplified Reissner theory of finite rotations, Problem S, or equivalently the integral equation* (2.19) *has a unique solution* $y(x)$ *for all* $S > 0$. *The solution is positive and* $y \in M_n$, $n = 0, 1, 2, \ldots$ [49].

For Problem H, this theorem only holds for $\nu = 0$, $H > 0$. The interpolated iteration (2.15) also applies to the simplified Reissner theory, rigorously for Problem S, that is, Theorem 2.9 remains valid. We have also tested (2.15) numerically in Problem H for $\nu \neq 0$ and found convergence for $H \geqslant 0$.

In the general case of Problem H, $\nu \neq 0$, equation (2.19) is a nonstandard integral equation because of the algebraic term multiplying $q_0 = \nu$. A complete solution of Problem H is due to Grabmüller and Pirner [19] and to Beck [3]. The uniqueness of positive solutions is obtained as follows. After multiplying by $x^3 w$, we integrate (2.12) over (0, 1) and integrate by parts to get for $w = y_1 - y_2$

$$w'(1)w(1) = \int_0^1 x^3[w'^2(x) + M(x)w^2(x)]\, \mathrm{d}x , \tag{2.24}$$

where $M(x)$ is defined by (2.23). Unless $w \equiv 0$, the right-hand side is strictly positive. In Problem S, $w(1) = 0$ and uniqueness follows. In Problem H, we have

$$w'(1) + w(1) - \nu[D(1, y_1(1)) - D(1, y_2(1))] = 0 . \tag{2.25}$$

Applying the mean-value theorem, we obtain $w'(1) + (1 - \nu\kappa)w(1) = 0$ where

$$0 < \kappa := \frac{y_\theta(1)}{[y_\theta^2(1) + k^2Q^2(1)]^{1/2}} \leqslant 1 , \tag{2.26}$$

$y_\theta = y_1 + \theta(y_2 - y_1)$ with an intermediate variable $\theta(x)$. It is seen that y_θ is bounded from below by $\min\{\min y_1(x), \min y_2(x)\} > 0$. This yields $w'(1)w(1) = -(1 - \nu\kappa)w^2(1) \leqslant 0$ since $0 \leqslant \nu \leqslant 1/2$, and hence $w \equiv 0$ by equation (2.24).

We proceed to outline the existence proof. To this end, an extended version of a theorem concerning positive solutions of integral equations due to Novak [18] is utilized. Letting $g(x) := y(x) - q$, the objective is to determine solutions g of

$$g(x)=\int_0^1 K(x,t)\tilde{f}(t,g(t)+q)\,\mathrm{d}t+q_0[(g(1)+q)^2+\Theta^2]^{1/2}\,, \tag{2.27}$$

satisfying $g(x)+q\geq 0$. Existence of solutions of (2.27) are obtained via Schauder's fixed point theorem. Consider the convex set M_δ defined in the Banach space $C[0,1]$

$$M_\delta=\Big\{g\in C[0,1]\,|\,\exists H\in C[0,1] \text{ with } 0\leq H(t)\leq k/\delta \text{ and } \\ g(x)=\int_0^1 K(x,t)H(t)\,\mathrm{d}t+q_0[\rho^2+\Theta^2]^{1/2},\ 0\leq\rho\leq\rho_0\Big\}\,. \tag{2.28}$$

Here $\delta>0$, to be chosen later, and $\rho_0>0$ is defined by

$$\rho_0=\frac{K_0+|q|+\Theta}{1-q_0}\,,\qquad K_0:=\frac{k}{\delta}\max_{0\leq x\leq 1}\int_0^1 K(x,t)\,\mathrm{d}t\leq\frac{8}{3\delta k}\,. \tag{2.29}$$

One can show that, if g is a solution of (2.27) satisfying $g(x)+q\geq\delta>0$ then $|g(x)|\leq\rho_0$ and $g\in M_\delta$. Hence the solution set of (2.27) is a subset of M_δ. Next one introduces an operator W acting on M_δ by

$$(Wg)(x):=\int_0^1 K(x,t)\tilde{f}(t,d_g(t))\,\mathrm{d}t+q_0[(g(1)+q)^2+\Theta^2]^{1/2}\,. \tag{2.30}$$

A similar operator and the cutoff function $d_g(x):=\max\{g(x)+q,\delta\}$, $0\leq x\leq 1$ had first been introduced by Novak [18] within the context of annular Föppl membrane problems. It is not difficult to prove the following

THEOREM 2.11 [18]. *Suppose there exists a constant $\delta>0$ such that for any $g\in M_\delta$ satisfying* $\min_{0\leq x\leq 1}[g(x)+q]\leq\delta$ *we have the property*

$$\exists x_0\in[0,1]\colon g(x_0)<(Wg)(x_0)\,. \tag{N}$$

Then equation (2.19) *has a solution $y(x)>\delta$.*

A sufficient condition for the existence of positive solutions of Problem H is now obtained by choosing H such that condition (N) is satisfied.

THEOREM 2.12 [18]. *Assume that either $H\geq 0$ or*

$$H<0 \quad \textit{and} \quad \int_0^1 K(1,t)\tilde{f}\Big(t,-H\,\frac{1+t^2}{2t^2}\Big)\,\mathrm{d}t+\nu\Theta>-H\,, \tag{2.31}$$

then there exists a number $\delta>0$ such that condition (N) *holds for Problem H.*

In the Föppl theory, solutions of Problem H exist for all real H (Theorem 2.8). This result does not carry over to the simplified Reissner theory. Indeed, a tensile solution y of Problem H necessarily satisfies $y(x)\geq y(1)>0$. But from (2.19) one has, if $Q^2(x)>0$,

$$0<(1-\nu)y(1)\leq H+\int_0^1 K(1,t)\,\frac{\mathrm{d}t}{t}+\nu\Theta\,, \tag{2.32}$$

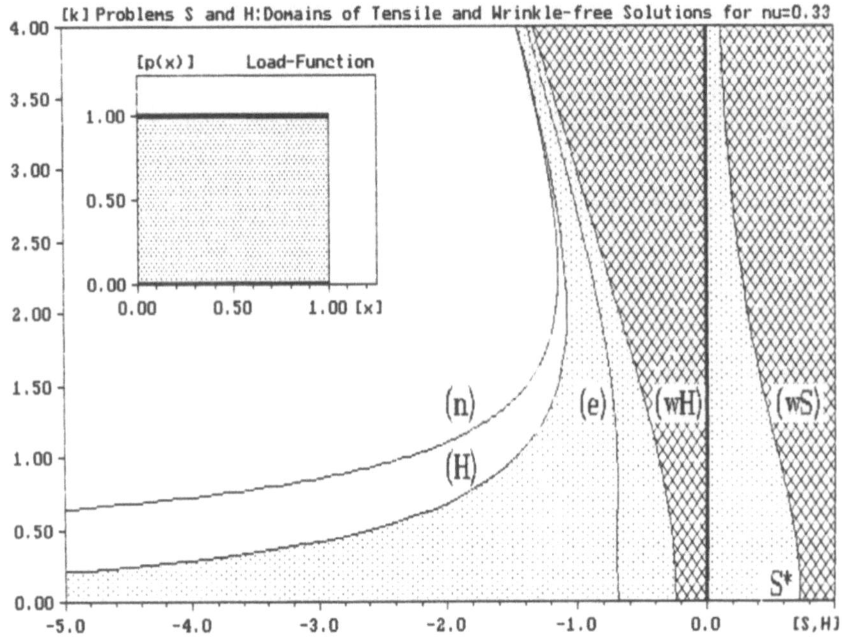

Fig. 1. The arcs (n) and (e) denote, respectively, the curves Γ_n and Γ_e which are approximations for the *separatrix* (H) between the domains of existence (dotted) and nonexistence (blank) of tensile solutions to Problem H. The domains of wrinkle-free solutions (crosshatched) which extend unboundedly to the right are bounded to the left by the arcs (wS) for Problem S and (wH) for Problem H. The load is uniform and Poisson's number is $\nu = 1/3$.

which provides a necessary condition for the existence of a solution to Problem H. Unfortunately the integral in (2.31) cannot be expressed in closed form, even in the case of uniform pressure $Q = 1$. But the important point is that (2.31) and (2.32) yield graphs Γ_e and Γ_n in the (H, k) plane bounding the domains of *existence* and *nonexistence* of tensile solutions of Problem H, respectively. For $Q = 1$, these graphs are shown in Fig. 1 (for $\nu = 1/3$).

It is clear from the above that there must be an arc Γ in the (H, k) plane, located between Γ_e and Γ_n in Fig. 1, which separates precisely the domains of existence and nonexistence. Indeed, since the solutions of Problem H are contained in those of Problem S for $S > 0$, we simply need to solve Problem S for $S = 0$ and determine the corresponding value of $H = H(k)$ from (2.16). As an analytic solution of (2.19) is not available, we solve the problem numerically. This was done by Weinitschke in [45], although an existence proof for Problem S, $S = 0$ was still lacking. Recently, Beck [3] has supplied such a proof, which is by no means simple. The separation curve Γ, whose existence and geometric properties were also proved in [3], is included in Fig. 1, again for the case $Q = 1$, $\nu = 1/3$.

3. Annular membranes

In this section we consider annular membranes of inner radius b and outer radius a under vertical pressure $p = p(r)$. At the edges, either the radial stress or the radial displacement is prescribed. The governing equations (1.6) and (1.8) are the same, and so are the dimensionless variables, except that the interval $(0, 1)$ is replaced by $(\varepsilon, 1)$ where $\varepsilon = b/a$. This also changes the lower limit in the integral for Q. Thus the annular membrane problems can be reduced to the differential equation

$$
\begin{aligned}
&Ly = f(x, y) := \begin{cases} (2/y^2)R^2(x, \varepsilon): & \text{Föppl theory}\,, \\ (2/kx)^2[1 - y/D(x, y)]: & \text{simplified Reissner theory}\,, \end{cases} \\
&0 < \varepsilon < x < 1\,, \qquad\qquad\qquad\qquad\qquad\qquad\qquad\qquad\qquad\qquad (3.1) \\
&R(x, \varepsilon) := \frac{2}{x^2}\int_\varepsilon^x t\bar{p}(t)\,\mathrm{d}t\,, \qquad D(x, y) := [y^2 + k^2x^2R^2(x, \varepsilon)]^{1/2}\,.
\end{aligned}
$$

The radial and circumferential stress components σ_r and σ_θ, and the radial and normal displacements u and w are related to y as in (2.2) and (2.17), provided $Q(x)$ is replaced by $R(x, \varepsilon)$ in all the formulas. If the surface load is *uniform*, then $R(x, \varepsilon) = 1 - \varepsilon^2/x^2$. A pair of boundary conditions at $x = \varepsilon$ and $x = 1$ is taken, respectively, from

$$
y(\varepsilon) = s \quad \text{or} \quad \varepsilon y'(\varepsilon) + (1 - \nu)y(\varepsilon) = h\,, \tag{3.2}
$$

at the inner edge and

$$
y(1) = S \quad \text{or} \quad \begin{cases} y'(1) + (1 - \nu)y(1) = H: & \text{Föppl}\,, \\ y'(1) + y(1) - \nu D(1, y(1)) = H: & \text{simplified Reissner}\,, \end{cases} \tag{3.3}
$$

at the outer edge. Introducing the notations (s, S), (s, H), (h, S) and (h, H) reference shall be made in an obvious manner to the four different boundary value problems arising from (3.1), (3.2) and (3.3).

The analysis of annular membrane problems defined by (3.1) and (3.2) for $s \geqslant 0$, $S \geqslant 0$, h and H real, and for an arbitrary load $\bar{p}(x)$ is more complicated than in the circular membrane problems since solutions $y(x)$ are permitted to become zero at the edges of the annulus. The *regular tensile solutions* of (3.1) (rt-solutions for short, which have to be understood as functions $y \in C^2(\varepsilon, 1) \cap C^1[\varepsilon, 1]$ satisfying $y(x) > 0$ for $\varepsilon \leqslant x \leqslant 1$) now have to be distinguished from *regular nonnegative solutions* (rn-solutions) which suffer a loss of regularity $y \in C^2(\varepsilon, 1) \cap C^0[\varepsilon, 1]$ and satisfy $y(x) > 0$ only for $\varepsilon < x < 1$. Furthermore, the solutions $y(x)$ are not necessarily monotone in the interval $(\varepsilon, 1)$, there is no counterpart of inequality (2.4) here. The four different boundary parameters s, h, S and H considerably enhance the variety of solution behavior.

As in Section 2, we begin by discussing results for the Föppl *small finite deflection theory*. The first solution of (3.1)–(3.3) for uniform pressure and $s = H = 0$ was given by Schwerin [39] in terms of a formal power series. In the absence of surface loads, he also found a closed form solution for a membrane subjected to axial edge load and fixed edges ($h = H = 0$). To simplify his calculations, he transformed the differential equation (3.1) into the form $U''(\xi) = -\xi^2/U^2$. This transformation turned out to be crucial for the theoretical analysis of rn-solutions for both Föppl and Reissner theory. We write the *Schwerin transformation* for our purposes as follows

$$
z = \frac{x^2 - \varepsilon^2}{1 - \varepsilon^2}\,, \qquad g(z) = \omega^4 x^2 y(x)\,, \quad \varepsilon \leqslant x \leqslant 1\,, \quad \omega := (1 - \varepsilon^2)^{-1/3}\,. \tag{3.4}
$$

This change of variables maps any rn-solution $y(x)$ of (3.1) into an rn-solution $g(z)$ of the differential equation

$$-\frac{d^2g}{dz^2} = F(z, g) := \begin{cases} 2P^2(z,\varepsilon)/g^2: & \text{Föppl}, \\ RS(z, g): & \text{simplified Reissner}, \end{cases} \qquad 0<z<1, \tag{3.5}$$

where the following notations are used:

$$P(z, \varepsilon) := \frac{1}{1-\varepsilon^2}\int_\varepsilon^{x(z)} t\bar{p}(t)\,dt, \qquad x(z) := [z(1-\varepsilon^2)+\varepsilon^2]^{1/2},$$

$$RS(z, g) := \left(\frac{1}{k\omega x(z)}\right)^2\left[1-\frac{g}{\sqrt{g^2+\Theta^2(z)}}\right], \qquad \Theta(z) := 2k\omega x(z)P(z, \varepsilon).$$

The boundary conditions (3.2), (3.3) can be expressed in terms of z and $g(z)$ by

$$g(0)=s \quad \text{or} \quad B_0[g]=h, \qquad g(1)=S \quad \text{or} \quad B_1[g]=H. \tag{3.6}$$

Here the new boundary parameters $s\geq 0$, $S\geq 0$, h and H real, are obtained by suitably stretching the old ones. The boundary operators $B_j[g]$ have the form

$$B_0[g] := \varepsilon^2\dot{g}(0)-\mu g(0), \qquad \mu := \tfrac{1}{2}(1-\varepsilon^2)(1-\nu),$$

$$B_1[g] := \begin{cases} \dot{g}(1)-\mu g(1): & \text{Föppl}, \\ \dot{g}(1)-\dfrac{1-\varepsilon^2}{2}[g(1)+\nu\sqrt{g^2(1)+\Theta^2(1)}]: & \text{simplified Reissner}. \end{cases}$$

Unless otherwise stated, a dot denotes differentiation with respect to z. Since $F(z, g)$ is nonnegative it is clear from (3.5) that any rn-solution $g(z)$ must be concave. Thus, $g(z)$ with $g(0)=s$ and $g(1)=S$ is supported by the linear function $q(z) := zS+(1-z)s$ in the sense that $g(z)-q(z)\geq 0$ holds.

The convergence of Schwerin's series solution of $\ddot{g}=-z^2/2g^2$, $(\bar{p}=1)$, was proved by Weinitschke in [48], but only for the case $s=0$ and S or H sufficiently large, not covering the situation $H=0$. An integral equation method analogous to (2.7) was also employed in [48] for the solution of Problem (s, H). However, the results obtained by applying the Banach fixed point theorem will not be discussed here, as they impose unacceptable restrictions on the parameters ε, ν, s and H, excluding, in particular, the case $s=H=0$. As for circular membranes, a constructive existence proof for an rt-solution (not covering a stress-free inner edge $s=0$) as a limit of $y_n(x)$ defined by (2.15), with an appropriate definition of T, has been given by Novak [33].

The solutions for a free inner edge $s=0$ and uniform load $\bar{p}=1$ can be expressed in terms of solutions of the circular membrane, as shown by Grabmüller and Weinitschke [21]. More precisely, we have the following

THEOREM 3.1. *Let $z(\xi)$ be the solution of* (2.1), (2.3), *Problem S with $z(1)=S_\varepsilon := (1-\varepsilon^2)^{-4/3}S>0$, $Q=1$. Then there exists a unique solution $y(x)$ of Problem (s, S), $s=0$, $\bar{p}=1$. This solution is positive and has the form*

$$y(x) = (1-\varepsilon^2)^{1/3}\left(1-\frac{\varepsilon^2}{x^2}\right)z(\xi), \qquad \xi=\left(\frac{x^2-\varepsilon^2}{1-\varepsilon^2}\right)^{1/2}. \tag{3.7}$$

Let $z(\xi)$ be the solution of (2.1), (2.3), *Problem H with* $z'(1)+(1-\nu_\varepsilon)z(1)=H_\varepsilon := (1-\varepsilon^2)^{-1/3}H$, *where* $\nu_\varepsilon := \nu - \varepsilon^2(1+\nu)$, $Q=1$. *Then there exists a unique solution of Problem* (s, H), $s=0$, $\bar{p}=1$. *This solution is positive and has the form* (3.7).

In particular, this theorem covers the physically interesting cases $s=0$, $S>0$ and $s=0$, $H\geqslant 0$, but only for uniform load. It also proves a formal result obtained by Schwerin [39] concerning the stress concentration factor. In addition, existence and uniqueness of rt-solutions for Problem (s, S), $s>0$, $S>0$ and for Problem (s, H) for $s>0$ and $s+H>0$, in terms of the new parameters defined in (3.6), for arbitrary load functions $\bar{p}(x)$, was proved in [21]. This was done by reducing these two problems to integral equations with antitone operators and applying Schauders fixed point theorem, along the lines described in Section 2.

Concerning the more difficult Problems (h, S) and (h, H), a new device was brought into the analysis of rt-solutions by E. Novak [33] who considered positive solutions of certain nonlinear integral equations. His method led to a considerable improvement of the known existence results. The standard method of using Green's function shows that rt-solutions $g(z)$, for example of Problem (s, S), coincide with the positive $C^0[0,1]$-solutions of the integral equation

$$g(z)=q(z)+\int_0^1 k(z,t)F[t,g(t)]\,dt\,,\quad 0\leqslant z\leqslant 1\,, \tag{3.8}$$

where $q(z)$ is defined above and where

$$k(z,t)=\begin{cases} t(1-z): & 0\leqslant t\leqslant z\leqslant 1\,,\\ z(1-t): & 0\leqslant z\leqslant t\leqslant 1\,.\end{cases}$$

A new dependent variable $f(z) := g(z)-q(z)$ is introduced, and instead of (3.8) the following integral equation is considered (see (2.30)):

$$f(z)=(Wf)(z) := \int_0^1 k(z,t)F[t,d_f(t)]\,dt\,,\quad 0\leqslant z\leqslant 1\,. \tag{3.9}$$

Here, for some $\delta>0$, the cutoff function $d_f(z) := \max\{f(z)+q(z),\delta\}$, $0\leqslant z\leqslant 1$, provides a lower bound for the solutions of (3.9). As in Section 2 a convex subset $M\subset C^0[0,1]$ is considered which consists of all functions f admitting an integral representation $f(z)=\int_0^1 k(z,t)H(t)\,dt$ for some $H\in C^0[0,1]$, $H\geqslant 0$. An application of the Schauder fixed point theorem then yields an analogon to Theorem 2.11, with $g(x)$ replaced by $f(z)$ and W defined by (3.9), proving that equation (3.9) has a $C^0[0,1]$-solution $f(z)$ with $f+q>\delta$.

It is clear that Novak's condition (N) must be void if $s>0$ and $S>0$ is considered. Thus, a new *existence* proof for rt-solutions to Problem (s, S) is obtained because it is seen from (3.9) that $g(z) := f(z)+q(z)$ solves the integral equation (3.8). In [17], the question of *uniqueness* of rt-solutions to each boundary value problem was fully resolved by a suitable application of Hopf's generalized maximum principle [36]. The above new integral equation method was also extended to Problems (s, H), (h, S) and (h, H). The particular constellation of Problem (s, H) again leads to a void condition (N). In summary, we have [17, 18]:

THEOREM 3.2. *Let* $P^2(1,\varepsilon)>0$, *then in the small finite deflection theory both Problem* (s, S) *and Problem* (s, H) *have a unique rt-solution for all* $s>0$, $S>0$ *and H real. Moreover, rt-solutions for Problems* (h, S) *and* (h, H) *are unique.*

Analogous existence results do not hold in the cases of Problems (h, S) and (h, H) for the Föppl theory. A discussion of the crucial condition (N) surprisingly shows that rt-solutions are absent in an unbounded, simply connected subset of the respective parameter ranges (h, S) and (h, H). This was first discovered by Grabmüller and Novak [18], who estimated the domains of existence and nonexistence of rt-solutions by simple analytical curves. Unfortunately, a gap was left in the parameter ranges where a definite statement on existence or nonexistence was impossible. We shall return to this point after having commented on Reissner's finite rotation theory for annular membrane problems.

Before 1980, a mathematical analysis of the boundary value problems posed by the simplified Reissner theory seemed to be nonexistent as far as annular membranes under a nonvanishing surface load are concerned. First existence results for a rather restricted range of the boundary data were obtained by Weinitschke [48], who extended the integral equation technique formerly used for the Föppl membrane equations. The discovery of Novak's device led to a marked improvement and extension of these results. The general analysis developed above covers also the membrane equations of Reissner's theory and thus provides the existence of rt-solutions to Problem (s, S) for each $s>0$, $S>0$. This result was established by Grabmüller and Pirner [19]. The uniqueness of rt-solutions was also proved in [19] for the whole set of physically meaningful boundary parameters $s>0$, $S>0$, h and H real, and for each of the problems defined by (3.1) and (3.3). In summary, there is the following

THEOREM 3.3. (a) *Let $P^2(1, \varepsilon)>0$, then Problem (s, S) has a unique rt-solution for all $s>0$ and $S>0$.*
(b) *Problems (s, H), (h, S) and (h, H) have at most one rt-solution for all $s>0$, $S>0$, h and H real.*

Problems (s, H) and (h, H) differ from those of the Föppl membrane model since the boundary operator $B_1[g]$ in (3.6) now is genuinely nonlinear. The integral equation (3.8) changes to the form

$$g(z) = q_j(z) + \int_0^1 k_j(z, t) RS[t, g(t)]\, dt + w_j(z)\sqrt{g^2(1) + \Theta^2(1)}\,, \quad 0 \leq z \leq 1\,, \tag{3.10}$$

with appropriate Green's functions $k_j(z, t)$ and with linear functions $q_j(z) := A_j z + B_j$, $w_j(z) := \nu(C_j z + D_j)$, dependent on the boundary parameters s, S, h and H. The subscripts $j = 1, 2, 3$ refer to Problems (s, H), (h, S) and (h, H), respectively. The operator W in (3.9) needs to be corrected by an additional term $w_j(z)\{[f(1) + q_j(1)]^2 + \Theta^2(1)\}^{1/2}$ and the subset $M \subset C^0[0, 1]$ must be suitably modified in order to show that Theorem 2.11 now holds for annular membranes within Reissner's theory. Details are elaborated in [19], and the results obtained from a discussion of condition (N) differ from those of the Föppl membrane model rather quantitatively than qualitatively. As before, the ranges of boundary parameters associated with Problems (h, S) and (h, H) are subdivided into the three simply connected subsets of existence, of nonexistence and a remainder of unknown relation to rt-solutions. Surprisingly, the behavior of Problem (s, H) alters significantly when the theory turns over from small finite deflections to finite rotations. Contrary to the statement of Theorem 3.2, the parameter range $s>0$, H real, now splits in the same manner as described above. An explanation stems from the concept of superfunctions which was more thoroughly discussed in [14].

A concave function $g_\infty \in C^2(0,1) \cap C^0[0,1]$ is said to be a *superfunction* of Problem (s, S) if for any rn-solution $g(z)$ the relations $g(j) = g_\infty(j)$ with $j = 0, 1$, and $g(z) \leq g_\infty(z)$ with $0 \leq z \leq 1$ hold. For $s \geq s_0$ and $S \geq S_0$, a superfunction $g_\infty(z)$ of Problem (s, S) is generated by $g_\infty(z) := (1-z)(s-s_0) + z(S-S_0) + g_{\infty,0}(z)$ where $g_{\infty,0}(z)$ denotes any superfunction of Problem (s_0, S_0). For example, suitable superfunctions of Problem (s = 0, S = 0) are provided by [14]

$$g_{\infty,0}(z) := \begin{cases} \sqrt{z(1-z)}: & \text{Föppl}\,, \\ -\dfrac{\omega^4}{k^2}\left[z \ln \varepsilon^2 + 2x^2(z)\ln\left(\dfrac{x(z)}{\varepsilon}\right)\right]: & \text{simplified Reissner}\,. \end{cases} \tag{3.11}$$

Pursuing an idea mentioned in Section 2, rt-solutions $g = g(z; s, S)$ of Problem (s, S) may be interpreted as rt-solutions of Problem (s, H) for a value H given by $B_1[g(\cdot\,; s, S)] = H$. The superfunction $g_\infty(z) := q(z) + g_{\infty,0}(z)$ with $g_{\infty,0}$ taken from (3.11) is appropriate to verify $q(z) \leq g(z; s, S) \leq g_\infty(z)$ and to show $B_1[g(\cdot\,; s, S)] \to +\infty$ as $S \to +\infty$. In addition, taking account of

$$\dot g_\infty(1) \leq \dot g(1; s, S) \leq \dot q(1) - \limsup_{z\to 0+} \frac{1}{z}\int_0^z k(1-z, t)F[t, g_\infty(t)]\,\mathrm{d}t\,, \tag{3.12}$$

an elementary calculation yields $\lim_{S\to 0+} B_1[g(\cdot\,; s, S)] = -\infty$ for Föppl's theory, while

$$\lim_{S\to 0+} B_1[g(\cdot\,; s, S)] \geq \dot g_{\infty,0}(1) - s - \tfrac{1}{2}\nu(1-\varepsilon^2)|\Theta(1)| =: H_l > -\infty$$

in case of Reissner's theory [16]. In this case the range of the mapping $(s, S) \mapsto B_1[g(\cdot\,; s, S)]$ does not cover the whole real axis. So rt-solutions must be absent at least for $H < H_l$.

First ideas of the mapping argument used above originate from Pirner's diploma thesis [35] where only the finite rotation case was treated. However, the ingenious mapping idea was appropriate to fill in the gap mentioned above between the domains of existence and nonexistence of rt-solution. The following strategy was successful.

Firstly, the natural domain $Ex(s, S) := \{(s, S): s \geq 0, S \geq 0\}$ of existence of rn-solutions to Problem (s, S) is introduced. Theorem 3.3 confirms that the unique *rt-solutions* $g := g(z; s, S)$ of Problem (s, S) are defined at every interior point (s, S) of $Ex(s, S)$. At every boundary point of $Ex(s, S)$ there exists a unique *rn-solution*. This was shown by Pirner [35] in the finite rotation case and by Grabmüller [14] within a more general setting which also covers the small finite deflection case, see also [20].

THEOREM 3.4. *Assume $P^2(1, \varepsilon) > 0$, and let (s, S) be any boundary point of the set $Ex(s, S)$.*

(a) The small finite deflection model: *There exists a unique rn-solution $g = g(z; s, S)$ of Problem (s, S) with a finite derivative $\dot g(0; s, S)$ but an unbounded derivative $\dot g(1; s, S) = -\infty$ if $S = 0$.*

(b) The finite rotation model: *There exists a unique rn-solution $g = g(z; s, S)$ of Problem (s, S) with additional regularity $g \in C^1[0, 1]$ for all $s \geq 0$, $S \geq 0$.*

Secondly, each boundary operator $B_j, j = 0, 1$, acting on the totality of rn-solutions $g(z; s, S)$ of Problem (s, S) is interpreted as a mapping $\beta_j(s, S) := B_j[g(\cdot\,; s, S)]$ from the domain $Ex(s, S)$ into the reals. If the images of the set $Ex(s, S)$ subject to the vectorial mappings $(s, \beta_1(s, S))$ and $(\beta_0(s, S), S)$ are, respectively, denoted by $Ex(s, H)$ and $Ex(h, S)$, it becomes evident that these sets can be expected to form the natural domains of existence of rn-solutions to Problems (s, H) and (h, S). The more involved Problem (h, H) will be discussed below.

Of major interest is the behavior of the semi-axes $s \geqslant 0$ and $S \geqslant 0$ under the mappings β_j since their images $\gamma_j(s) := \beta_j(s, 0)$ and $\Gamma_j(S) := \beta_j(0, S)$ should be expected to separate the domain of existence of rt-solutions from that of nonexistence.

The development of the necessary analysis initiated by Pirner [35] was concerned with the simplified Reissner model. Pirner's results were substantially based on the following theorem which summarizes the main properties of the mappings β_j.

THEOREM 3.5 [The finite rotation model]. *Let $s_0 \geqslant 0$ and $S_0 \geqslant 0$ be fixed.*

(a) *The projections $\beta_j(s_0, \cdot)$: $\bar{\mathbb{R}}_+ \to \mathbb{R}$ are strictly increasing and continuous, and so must be Γ_j. The ranges of $\beta_j(s_0, \cdot)$ are the segments $[\gamma_j(s_0), +\infty)$. Asymptotic forms of Γ_j are provided by*

$$\left.\begin{aligned} &0 \leqslant \Gamma_0(S) - \varepsilon^2 S = o(1) \\ &0 \leqslant (1-\mu)S - \Gamma_1(S) = o(1) \end{aligned}\right\} \quad \text{as } S \to +\infty . \tag{3.13}$$

(b) *The projections $\beta_j(\cdot\,, S_0)$: $\bar{\mathbb{R}}_+ \to \mathbb{R}$ are strictly decreasing and continuous, and so must be γ_j. The ranges of $\beta_j(\cdot\,, S_0)$ are the segments $(-\infty, \Gamma_j(S_0)]$. Asymptotic forms of γ_j are provided by*

$$\left.\begin{aligned} &0 \leqslant \gamma_0(s) + (\varepsilon^2 + \mu)s = o(1) \\ &0 \leqslant -(1-\varepsilon^2)\nu|\Theta(1)|/2 - s - \gamma_1(s) = o(1) \end{aligned}\right\} \quad \text{as } s \to +\infty . \tag{3.14}$$

A proof of the continuity and monotonicity statements in Theorem 3.5, which is by no means simple, was supplied in [20]. The asymptotic forms (3.13) and (3.14) are easily derived from the integral representation (3.8) using

$$\left.\begin{aligned} \dot{g}(0; s, S) &= S - s + \int_0^1 (1-t)RS[t, g(t)]\,\mathrm{d}t , \\ \dot{g}(1; s, S) &= S - s - \int_0^1 tRS[t, g(t)]\,\mathrm{d}t , \end{aligned}\right\} \tag{3.15}$$

and observing that $RS[z, q(z)] \to 0$ as $s \to +\infty$ or $S \to +\infty$. The last relation holds pointwise for each $z \in (0, 1)$ and for $q(z) := zS + (1-z)s$ supporting $g(z)$.

By Theorem 3.5, an rn-solution $g(z; s, S)$ corresponding to $(s, S) \in Ex(s, S)$ induces the displacements $h := \beta_0(s, S) \leqslant \Gamma_0(S)$ at the inner edge of the annulus, and $H := \beta_1(s, S) \geqslant \gamma_1(s)$ at the outer edge. Since the mappings $\beta_j(s, \cdot)$ and $\beta_j(\cdot\,, S)$ are one-to-one, the representations

$$Ex(s, H) = \{(s, H) \colon H \geqslant \gamma_1(s), s \geqslant 0\} , \qquad Ex(h, S) = \{(h, S) \colon h \leqslant \Gamma_0(S), S \geqslant 0\}$$

become evident which in turn allow to be interpreted as an existence result for both Problems (s, H) and (h, S).

THEOREM 3.6 [The existence statement of Problems (s, H) and (h, S) for the finite rotation theory]. *Assume* $P^2(1, \varepsilon) > 0$. *Let, respectively,* $g_0(z; S)$ *and* $g_1(z; s)$ *denote the unique rn-solutions of Problems* $(s = 0, S)$ *and* $(s, S = 0)$, *and define the continuous curves*

$$\gamma_1(s) := \dot{g}_1(1; s) - \frac{1-\varepsilon^2}{2}\,\nu|\Theta(1)|\,, \quad s \geqslant 0\,, \qquad \Gamma_0(S) := \varepsilon^2 \dot{g}_0(0; S)\,, \quad S \geqslant 0\,. \tag{3.16}$$

Then the parameter ranges of Problems (s, H) *and* (h, S) *are decomposed, respectively, by* γ_1 *and* Γ_0 *into complementary subsets of existence and nonexistence as follows*:

(a) *Problem* (s, H) $((h, S)$, *resp.) has a unique rt-solution* $g(z)$ *if and only if* $s > 0$ *and* $H > \gamma_1(s)$ $(S > 0$ *and* $h < \Gamma_0(S)$, *resp.).*

(b) *At any boundary point* $(s \geqslant 0, H = \gamma_1(s))$ *a unique rn-solution* $g(z)$ *of Problem* (s, H) *is given by* $g_1(z; s)$ *and thus satisfies* $g(0) = s$ *and* $g(1) = 0$. *At any boundary point* $(s = 0, H)$, $H > H_0 := \gamma_1(0)$, *there exists a unique rn-solution* $g(z)$ *which satisfies* $g(0) = 0$ *and* $g(1) > 0$.

(c) *At any boundary point* $(h = \Gamma_0(S), S \geqslant 0)$ *a unique rn-solution* $g(z)$ *of Problem* (h, S) *is given by* $g_0(z; S)$ *and thus satisfies* $g(0) = 0$ *and* $g(1) = S$. *At any boundary point* $(h, S = 0)$, $h < h_0 := \Gamma_0(0)$, *there exists a unique rn-solution* $g(z)$ *which satisfies* $g(0) > 0$ *and* $g(1) = 0$.

(d) *Outside the sets* $Ex(s, H)$ *and* $Ex(h, S)$ *rn-solutions cannot exist.*

It is worth noting that the relations (3.13) and (3.14) supplemented by the superfunction (3.11) provide lower and upper bounds for the *separatrices* $\gamma_1(s)$ and $\Gamma_0(S)$. Indeed, letting $\gamma_1^*(s) := -s - (1-\varepsilon^2)\nu|\Theta(1)|/2$, a simple calculation yields

$$\left.\begin{aligned} &\gamma_1^*(s) + \dot{g}_{\infty,0}(1) \leqslant \gamma_1(s) \leqslant \gamma_1^*(s)\,, \quad s \geqslant 0\,, \\ &\varepsilon^2 S \leqslant \Gamma_0(S) \leqslant \varepsilon^2(S + \dot{g}_{\infty,0}(0))\,, \quad S \geqslant 0\,. \end{aligned}\right\} \tag{3.17}$$

The bounds (3.17) are simpler, but slightly coarser than those given in the paper [19]. The domains of existence of tensile solutions of Problem (h, S) for various k are illustrated in Fig. 2.

The analysis of Problem (h, H) is more complex because the images of the semi-axes $s \geqslant 0$ and $S \geqslant 0$ in (h, H)-plane now are *parametrized* arcs

$$\begin{aligned} \Sigma' &:= \{(h, H) \colon h = \gamma_0(s),\ H = \gamma_1(s),\ s \geqslant 0\}\,, \\ \Sigma'' &:= \{(h, H) \colon h = \Gamma_0(S),\ H = \Gamma_1(S),\ S \geqslant 0\}\,, \end{aligned}$$

which form a connected continuous arc $\Sigma := \Sigma' \cup \Sigma'' \subset (h, H)$. To make the representation of Σ *explicit*, Theorem 3.5 is utilized. The functions

$$\rho_1(h) := \gamma_1[\gamma_0^{-1}(h)]\,, \quad h \leqslant h_0 \quad \text{and} \quad \rho_2(h) := \Gamma_1[\Gamma_0^{-1}(h)]\,, \quad h \geqslant h_0\,,$$

are properly defined and map onto the segments $(-\infty, H_0]$ and $[H_0, +\infty)$, respectively. This shows the (h, H)-plane is complementarily subdivided by the arc

$$\Sigma = \{(h, H) \colon H = \rho_1(h) \text{ for } h \leqslant h_0 \text{ and } H = \rho_2(h) \text{ for } h > h_0\}$$

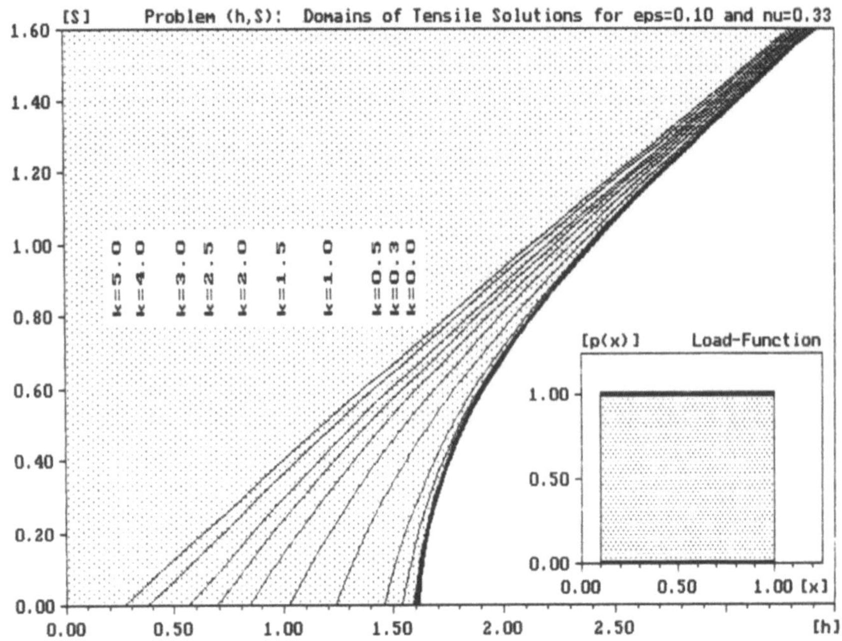

Fig. 2. Tensile solutions $y(x)$ of Problem (h, S) only exist for boundary parameters h and S within the dotted domain which extends unboundedly to the left. The diagram shows how the *separatrix* $\Gamma_0(S)$ varies with k. For $k = 0$ (*Föppl model*) the domain of existence is maximal. The surface load is uniform, $\nu = 1/3$, and $\varepsilon = 0.1$.

into two subdomains, which clearly are the domains of existence and nonexistence of rt-solutions to Problem (h, H). Via an implicit-function argument the following set is recognized as the domain of *existence*:

$$Ex(h, H) := \{(h, H): H \geqslant \rho_1(h) \text{ for } h \leqslant h_0,\ H \geqslant \rho_2(h) \text{ for } h > h_0\}\,. \tag{3.18}$$

Here, the details are omitted and reference is made to [20]. Again, the results can be interpreted as an existence theorem.

THEOREM 3.7 [The existence statement of Problem (h, H) for the finite rotation theory]. *In addition to the assumptions of Theorem* 3.6, *define*

$$\left.\begin{aligned} &\gamma_0(s) := \varepsilon^2 \dot{g}_1(0; s) - \mu s\,, \quad s \geqslant 0\,, \\ &\Gamma_1(S) := \dot{g}_0(1; S) - \frac{1-\varepsilon^2}{2}\,[S + \nu\sqrt{S^2 + \Theta^2(1)}]\,, \quad S \geqslant 0\,. \end{aligned}\right\} \tag{3.19}$$

Then the parameter range of Problem (h, H) *is decomposed by* Σ *into complementary subsets of existence and nonexistence as follows*:

(a) *A unique rt-solution* $g(z)$ *exists if and only if*

$$H > \begin{cases} \rho_1(h) := \gamma_1[\gamma_0^{-1}(h)]\,, & h \leqslant h_0 := \gamma_0(0)\,, \\ \rho_2(h) := \Gamma_1[\Gamma_0^{-1}(h)]\,, & h > h_0\,. \end{cases}$$

(b) *At any boundary point* $(h \leqslant h_0,\ H = \rho_1(h))$ *a unique rn-solution* $g(z)$ *is given by*

$g_1(z; s)$ *and thus satisfies* $g(0) = s = \gamma_0^{-1}(h) > 0$ *and* $g(1) = 0$. *At any boundary point* $(h > h_0,\ H = \rho_2(h))$, *a unique rn-solution* $g(z)$ *is given by* $g_0(z; S)$ *and thus satisfies* $g(0) = 0$ *and* $g(1) = S = \Gamma_0^{-1}(h) > 0$.

(c) *Outside the set* $Ex(h, H)$ *rn-solutions cannot exist.*

The domains of existence of tensile solutions are given in Fig. 3, for various values of k.

Lower and upper bounds for the separatrix Σ are derived from a fundamental relation between rn-solutions $g(z)$ and any superfunction $g_\infty(z)$ of Problem (s, S). Notice that $q(z) \leqslant g(z) \leqslant g_\infty(z)$ holds. Taking finite differences at $z = 0$ and $z = 1$, in the limit the derivatives are subjected to the following inequalities:

$$\dot{g}_\infty(0) \geqslant \dot{g}(0) \geqslant S - s \geqslant \dot{g}(1) \geqslant \dot{g}_\infty(1)\,. \tag{3.20}$$

Using this and the function $\gamma_1^*(s)$ defined above one obtains straightforwardly

$$\left.\begin{aligned} &\dot{g}_{\infty,0}(0) \geqslant \frac{1}{\varepsilon^2} B_0[g] + \left(1 + \frac{\mu}{\varepsilon^2}\right) s - S \geqslant 0\,, \\ &B_1[g] \geqslant (1-\mu)S + \dot{g}_{\infty,0}(1) + \gamma_1^*(s)\,. \end{aligned}\right\} \tag{3.21}$$

From these relations and from (3.17) the following bounds are obtained for the strictly increasing arc $\rho_1(h)$, where $\rho_1^*(h) := H/(\varepsilon^2 + \mu) - (1 - \varepsilon^2)\nu|\Theta(1)|/2$ will be used

$$\rho_1^*(h) + \dot{g}_{\infty,0}(1) - \frac{\varepsilon^2}{\varepsilon^2 + \mu}\, \dot{g}_{\infty,0}(0) \leqslant \rho_1(h) \leqslant \rho_1^*(h)\,, \quad h \leqslant h_0\,. \tag{3.22}$$

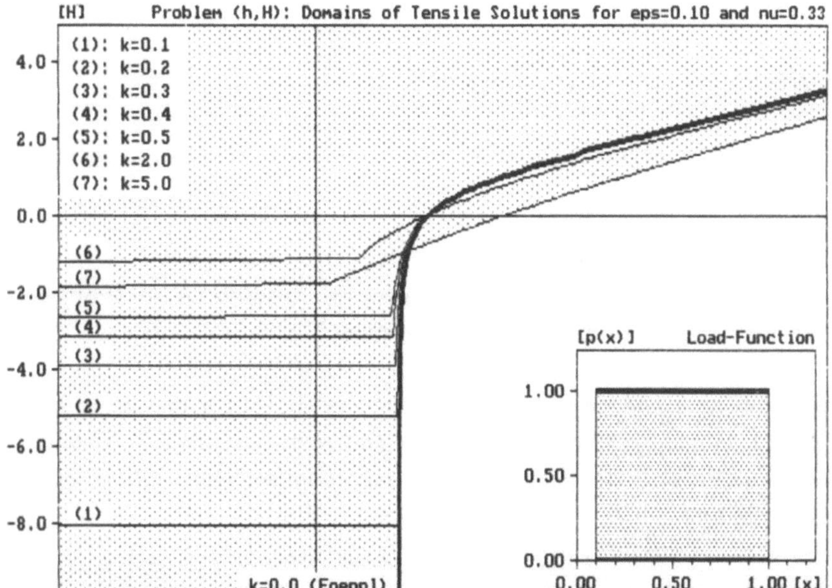

Fig. 3. Domains of existence (dotted) of tensile solutions $y(x)$ to Problem (h, H). The boundary manifold Σ defined in Theorem 3.7 depends on k. For $k = 0$ (*Föppl model*) existence extends to all values $H < 0$. The surface load is uniform, $\nu = 1/3$, and $\varepsilon = 0.1$.

Analogues estimates for the arc $\rho_2(h)$ are somewhat more complicated but principally derivable from (3.17) and (3.22).

An extension of Pirner's mapping argument to Föppl's small finite deflection theory needs a few changes in the above analysis and has been elaborated in [16, 22] within a more general setting of two-point nonlinear boundary value problems. As was seen in the discussion of Problem (s, H), the set $Ex(s, H)$ alters significantly when the theory changes from Föppl's model to Reissner's model. This was a consequence of rn-solutions $g(z; s, S)$ to Problem (s, S) getting an *unbounded* derivative $\dot{g}(1; s, S) \to -\infty$ as $S \to 0+$ in Föppl's theory, which was seen by virtue of (3.12). The question whether a similar behavior occurs at $z = 0$ has a negative answer. For a proof, the superfunction $g_\infty(z)$ of Problem (s, S), $s = 0$, derived from (3.11) has to be replaced by a more appropriate one, for example by

$$g_\infty(z) := zS + \frac{2}{S^2} \int_0^1 k(z, t) \frac{P^2(t, \varepsilon)}{t^2} \, dt ,$$

which possesses a *bounded* derivative

$$\dot{g}_\infty(0) = S + \frac{2}{S^2} \int_0^1 (1 - t) \frac{P^2(t, \varepsilon)}{t^2} \, dt < +\infty .$$

Using (3.20) it becomes evident that $g(z; s, S)$ has a bounded derivative at $z = 0$, $s = 0$. As a consequence, the existence statement of Problem (h, S) expressed by Theorem 3.6 does not change substantially if the small finite deflection theory is considered.

However, Theorem 3.7 needs some modifications since the arc $\gamma_1(s)$ degenerates to $-\infty$ because $\dot{g}_1(1; s)$ now becomes unbounded for each $s \geq 0$, and because $\Gamma_1(S) \to -\infty$ as $S \to 0+$. Thus the separatrix Σ consists of the arc Σ'' alone. A suitable explicit representation of Σ'' is provided by

$$\Sigma'' = \{(h, H) : h = \rho_2^{-1}(H) = \Gamma_0[\Gamma_1^{-1}(H)],\ H \text{ real}\} .$$

The range of ρ_2^{-1} is the segment $(h_0, +\infty)$ with $h_0 := \Gamma_0(0)$. Therefore, the set

$$Ex(h, H) := \{(h, H) : h \leq \rho_2^{-1}(H),\ H \text{ real}\}$$

has to be recognized as the domain of existence of rn-solutions to Problem (h, H). In summary, the following final existence theorem holds.

THEOREM 3.8 [The existence statement of Problem (h, H) for the small finite deflection theory]. *Let $g_j(z; \cdot)$ and $\Gamma_j(S)$ be defined as in Theorems* 3.6 *and* 3.7. *Then the parameter range of Problem (h, H) is decomposed by Σ'' into complementary subsets of existence and nonexistence as follows*:

(a) *A unique rt-solution $g(z)$ exists if and only if $h < \rho_2^{-1}(H) := \Gamma_0[\Gamma_1^{-1}(H)]$ and $H \in \mathbb{R}$. The continuous arc ρ_2^{-1} is strictly increasing with $\lim_{H \to -\infty} \rho_2^{-1}(H) = h_0 := \Gamma_0(0)$ and $\lim_{H \to +\infty} \rho_2^{-1}(H) = +\infty$.*

(b) *At any boundary point $(h = \rho_2^{-1}(H), H)$ a unique rn-solution $g(z)$ is given by $g_0(z; S)$ and thus satisfies $g(0) = 0$ and $g(1) = S = \Gamma_1^{-1}(H)$.*

(c) *Outside the set $Ex(h, H)$ rn-solutions cannot exist.*

4. Wrinkle-free solutions of circular and annular membranes

Stability considerations limit the applicability of tensile and rn-solutions to engineering problems. According to Stein and Hedgepeth [42], buckling of a stretched membrane is termed 'wrinkling', and the criterion adopted in [42] for wrinkling is that in a membrane compressive principal stresses cannot occur. Hence the vanishing of the minimum principal stress is taken to be the condition for incipient wrinkling. Recently this criterion has been demonstrated by a stability analysis, even for physically nonlinear membrane theory, by Steigmann [41]. Since $\sigma_r \geqslant 0$ in the solutions discussed in the preceding sections, the condition of wrinkling is the vanishing of σ_θ. The same criterion was used also by Jahsman et al. [24] and by Nachbar [32] in their membrane analyses under point loads.

In his work on the Föppl circular membrane under uniform pressure, Dickey [10] observed from his numerical results that $\sigma_\theta(r)$ is monotone decreasing in $0 \leqslant r \leqslant a$. Hence, $\sigma_\theta(a) = 0$ is the wrinkling criterion. He found numerically that $\sigma_\theta(a) = 0$ for $S = 0.7292$, which implies that for $S > 0.7292$ the membrane is entirely in tension, while for $S < 0.7292$ circumferential compression occurs in some annulus $0 < c < r \leqslant a$. For the fixed edge problem $H = 0$, he found that $S \geqslant 0.7292$ for all ν, $0 \leqslant \nu < 1/2$.

It is seen that the values S and H separating stable from unstable solutions can easily be determined by solving the differential equations (2.1) or (2.16) (the ν-term omitted) for the boundary condition $y'(1) + y(1) = 0$, provided that $\sigma_\theta'(x) \leqslant 0$ holds for all $x \in [0, 1]$. This monotonicity was first proved for the circular membrane under certain variable loads $\bar{p}(x)$ by Weinitschke [45], both within the Föppl and the simplified Reissner theory. Set $z(x) := xy' + y$, then from (2.16)

$$x^3 z' = -x^3 y' - \frac{4}{k^2} x^2 \tilde{f}(x, y) , \tag{4.1}$$

from which

$$x^3 z' = h(x) - xh'(x) , \qquad h(x) = \frac{4}{k^2} \int_0^x t^2 \tilde{f}(t, y(t)) \, \mathrm{d}t \tag{4.2}$$

is easily derived. An elementary argument then shows that $z' \leqslant 0$ provided that $Q(x) > 0$ and $[x\bar{p}(x)]' \geqslant 0$. In that case z and therefore σ_θ is monotone decreasing. The same argument goes through for the Föppl membrane. The resulting stability limit curves are shown in Fig. 1.

A special result for annular membranes, but within the Föppl theory only, has also been established in [45]. It concerns the case $\bar{p} = 1$ and $s = 0$, for which we have the representation theorem (3.7). It is easy to derive a formula for $z = xy' + y$ from (3.7) which shows that z is monotone decreasing for all $x \in [\varepsilon, 1]$. Hence, the limit curves $S(\varepsilon)$ and $H(\varepsilon)$ separating the wrinkle-free solutions from the unstable ones are given by the condition $y'(1) + y(1) = 0$.

It was a major task to generalize the above results to circular and annular membranes under general boundary conditions, and to remove restrictions such as $(x\bar{p})' \geqslant 0$ or $\bar{p} = 1$ in the above results. This has been achieved in 1990 by Grabmüller [15] for the annular membrane and in 1991 by Beck and Grabmüller [4] for the circular membrane. We proceed to outline these new results.

We first observe that the transformation (3.4) can also be used in the case $\varepsilon = 0$. If we let $t := x^2$ and $g(t) := x^2 y(x)$, then Problem S reduces to

$$-\ddot{g}(t) = F(t, g), \quad 0<t<1, \qquad g(0)=0, \qquad g(1)=S, \tag{4.3}$$

where a dot denotes differentiation with respect to t in this section. $F(t, g)$ is obtained from (3.5) by setting $\varepsilon = 0$. Since F is nonnegative, any rn-solution $g(t)$ of (4.3) must be concave. Furthermore, $g(t)$ satisfies an integral equation of the form (3.8), with $q(t) = St$, which can be utilized to extend differentiability of $g(t)$ to the end points $t = 0$ and $t = 1$. In particular, it can be shown that any solution of (4.3) satisfies $\dot{g}(0) = y(0) > 0$ and implies the correct boundary condition $y'(0) = 0$.

Let $g_j(t)$ be solutions of Problem S for $S = S_j$, $j = 1, 2$ with $S_2 > S_1 \geqslant 0$, and let $z(t) := g_2(t) - g_1(t)$. Then the integral representations for $g_j(t)$ can be used to prove that sign $\ddot{z}(t) =$ sign $z(t)$ and that $z(t) \geqslant 0$ and $\dot{z}(t) > 0$ holds for all $t \in [0, 1]$. In terms of the variables t and g, the circumferential stress can be written as

$$\sigma_\theta(t) = \frac{1}{4} Ek^2\left[2\dot{g}(t) - \frac{1}{t} g(t)\right], \quad 0 \leqslant t \leqslant 1. \tag{4.4}$$

It is seen that the rn-solution $g(t)$ of Problem S for $S = 0$ is not wrinkle-free. Indeed, by concavity we have $\dot{g}(1) < 0$ while $g(1) = 0$, implying $\sigma_\theta(1) < 0$. Consequently, a wrinkle-free solution must be an rt-solution satisfying $\dot{g}(t) > 0$ for all $t \in [0, 1]$. The function $\rho(t) := 2t\dot{g}(t) - g(t)$ obviously preserves the positivity of σ_θ except at $t = 0$. Let $g_j(t)$ be defined as above, then it follows from the monotonicity properties of z, $\dot{z}$ and $\ddot{z}$ that the corresponding functions $\rho_j(t)$ satisfy the inequality

$$\rho_2(t) - \rho_1(t) = t\dot{z}(t) + \int_0^t \tau\ddot{z}(\tau)\,d\tau > 0, \quad 0 < t \leqslant 1. \tag{4.5}$$

Therefore, if $g(t)$ is wrinkle-free at a point (k, S), $S > 0$, then this property is preserved at any point (k, S_1) with $S_1 > S$. As the set

$$W := \{(k, S): \inf_{0<t\leqslant 1} \rho(t; k, S) \geqslant 0\}, \tag{4.6}$$

is the domain of existence of wrinkle-free solutions, its boundary $\Gamma(k) = \inf_{S\geqslant 0} W$, $k > 0$ is well-defined.

In order to study $\Gamma(k)$, two maximum principles are needed. They can be proved under the following assumptions on the load: $p(t) \geqslant 0$ (or $p(t) \leqslant 0$) is measurable and bounded, $p > 0$ (or $p < 0$) holds on a subset of $[0, 1]$ of positive measure, $Q(t) \neq 0$ for all $t \in [t_0, 1]$, $0 < t_0 < 1$, and $(d/dt)Q^2(t)$ is piece-wise continuous for $t \in (0, 1)$, where $Q(t) = \int_0^{\sqrt{t}} \tau\bar{p}(\tau)\,d\tau$ [15].

THEOREM 4.1. *Let $g(t)$ denote an rt-solution of Problem S, $S > 0$, then a local minimum $\rho(t_1) \leqslant 0$ cannot be attained at an interior point $t_1 \in (0, 1)$, unless $\dot{g}(1) \leqslant 0$.*

THEOREM 4.2. *Given $k_2 > k_1 > 0$, denote by $g_j(t) := g(t; k_j)$ the rn-solutions of Problem S for $S = S_j \geqslant 0$, $j = 1, 2$. Let $z(t) := g_2(t) - g_1(t)$. Then a local maximum $z(t_1) > 0$ cannot be attained at an interior point $t_1 \in (0, 1)$.*

According to Theorem 4.1, the stress component $\sigma_\theta(t)$ of a wrinkle-free solution $g(t)$ cannot vanish at an interior point of the membrane. Thus, positivity of $\sigma_\theta(t)$ is controllable via the positivity of the boundary value $\rho(1)$. This generalizes a similar result derived from (4.1) and (4.2), under the more restrictive assumptions $Q(t) > 0$ and $(d/dt)[t\bar{p}(t)] \geqslant 0$. Integral representations of $g(t)$ and $\dot{g}(t)$ give rise to the formula

$$\rho(1) \equiv \rho(1; k, S) = 2\dot{g}(1) - S = S - \frac{2}{k^2} \int_0^1 R[\tau, g(\tau)] \, d\tau , \tag{4.7}$$

where

$$R(t, g) := 1 - \frac{g}{[g^2 + k^2 P^2(t)]^{1/2}} , \qquad P(t) := 2\sqrt{t} \int_0^{\sqrt{t}} \tau \bar{p}(\tau) \, d\tau .$$

R is related to F of (4.3) by $F(t, g) = R(t, g)/(k^2 t)$. It is seen from (4.7) that $\rho(1; k, S) \to +\infty$ for each fixed $k > 0$, as $S \to +\infty$. Since $\rho(1; k, 0) < 0$, the segment $[0, +\infty)$ is contained in the range of the mapping $S \mapsto \rho(1; k, S)$, for $S \geqslant 0$, $k > 0$. Hence the boundary $\Gamma(k)$ defined above can be obtained from the nonempty set

$$\Gamma := \{(k, S), k > 0, S \geqslant 0 : \rho(1; k, S) = 0\} , \tag{4.8}$$

by applying an implicit function theorem to $\rho(1; k, S) = 0$. At this point the continuity and strict monotonicity (4.5) of $\rho(1)$ is used. The set (4.8) then is the graph of a uniquely defined function $\Gamma: \mathbb{R}_+ \to \mathbb{R}$ satisfying $\rho(1; k, \Gamma(k)) = 0$ for all $k > 0$. The function $\Gamma(k)$ constitutes the finite boundary part of W provided $\dot{g}(1) > 0$. In view of (4.7), this condition holds for $S \geqslant \Gamma(k)$, $k > 0$.

Now Theorem 4.2 is applied to prove

THEOREM 4.3. *The mapping* $\Gamma(k)$ *is strictly decreasing and continuous*, $\lim_{k\to\infty} \Gamma(k) = 0$ *and* $\lim_{k\to 0+} \Gamma(k) = S^* > 0$ *exist.*

The following representation is an immediate consequence of (4.7)

$$\Gamma(k) = \frac{2}{k^2} \int_0^1 R[\tau, g(\tau)] \, d\tau , \quad k > 0 , \tag{4.9}$$

where $g(t)$ is the rt-solution satisfying $\rho(1) = 0$. Equation (4.9) shows that $\Gamma(k) \to 0$ as $k \to +\infty$. The limit for $k \to 0+$ can be obtained by applying l'Hospital's rule to (4.9). The result is

$$\Gamma(0) = S^* = \int_0^1 \frac{P^2(\tau)}{g^2(\tau)} \, d\tau > 0 . \tag{4.10}$$

S^* is the boundary of the Föppl model, so that rt-solutions are wrinkle-free for $S \geqslant S^*$. For uniform load $S^* = 0.7292$, as computed by Dickey [10].

It remains to consider the displacement problem. In terms of the variables t and g, the boundary conditions given in (2.3) and (2.16) for Problem H are

$$\begin{aligned} &2\dot{g}(1) - (1 + \nu) g(1) = H : &&\text{Föppl} , \\ &2\dot{g}(1) - g(1) - \nu [g^2(1) + k^2 P^2(1)]^{1/2} = H : &&\text{simplified Reissner} . \end{aligned} \tag{4.11}$$

Introducing $\rho(1)=0$ and making use of (4.7) the domain of wrinkle-free solutions for Problem H is bounded by the graph of

$$H(k) = -\nu S^*\,, \qquad k=0: \quad \text{Föppl}\,,$$
$$H(k) = -\nu[\Gamma^2(k) + k^2P^2(1)]^{1/2}\,, \quad k>0: \quad \text{simplified Reissner}\,. \tag{4.12}$$

An asymptotic form of $H(k)$ is $0 \leqslant -H(k) - \nu P(1)k = o(1)$ as $k \to \infty$, which follows from Theorem 4.3. The limit curves $\Gamma(k)$ and $H(k)$ are presented in Fig. 1 for the case of uniform load and $\nu = 1/3$. It is worth noting that there is a strong dependence on ν. In fact, (4.12) shows that $H(k)=0$ for all $k \geqslant 0$ if $\nu = 0$.

We now discuss the problem of wrinkle-free solutions for *annular membranes*, solved by Grabmüller [15] under similar assumptions on the surface load as for the circular membrane. The basic boundary value problems have been formulated in equations (3.1)–(3.3). The Schwerin transformation (3.4) is used to allow the application of concavity and monotonicity arguments. A maximum principle is then established which shows that the positivity of $\sigma_\theta(x)$ is controllable via the values taken at the boundary $x=\varepsilon$ and $x=1$. We then examine in the set of boundary data $s \geqslant 0$, $S \geqslant 0$ the manifolds $\sigma_\theta(\varepsilon)=0$ and $\sigma_\theta(1)=0$ (keeping the parameters k and ε fixed). They determine the boundary of the domain $W(s,S)$ of wrinkle-free solutions of Problem (s, S).

In terms of the variables $t=(x^2-\varepsilon^2)/(1-\varepsilon^2)$, $g(t)=\omega^4x^2y(x)$ we have the annular membrane boundary value problems (3.5) and (3.6), replacing z by t, as above. Let $g_j(t)$ denote any two rn-solutions of Problem (s_j, S_j), $j=1,2$, and let $z(t) := g_2(t) - g_1(t)$, then we have again monotonicity $z(t) \geqslant 0$, $t \in [0,1]$, $\ddot{z}(t) \geqslant 0$, $t \in (0,1)$, provided that $s_2 \geqslant s_1 \geqslant 0$ and $S_2 \geqslant S_1 \geqslant 0$. Furthermore, $\dot{z}(t) \geqslant 0$, $t \in [0,1]$, provided that $s_2 > s_1 \geqslant 0$ and $S_2 = S_1 \geqslant 0$. The function

$$\rho(t) := \omega(t+\gamma)\sigma_\theta(t)(4/Ek^2) = 2(t+\gamma)\dot{g}(t) - g(t)\,, \tag{4.13}$$

where $\gamma = \varepsilon^2/(1-\varepsilon^2)$, preserves both the regularity and positivity of σ_θ, thus

$$W(s,S) = \{(s,S) \colon \inf_{0<t\leqslant 1} \rho(t; s, S) \geqslant 0\}\,. \tag{4.14}$$

As in the case of the circular membrane, it is observed that an rn-solution $g(t)$ of Problem (s, S) is not wrinkle-free if $s=S=0$. Clearly a wrinkle-free solution $g(t)$ must satisfy $\dot{g}(t) > 0$ for all $t \in [0,1]$, which follows from (4.13) noting that $g(1) > 0$, and that $\dot{g}(0) \geqslant \dot{g}(t) \geqslant \dot{g}(1)$ by concavity.

In order to derive monotonicity properties for $\rho(t)$, the integral equation (3.8) is employed. Together with the monotonicity of $z = g_2 - g_1$, the following inequalities analogous to (4.5) are obtained for $\rho_j(t) = \rho(t; s_j, S_j)$, $j=1,2$

$$\rho_2(t) - \rho_1(t) = (t+2\gamma)\dot{z}(t) + \int_0^t \tau\ddot{z}(\tau)\,\mathrm{d}\tau \geqslant 0\,, \quad t \in [0,1]\,, \tag{4.15}$$

provided that $s_2 = s_1 \geqslant 0$ and $S_2 > S_1 \geqslant 0$. On the other hand, if $s_2 > s_1 \geqslant 0$ and $S_2 = S_1 \geqslant 0$, then

$$\rho_2(t) - \rho_1(t) = 2(t+\gamma)\dot{z}(t) - z(t) \leqslant 0\,, \quad t \in [0,1]\,. \tag{4.16}$$

Next the structure of the boundary $\Gamma(s) = \inf_S W(s, S)$, $s \geq 0$, is studied under the assumption that $p(x)$ is measurable and positive on $\varepsilon \leq x \leq 1$ in the sense defined above for the circular membrane. Theorem 4.1 holds for any rn-solution of Problem (s, S), $S > 0$ ($0 \leq t \leq 1$ represents the interval $\varepsilon \leq x \leq 1$). Thus, positivity of σ_θ is here controlled by the boundary values $\rho(0)$ *and* $\rho(1)$, given from (4.13) by

$$\rho(0) = 2\gamma\dot{g}(0) - s\,, \qquad \rho(1) = 2(1+\gamma)\dot{g}(1) - S\,. \tag{4.17}$$

If both $\rho(0; s, S)$ and $\rho(1; s, S)$ are zero for $s = s_p$, $S = S_p$, then (s_p, S_p) is called a *switch point*. The integral representation for $\rho(t)$, evaluated at $t = 0$ and $t = 1$, supplies two linear algebraic equations for calculating switch points. The result is

$$\begin{aligned} s_p &= 2\gamma \int_0^1 (1 + 2\gamma + \tau) F[\tau, g(\tau)]\, d\tau > 0\,, \\ S_p &= 2(1+\gamma) \int_0^1 (2\gamma + \tau) F[\tau, g(\tau)]\, d\tau > 0\,. \end{aligned} \tag{4.18}$$

It can be shown that there exists exactly one switch point for any given $k > 0$ and $\varepsilon > 0$. Here again the monotonicity of the function $z(t)$ is crucial in the proof. It is easy to compute the switch point numerically. One simply solves the differential equation $-\ddot{g}(t) = F(t, g)$ subject to the boundary conditions (4.17) setting $\rho(0) = \rho(1) = 0$ and $s = g(0)$, $S = g(1)$.

Physically, one expects wrinkling to occur if s at the inner edge is sufficiently large. On the other hand, no wrinkling should occur if S at the outer edge is sufficiently large. Indeed, from an integral representation for $\rho(t)$ one can show rigorously that $\rho(0; s, S) \to -\infty$ for each $S \geq 0$ as $s \to +\infty$, and $\rho(1; s, S) \to +\infty$ for each $s \geq 0$ as $S \to +\infty$. Hence the ranges of the mappings $s \mapsto \rho(0; s, S)$ and $S \mapsto \rho(1; s, S)$ are such that the nonempty sets

$$\Gamma_j = \{(s, S)\colon \rho(j; s, S) = 0\}\,, \quad j = 0, 1 \tag{4.19}$$

give rise, via the implicit function theorem, to a pair of uniquely defined functions $\Gamma_0^{-1}\colon \bar{\mathbb{R}}_+ \to \mathbb{R}$ and $\Gamma_1\colon \bar{\mathbb{R}}_+ \to \mathbb{R}$ satisfying

$$\rho(0; \Gamma_0^{-1}(S), S) = 0 \quad \text{for } S \geq 0 \quad \text{and} \quad \rho(1; s, \Gamma_1(s)) = 0 \quad \text{for } s \geq 0\,. \tag{4.20}$$

With $s^* := \Gamma_0^{-1}(0)$ and $S^* := \Gamma_1(0)$, it follows that $D(\Gamma_0) = \mathrm{Range}(\Gamma_0^{-1}) = [s^*, +\infty)$ and $\mathrm{Range}(\Gamma_1) = [S^*, +\infty)$ because the mapping $\Gamma_0^{-1}\colon \bar{\mathbb{R}}_+ \to D(\Gamma_0)$ is one-to-one and the mappings $\Gamma_0\colon D(\Gamma_0) \to \mathbb{R}$ and $\Gamma_1\colon \bar{\mathbb{R}}_+ \to \mathbb{R}$ are strongly increasing and continuous; furthermore, $\lim_{s\to\infty} \Gamma_j(s) = +\infty$, $j = 0, 1$.

The domain $W(s, S)$ of wrinkle-free solutions of Problem (s, S) can now be determined. In view of (4.17), the condition $\dot{g}(1) > 0$ holds for $S \geq \Gamma_1(s)$ and $s \geq 0$. Thus Theorem 4.1 applies which shows that the boundary Γ of $W(s, S)$ is part of the arcs Γ_0 and Γ_1. More precisely, we have:

THEOREM 4.4. *There exists a unique switch point* (s_p, S_p). *The parameter domain of wrinkle-free solutions of Problem* (s, S) *is the set (see Fig.* 4)

$$W(s, S) = \{(s, S)\colon S \geq \Gamma_1(s) \text{ for } 0 \leq s \leq s_p,\ S \geq \Gamma_0(s) \text{ for } s_p \leq s < \infty\}\,.$$

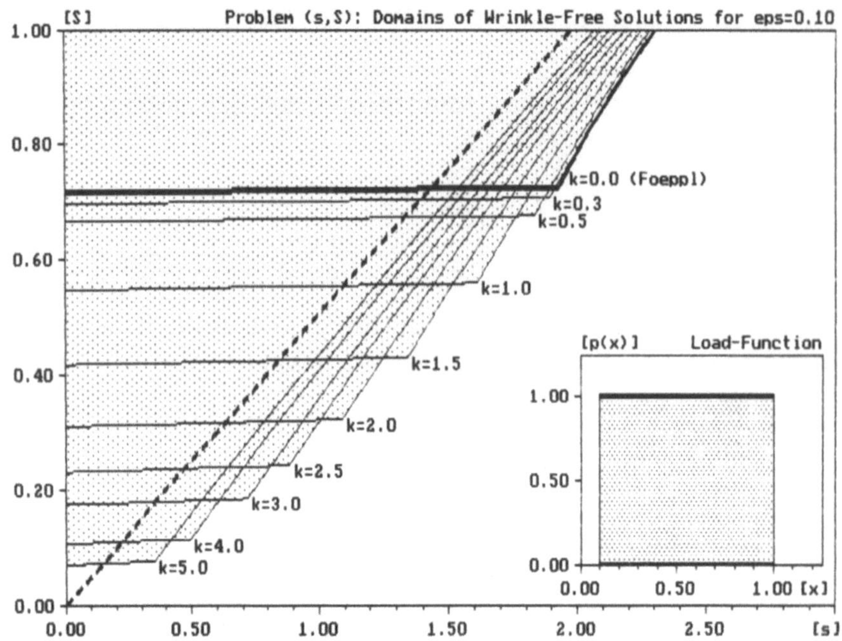

Fig. 4. Wrinkle-free solutions of Problem (s, S) only exist within the dotted domains. At the boundary curves that asymptotically approach the straight line $S=(1+\varepsilon^2)s/2$ (bold dotted), the circumferential stress σ_θ vanishes at one of the edges of the annulus, i.e., at the outer edge on the horizontal arcs, and at the inner edge on the inclined arcs. The surface load is uniform, $\nu=1/3$, and $\varepsilon=0.1$.

Next we consider problems involving displacement data in the boundary conditions. Following Theorem 3.4, the images of mappings defined via the displacement boundary operators B_j, $j=0,1$, were seen to be the domains where rn-solutions of the respective Problems (s, H), (h, S) and (h, H) exist. Now it becomes obvious that the subdomain $W(s,S)\subseteq Ex(s,S)$ of wrinkle-free solutions is mapped onto a corresponding subdomain in the boundary-parameter ranges (s,H), (h,S) and (h,H). In particular, we are interested in finding the images of the arcs $\Gamma_0(s)$ and $\Gamma_1(s)$. To this end we simply have to combine (4.17), setting $\rho(0)$ and $\rho(1)$ equal zero, with the displacement boundary operators B_j defined following (3.6). Hence, we have, with $\tilde{\nu}:=(1-\varepsilon^2)\nu/2$ and $\Theta_1:=\Theta(1)$

$$\text{on } \Gamma_0\text{:}\quad h_0(s)=\varepsilon^2\dot{g}(0)-\mu s=\frac{\varepsilon^2}{2\gamma}\,s-\mu s=-\tilde{\nu}s\,, \tag{4.21}$$

$$\text{on } \Gamma_1\text{:}\quad H_1(s)=\begin{cases}\dot{g}(1)-\mu g(1)=\dfrac{1}{2(1+\gamma)}\,S-\mu S=-\tilde{\nu}\Gamma_1(s)\,,\\[2ex] \dot{g}(1)-\dfrac{1-\varepsilon^2}{2}\,[g(1)+\nu\sqrt{g^2(1)+\Theta_1^2}]=-\tilde{\nu}(\Gamma_1^2(s)+\Theta_1^2)^{1/2}\,,\end{cases} \tag{4.22}$$

for the Föppl and simplified Reissner theory. If the boundary condition $B_1[g]=H$ is rewritten in terms of $\Gamma_0(s)$ we find

$$H_0(s)=\begin{cases}D_1(s)-\mu\Gamma_0(s)\text{:} & \text{Föppl}\,,\\[2ex] D_1(s)-\mu\Gamma_0(s)-\tilde{\nu}\Gamma_0(s)[(1+\Theta_1^2/\Gamma_0^2(s))^{1/2}-1]\text{:} & \text{simpl. Reissner}\,,\end{cases} \tag{4.23}$$

where $D_1(s)=\dot{g}(1)$, $g(t)$ being the solution of Problem (s, S) for $S=\Gamma_0(s)$.

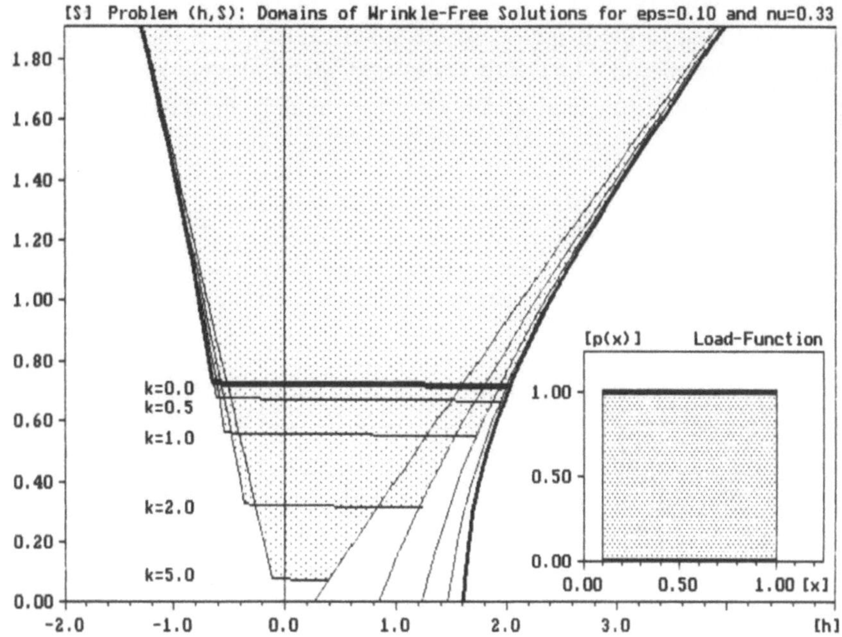

Fig. 5. The domains (dotted) of wrinkle-free solutions $y(x)$ of Problem (h, S) for various k. Only minor differences are found between the results for $k = 0$ (*Föppl model*) and $k > 0$ (*Reissner model*). The surface load is uniform, $\nu = 1/3$, and $\varepsilon = 0.1$.

The set $W(s, H)$ of wrinkle-free solutions of Problem (s, H) can thus be written as follows:

$$W(s, H) = \{(s, H): H \geq H_1(s) \text{ for } 0 \leq s \leq s_p,\ H \geq H_0(s) \text{ for } s_p \leq s < \infty\} .$$

We note that $H_1(s) = 0$ for all k if $\nu = 0$. Therefore, no wrinkle-free solution can exist for $H < 0$.

Changing to a parametrization with respect to S, the domain $W(h, S)$ of wrinkle-free solutions of Problem (h, S) can similarly be determined. Clearly (4.21) can be written as $h_0(S) = -\tilde{\nu}\Gamma_0^{-1}(S)$, in (4.22) $\Gamma_1(s)$ is simply replaced by S. The domains of wrinkle-free solutions of Problem (h, S) are shown in Fig. 5. In the case of Problem (h, H) the boundary of the corresponding domain $W(h, H)$ decomposes into two connecting arcs which can be expressed via the function H_0, H_1 defined in (4.22) and (4.23) and a function $h_1(S)$ defined for describing $W(h, S)$; for details see [15].

5. Curved membranes of revolution

We begin by discussing some recent work for shallow membranes for small finite deflections (Föppl theory), obtained by Baxley [2] and Dickey [11]. The governing equation is (1.7). For a shallow spherical cap under normal pressure $p_s = 0 = p_H$ and $z \doteq z_0(1 - r^2)$, approximately. Since $z' = \sin\varphi$, this implies $1 - \cos\varphi \doteq 2z_0 r^2$ in (1.7). Introducing dimensionless variables as for the circular membrane, equation (1.7) can be reduced to the form

$$Ly = \frac{2}{y^2} Q^2(x) - \lambda_0^2 , \qquad 0 < x < 1 , \qquad y'(0) = 0 , \tag{5.1}$$

where $x = r/a$, L and Q are defined in (2.1), and $\lambda_0 = (4Ea^2d/p_0)^{1/3}/R\sqrt{2}$ is a geometry-load parameter proportional to the height z_0 of the spherical cap, R is the radius of the sphere. Applying the transformation (3.4) with $t = x^2$ and $g(t) = x^2y(x)$, we obtain the problem formulated in [2] for uniform load and fixed edge, namely Problem H for $H = 0$, that is (apart from notation)

$$\begin{aligned} &\ddot{g}(t) = -\frac{t^2}{g^2} + \lambda^2, \qquad 0 < t < 1, \\ &g(0) = 0, \qquad 2\dot{g}(1) - (1+\nu)g(1) = 0. \end{aligned} \tag{5.2}$$

It is observed in [2] that the right-hand side of the differential equation is nondecreasing in g, at least for tensile solutions $g > 0$, but that the coefficients in the boundary condition at $t = 1$ do not have the right signs for a classical existence theorem to apply. However, it was shown in [17], how this difficulty can be overcome by the well-known maximum principle of Hopf. Hence, applying the technique of [17], the *uniqueness* of positive solutions of the boundary value problem (5.2) follows, even for variable load and more generally for Problem S and Problem H, in our terminology. In order to prove *existence*, Baxley substitutes $u = 1/t$, reduces (5.2) to a boundary value problem on $1 \leq u < \infty$, and applies an existence theorem valid for problems of the type

$$\begin{aligned} &w''(u) = f(u, w, w'), \qquad a \leq u < \infty, \\ &a_0w(a) - a_1w'(a) = A, \qquad a_0 > 0, \quad a_1 \geq 0, \end{aligned}$$

where f must satisfy a number of smoothness and monotonicity conditions. We summarize the results as

THEOREM 5.1 (Baxley). *Suppose* $(1+\nu)/2 < \rho < 1$, *then Problem* (5.2) *has at most one solution satisfying* $g(t)/t^\rho \to 0$ *as* $t \to 0+$. *A positive solution* $g(t)$ *exists for all* λ, *which has the properties*

$$\frac{g(t)}{t} < \frac{1}{2\lambda}, \qquad g(t) \geq \alpha t(\beta - t), \qquad \frac{g(t)}{t} \text{ decreasing}, \tag{5.3}$$

for $0 < t \leq 1$ *and some constants* $\alpha > 0$, $\beta > 1$. *Furthermore*, $\lim(g(t)/t)$ *exists for* $t \to 0+$.

It appears that the existence part of the theorem can also be proved by transforming (5.2) into an integral equation and applying Schauder's fixed point theorem. Indeed, this is the technique employed by Dickey [11] in his work on shallow shells for surfaces generated by rotating the curve

$$z = z(r) = C\left(1 - \left(\frac{r}{a}\right)^\gamma\right), \qquad 0 \leq r \leq a, \tag{5.4}$$

for a constant $C > 0$ and a shape factor $\gamma > 1$. The case $\gamma = 2$ and C sufficiently small corresponds to the spherical cap (5.1). Given the radial stress or the radial displacement at the edge $r = a$, we obtain Problems S and H in the dimensionless form

$$Ly = \frac{2}{y^2} Q^2(x) - \lambda_0^2 x^{2\gamma-4}, \qquad \lambda_0 = \frac{1}{\sqrt{2}} \left(\frac{4EC^2\gamma^3 d}{p_0 a^4} \right)^{1/3},$$
$$y'(0) = 0, \qquad y(1) = S \quad \text{or} \quad y'(1) + (1-\nu)y(1) = H. \tag{5.5}$$

The technique of solving (5.5) employed in [11] is quite similar to the case $\lambda_0 = 0$, as discussed in detail in Section 2. The boundary value problem is transformed into an integral equation of type $y = Ty$, then the operator T is shown to have properties such that the Schauder fixed point theorem can be applied. The result is, proved in [11] for $Q = 1$ (uniform load):

THEOREM 5.2 (Dickey). *Set $D = \lambda_0^2 4^{-5/3}/[\gamma(\gamma-1)]$ and assume $\gamma > 1$, $S > D$. Then Problem S has a solution $y(x)$, which is positive for $x \in [0, 1]$.*

Although uniqueness is not discussed in [11], it is obvious that a standard uniqueness argument applies to (5.5) for any $S > 0$, so that positive solutions of Problem S are unique. Problem H, $H = 0$, is solved by interpolation as in [10], showing that $u(a)$ changes sign for solutions of Problem S in a range of S covered by Theorem 5.2. In addition, D must be sufficiently small. The result is, again for $Q = 1$:

THEOREM 5.3 (Dickey). *Assume $\gamma > 1$ and D is sufficiently small, then Problem H, for $H = 0$, has a positive solution $y(x)$.*

In the special case $\gamma = 2$, Theorem 5.1 shows that a smallness assumption on D is unnecessary. However, it is proved in [11] that if $\gamma = 4/3$, Problem H for $H = 0$ has no positive solution unless D is sufficiently small. This particular case is amenable to a phase plane analysis, from which the following conclusions are drawn in [11]. If $\lambda^2 \leq (64/3)^{1/3}$, where $\lambda^2 = c\lambda_0^2$, then Problem S has a unique rt-solution $y(x)$ for all $S > 0$, and Problem H has a unique rt-solution $y(x)$ for all H. If $\lambda^2 > (64/3)^{1/3}$ then Problem S has no tensile solution unless $S > S_0$, and Problem H has no rt-solution unless $H > H_0$ where S_0 and H_0 are positive numbers. In particular, there is no (rotationally symmetric) solution of Problem H for $H = 0$ in the case $\gamma = 4/3$ and $\lambda^2 > (64/3)^{1/3}$.

Finally, we present some work from the Thesis of J. Arango [1], which concerns deformation of curved membranes with finite rotations under uniform *normal pressure* $p_n = q$. The basic equation is (1.6), where $p_s = 0$. From equations (1.4) we have $p_H = q \sin \Phi$, which means that V in (1.6) is not a given quantity, but rather depends on the solution Φ. In fact, equilibrium in the z-direction implies

$$\frac{\mathrm{d}(rV)}{\mathrm{d}s} + rp_V = 0 = \frac{\mathrm{d}(rV)}{\mathrm{d}s} - r(s)q \cos \Phi. \tag{5.6}$$

Following the notation in Clark et al. [8], we substitute

$$M = rS_\phi \cos \Phi, \qquad N = rS_\phi \sin \Phi, \qquad H = rS_\theta, \tag{5.7}$$

where S_ϕ and S_θ are pseudo stress resultants which measure the tension in the deformed membrane per unit length of the undeformed membrane. Then the basic equations (1.6) and (5.6) can also be written as an equivalent first order system of two equilibrium equations and one compatibility equation [8]

$$r\frac{dM}{ds} - H + r^2 q \sin\Phi = 0\,, \qquad r\frac{dN}{ds} - r^2 q \cos\Phi = 0\,,$$
$$r\frac{dH}{ds} - M - Edr(\cos\Phi - \cos\varphi) = 0\,. \tag{5.8}$$

Assuming $0 < s \leq L$, tensile solutions must satisfy $S_\phi(s) > 0$ for $0 < s \leq L$. In addition, rotations are restricted to $0 \leq \Phi(s) < \pi/2$, so that $M > 0$ and $N > 0$ for $0 < s \leq L$. For wrinkle-free solutions we also have $H \geq 0$. Observing the relations

$$\cos\Phi = M/(M^2 + N^2)^{1/2}\,, \qquad \sin\Phi = N/(M^2 + N^2)^{1/2}\,,$$

the system (5.8) can be re-written in the dimensionless form

$$x'(t) = z(t)/\rho(t) - \rho(t)y(t)/Q(x, y)\,,$$
$$y'(t) = \rho(t)x(t)/Q(x, y)\,, \quad 0 < t < 1\,, \tag{5.9}$$
$$z'(t) = x(t)/\rho(t) + e[-\rho'(t) + x(t)/Q(x, y)]\,,$$

where $t = s/L$, $\rho = r/L$, $(x, y, z) = (M, N, H)/(L^2 q)$, $Q(x, y) = (x^2 + y^2)^{1/2}$ and $e = Ed/qL$. If the membrane is closed at the apex, one has $\rho(0) = 0$ and $\rho'(0) = 1$. Any regular solution of (5.9) must satisfy $x(0) = y(0) = z(0) = 0$. If the membrane has a circular inner edge at $r = r_0 > 0$, then $\rho(t) > 0$ for $0 \leq t \leq 1$ and $x(0)$, $y(0)$ or $x(0)$, $z(0)$ can be prescribed at the inner edge. In both cases the geometry is restricted by the assumption $\rho'(t) > 0$ for $0 \leq t \leq 1$. At the outer edge, a boundary condition

$$B(x(1), y(1), z(1)) = m \tag{5.10}$$

is imposed, where $B\colon \mathbb{R}^3 \to \mathbb{R}$ is a given function and m is a given number. The problem is then to find sufficient conditions on the boundary conditions, for certain physically meaningful B, such that equations (5.9), (5.10) have a tensile solution, and also to formulate conditions for uniqueness, and for wrinkle-free solutions.

The method of proof in [1] is essentially the shooting method. A set of *admissible solutions* $X(t, r) = (x(t, r), y(t, r), z(t, r))$, $0 \leq t \leq 1$, is introduced that satisfy the conditions at $t = 0$ stated above and the differential equations (5.9), and that depend on a shooting parameter r. The existence of solutions of (5.9) and (5.10) is then equivalent with the existence of zeros of the shooting function

$$F(r) := B(X(1, r)) - m\,.$$

Under appropriate assumptions on $F(r)$ there is a unique zero which yields uniqueness of the solution of the boundary value problem. It is clear that one of the main difficulties is to establish the existence of admissible solutions $X(t, r)$ in the interval $0 \leq t \leq 1$. Fortunately, the nonlinear nonnegative terms x/Q and y/Q in (5.9) are bounded by unity. This fact, together with some monotonicity properties of the solutions $X(t, r)$, allows one to extend the local existence of $X(t, r)$ near $t = 0$, which is guaranteed by classical theory, to a global existence in the whole interval $0 \leq t \leq 1$. The details are quite involved and cannot be discussed here. Some representative results of [1] for a membrane of revolution closed at the apex are contained in the next two theorems. In analogy to flat and shallow membranes two boundary value problems are defined by prescribing the radial stress or the radial displacement at the boundary $t = 1$ (see (5.7))

$$\begin{aligned} &\text{Problem S:} \quad B(X) = x(1)\,, \\ &\text{Problem H:} \quad B(X) = z(1) - \nu[x^2(1) + y^2(1)]^{1/2}\,. \end{aligned} \tag{5.11}$$

Let

$$\bar{\rho}(t) = -\int_1^t \frac{1}{\rho(\tau)}\,\mathrm{d}\tau\,, \qquad \rho'(t) = 1 - t\hat{\rho}(t)\,, \quad t > 0\,,$$

then the assumptions on $\rho(t)$ imply that $\bar{\rho} > 0$, $\hat{\rho} > 0$ for $0 < t \leq 1$.

THEOREM 5.4 (Arango). *Problem S has a tensile solution for all $m > 0$ satisfying*

$$m > m_0 := e \int_0^1 \hat{\rho}(t) t \sinh(\bar{\rho}(t))\,\mathrm{d}t > 0\,.$$

Problem H has a tensile solution for all $m > 0$ satisfying

$$m > m_1 := \sup\left\{ e \int_0^1 \hat{\rho}(t) t \cosh(\bar{\rho}(t))\,\mathrm{d}t,\ \int_0^1 [\cosh(\bar{\rho}(t)) - \sinh(\bar{\rho}(t))][\rho(t) + et\hat{\rho}(t)]\,\mathrm{d}t \right\} > 0\,.$$

These solutions of Problems S and H are unique if

$$1 + (\rho'(t))^2 + \rho(t)\rho''(t) \geq 0\,, \quad 0 \leq t \leq 1\,. \tag{5.12}$$

The conditions $m > m_0$ and $m > m_1$ are perhaps overly restrictive. On the other hand, Theorem 5.4 is valid for quite arbitrary membranes of revolution. The results of Dickey show that some particular geometries do require restrictions on m, excluding $m = 0$ in Problem H. The above results simplify considerably for the circular membrane problem under *normal* pressure, which in the Reissner theory of finite rotations is different from the corresponding problem under *vertical* pressure.

THEOREM 5.5 (Arango). *In the case of a circular membrane $\rho'(t) \equiv 1$, Problem S has a unique tensile solution for all $m \geq 0$. Problem H has a unique tensile solution for all $m \geq 1/2$.*

While the result for Problem S is best possible, Problem H for a fixed edge ($m = 0$) is not covered by Theorem 5.5.

Existence and uniqueness results for membranes of revolutions with a circular opening have also been obtained in [1]. Depending on which boundary data are prescribed at $t = 0$ and $t = 1$, there results a variety of boundary value problems. As an example, we briefly discuss results corresponding to Problems (s, S) and (s, H) defined in Section 3. Suppose $x_0 \geq 0$, $y_0 \geq 0$ are such that $x_0^2 + y_0^2 > 0$, then one has

THEOREM 5.6 (Arango). *Let* $B(X) = x(1)$ *and* $m \geqslant 0$. *Then the boundary value problem* (5.9), (5.10) *has a tensile solution. This solution is unique, if* $\rho(t)$ *satisfies* (5.12). *Let* $B(X)$ *be defined as in* (5.11), *Problem H, and* $m \geqslant \rho^2(1)$, *then* (5.9), (5.10) *has at least one tensile solution.*

In the case of a flat annular membrane $\rho(t) = t$, the condition $m \geqslant \rho^2(1)$ can be improved to $m \geqslant 1/3$.

The above conditions do not cover the whole range of physically meaningful boundary data. Separation curves, as in Section 3, separating exactly the domains of existence and nonexistence of tensile solutions or establishing the domain of wrinkle-free solutions for the various boundary value problems have yet to be found. For the circular and annular membrane it is known that $S_\theta(t) \geqslant 0$ can be controlled by the boundary data as in Section 4. The following result was proved in [1]: An admissible solution of equations (5.9) is wrinkle-free if and only if $z(0) \geqslant 0$ and $z(1) \geqslant 0$.

The problem of curved membranes under variable *vertical* load has apparently not been investigated as yet (see a forthcoming thesis by A. Beck). In closing, we would also like to stress that the results of Sections 2–4 concerning finite rotation problems have all been derived under the simplifying assumption that the term $\nu r p_s$ in (1.6) is ignored. It remains to be seen whether the qualitative results concerning existence, uniqueness and wrinkle-free solutions change significantly if that term is included in the analysis. Numerical calculations in [45] show that quantitative differences are not negligible for larger values of k.

Acknowledgement

Work on this paper was started by H.J.W. in 1990, while he was a Visiting Professor of Mathematics at the University of British Columbia. The work was supported in part by a Canadian NSERC Grant No. 8–4559.

References

1. J. Arango, Existenz und Eindeutigkeitsaussagen über Lösungen von Randwertproblemen der nichtlinearen Membrantheorie, Dissertation (Ph.D. Thesis), Faculty of Science I, Univ. Erlangen-Nürnberg, July 1989, 121 pp.
2. J.V. Baxley, A singular nonlinear boundary value problem: Membrane response of a spherical cap. *SIAM J. Appl. Math.* 48 (1988) 497–505.
3. A. Beck, Die Reissnersche Kreismembran: Existenz und Eindeutigkeit von positiven Lösungen, Diploma-Thesis, Faculty of Science I, University of Erlangen-Nürnberg, April 1989, 63 pp.
4. A. Beck and H. Grabmüller, Wrinkle-free solutions of circular membrane problems, *J. Appl. Math. and Phys.* (ZAMP) (in press).
5. E. Bromberg and J.J. Stoker, Non-linear theory of curved elastic sheets. *Quart. Appl. Math.* 3 (1945) 246–265.
6. A.J. Callegari and E.L. Reiss, Nonlinear boundary value problems for the circular membrane. *Arch. Rat. Mech. Anal.* 31 (1968) 390–400.
7. P. Ciarlet, *Plates and Functions in Elastic Multi-Structures, an Asymptotic Analysis.* Masson, Paris and Springer-Verlag, Berlin (1990).
8. R.A. Clark and O.S. Narayanaswamy, *Nonlinear membrane problems for elastic shells of revolution.* In: D. Muster (ed.), *Proc. Sympos. Theory of Shells* (L.H. Donnell Anniversary Volume). Univ. of Houston Press, Houston (1967) pp 80–110.
9. L. Collatz, *Funktionalanalysis und Numerische Mathematik.* Springer, Berlin–Göttingen–Heidelberg (1964).
10. R.W. Dickey, The plane circular elastic surface under normal pressure. *Arch. Rat. Mech. Anal.* 26 (1967) 219–236.

11. R.W. Dickey, Rotationally symmetric solutions for shallow membrane caps. *Quart. Appl. Math.* 47 (1989) 571–581.
12. L.H. Donnell, A new theory for the buckling of thin cylinders under axial compression and bending. *Trans. ASME* 56 (1934) 795–806.
13. A. Föppl, *Vorlesungen über Technische Mechanik*, Vol. 5. R. Oldenbourg, München (1907).
14. H. Grabmüller, Monotone methods for scalar second order boundary value problems with decreasing discontinuous nonlinearities and applications to annular membrane problems, Inst. Angew. Math., Univ. Erlangen-Nürnberg, Report 129, April 1988.
15. H. Grabmüller, Wrinkle-free solutions in the theory of annular elastic membranes. *J. Appl. Math. and Phys.* (ZAMP) 42 (1991) 783–805.
16. H. Grabmüller, Positive solutions and monotone methods for second order differential equations with discontinuous nonlinearity and one-sided nonlinear boundary condition, Inst. Angew. Math., Univ. Erlangen-Nürnberg, Report 130, April 1988.
17. H. Grabmüller and E. Novak, Nonlinear boundary value problems for the annular membrane: a note on uniqueness of positive solutions. *J. Elasticity* 17 (1987) 279–284.
18. H. Grabmüller and E. Novak, Nonlinear boundary value problems for the annular membrane: new results on existence of positive solutions. *Math. Meth. Appl. Sci.* 10 (1988) 37–49.
19. H. Grabmüller and R. Pirner, Positive solutions of annular elastic membrane problems with finite rotations. *Studies in Appl. Math.* 77 (1987) 223–252.
20. H. Grabmüller and R. Pirner, Existence theorems for some boundary value problems in the nonlinear theory of annular elastic membranes, *European J. of Appl. Math.* (in press).
21. H. Grabmüller and H.J. Weinitschke, Finite displacements of annular elastic membranes. *J. Elasticity* 16 (1986) 135–147.
22. H. Grabmüller, Monotone methods for second order differential equations with decreasing nonlinearity and nonlinear boundary conditions, SFB 123 Stochastische Mathematische Modelle, Universität Heidelberg, Preprint Nr. 467, June 1988.
23. H. Hencky, Über den Spannungszustand in kreisrunden Platten. *J. Math. Phys.* 63 (1915) 311–317.
24. W.E. Jahsman, F.A. Field and A.M.C. Holmes, Finite deformations in a prestressed, centrally loaded, circular elastic membrane. In: *Proc. 4th U.S. National Congress of Applied Mechanics*, ASME, New York (1962).
25. R. Kao and N. Perrone, Large deflections of axisymmetric circular membranes. *Int. J. Solids & Structures* 7 (1971) 1601–1612.
26. Th. von Kármán, *Festigkeitsprobleme im Maschinenbau*. Encyklopädie der Mathematischen Wissenschaften, Vol. 4/4 (1910) pp. 311–385.
27. H.B. Keller and E.L. Reiss, Iterative solutions of the non-linear bending of circular plates. *Comm. Pure Appl. Math.* 11 (1958) 273–292.
28. G.R. Kirchhoff, Über das Gleichgewicht und die Bewegung eines unendlich dünnen elastischen Stabes. *J. Reine u. Angew. Math.* 56 (1859) 285–313 (also *Vorlesungen über Math. Physik, Mechanik*, 2nd edn, Leipzig (1877)).
29. W.T. Koiter, The intrinsic equations of shell theory with some application. In: S. Nemat-Nasser (ed.), *Mechanics Today* 5 (E. Reissner Anniversary Volume). Pergamon Press (1980) pp. 139–154.
30. A. Libai and J.G. Simmonds, *The Nonlinear Theory of Elastic Shells: One Spatial Dimension*. Academic Press Inc., Boston (1988).
31. K. Marguerre, Zur Theorie der gekrümmten Platte großer Formänderung. In: *Proc. 5th Intern. Congr. Appl. Mech.* (1938) pp. 93–101.
32. W. Nachbar, Finite deformation of membranes and shells under localized loading. In: F.I. Niordson (ed.), *Theory of Thin Shells*. IUTAM Symposium, Copenhagen 1967. Springer-Verlag, Berlin/Heidelberg/New York (1969) pp. 176–211.
33. E. Novak, On the convergence of interpolated iteration methods. *SIAM J. Math. Anal.* 19 (1988) 1174–1182.
34. A. Ostrowski, *Solution of Equations in Euclidean and Banach Spaces*. Academic Press, New York/London (1973).
35. R. Pirner, Randwertprobleme der nichtlinearen Membrantheorie: Über Existenz und Nichtexistenz positiver Lösungen. Diploma-Thesis, Faculty of Science I, Univ. Erlangen-Nürnberg, April 1987, 82 pp.
36. M.H. Protter and H.F. Weinberger, *Maximum Principles in Differential Equations*. Prentice-Hall, Inc., Englewood Cliffs, N.J. (1967).
37. E. Reissner, On axisymmetrical deformations of thin shells of revolution. In: *Proc. Symp. Appl. Math.*, Vol. III, McGraw-Hill (1950) pp. 27–52.
38. E. Reissner, On the equations for finite symmetrical deflections of this shells of revolution. In: D.C. Drucker (ed.), *Progress in Mechanics* (Prager Anniversary Volume). Macmillan, New York (1963) pp. 171–178.
39. E. Schwerin, Über Spannungen und Formänderungen kreisringförmiger Membranen. *Z. Tech. Phys.* 12 (1929) 651–659.
40. J.G. Simmonds, Rigorous expunction of Poisson's ratio from the Reissner-Meissner equations. *Int. J. Solids & Structures* 11 (1975) 1051–1056.

41. D. Steigmann, Proof of a conjecture in elastic membrane theory. *J. Appl. Mech.* 53 (1988) 955.
42. M. Stein and J.M. Hedgepeth, Analysis of partly wrinkled membranes, NASA TN D-813, July, 1961.
43. C.A. Stuart, Integral equations with decreasing nonlinearities and applications. *J. Differential Equations* 18 (1975) 202–217.
44. H.J. Weinitschke, Some mathematical problems in the nonlinear theory of elastic membranes, plates and shells. In: G. Fichera (ed.), *Trends in Applications of Pure Mathematics to Mechanics*. Pitman, London (1976) pp. 409–429.
45. H.J. Weinitschke, Stable and unstable axisymmetric solutions for membranes of revolution. *Appl. Mech. Rev.* 42, No. 11, Part 2 (1989) S289–S294.
46. H.J. Weinitschke, *Zur mathematischen Theorie der endlichen Verbiegung elastischer Platten*. Habilitationsschrift, Universität Hamburg (1965).
47. H.J. Weinitschke, Existenz- und Eindeutigkeitssätze für die Gleichungen der kreisförmigen Membran. *Meth. u. Verf. d. Math. Physik* 3 (1970) 117–139.
48. H.J. Weinitschke, On axisymmetric deformations of nonlinear elastic membranes. In: S. Nemat-Nasser (ed.), *Mechanics Today* 5 (E. Reissner Anniversary Volume). Pergamon Press, Oxford (1980) pp. 523–542.
49. H.J. Weinitschke, On finite displacements of circular elastic membranes. *Math. Meth. Appl. Sci.* 9 (1987) 76–98.

Journal of Engineering Mathematics **26**: 195–210, 1992.
H.K. Kuiken and S.W. Rienstra (eds), Problems in Applied, Industrial and Engineering Mathematics.

The emission of sound by statistically homogeneous bubble layers

L. van WIJNGAARDEN and J. BUIST*

*University of Twente, P.O Box 217, 7500 AE Enschede, The Netherlands; *present address: Ultracentrifuge Nederland, P.O. Box 158, 7600 AD Almelo, The Netherlands*

Abstract. This paper is concerned with the flow of a bubbly fluid along a wavy wall, which is one Fourier component of a linearized hydrofoil. The bubbles are dispersed, not throughout the whole of the liquid, but only over a certain distance from the wall, as occurs in practice with cavitation bubbles. Outside the bubbly regime there is pure liquid.

The interface between the bubbly fluid and pure liquid fluctuates for various reasons. One of these is the relative motion between bubbles and liquid. This is considered here in detail. A calculation is made of the sound emitted by the bubbly layer into pure liquid as a result of this stochastic motion of the interface.

1. Introduction

Cavitating ship propellers emit sound. This may reach in the region around 1 KHz an intensity of 150 dB in the far field spectrum. The dangers associated with such a strong sound emission are obvious for navy vessels, but also with merchant ships the strong sound emission causes great discomfort inboard. The spectral distribution of the emitted sound looks like that sketched in Fig. 1, taken from the thesis by Buist [1]. Such distributions can be measured on model scale. It is, at the present time, not possible to predict from such measurements what the full scale emission is, in magnitude and in spectral distribution. A study on cavitation sound prediction is being made by us in cooperation with the Maritime Research Institute in the Netherlands (MARIN).

It has been known for a long time that the sound stems from the collapse of cavitation bubbles. Recently, however, evidence was gained that only at high frequencies does the sound come from individual bubbles, while at lower frequencies interaction between bubbles is important.

The work by Omta [2] and Buist [1] showed in particular, that a broad plateau at 1 KHz is generated by clouds of collapsing bubbles. At MARIN experiments were carried out in this connection. A hydrofoil (for a detailed description see Buist [1]) was positioned in a cavitation tunnel. Small gas bubbles were injected in the flow adjacent to the hydrofoil. In this way a bubbly flow passed closely near the foil. Under influence of the pressure distribution associated with the flow along the hydrofoil, the bubbles grew and collapsed.[†] Pressures in the water flow, resulting from this interaction between the bubbly layer and the outer (pure liquid) flow, were measured. Bubbles could, at the injection site, be produced intermittently or continually. In both cases the measured sound intensity, as derived from the pressure fluctuations, agreed well with the theory in Omta [2] and Buist [1]. In these theories the hydrodynamics of collapsing clouds of bubbles is considered. For the phenomena inside a cloud, equations are used which were derived in Biesheuvel & van Wijngaarden

[†]In practice bubbles contain both gas and vapour. The difference in behaviour with gas filled bubbles is quantitative rather than qualitative.

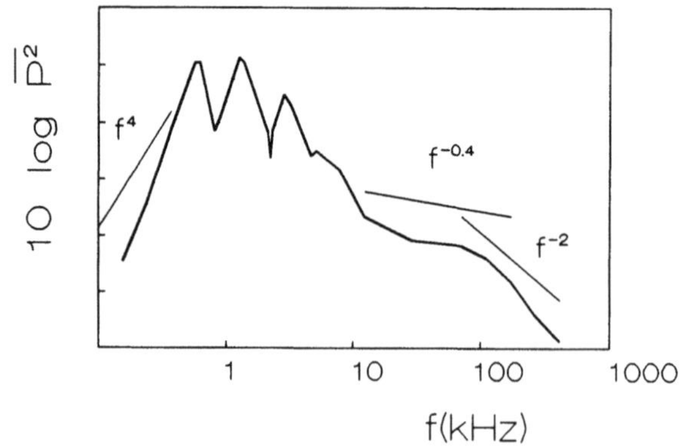

Fig. 1. Schematical representation of the spectral distribution of the sound emitted by a cavitating ship propeller. From Buist [1].

[3]. These equations are established by averaging on the mesoscale. The latter is a scale in between the microscale (in the case of a bubble cloud represented by the inter bubble distance) and the macroscale (for example the chord length of the hydrofoil). Such average equations predict a steady flow, both in the bubble layer and in the outer flow, when external parameters as incidence, rate of injection of bubbles and incoming flow velocity, are kept constant.

Such a flow, by its steady nature, does not emit sound. Nevertheless, sound emission is observed. This has two causes. Firstly, bubbles cluster together and form clouds, which is an unsteady process. Secondly, fluctuations around the average quantities contribute to sound emission. These fluctuations are suppressed by the averaging process. It is important to estimate the contribution by the fluctuations, because, if they are important, the use of average equations would be dubious or would, at the least, not provide a sufficiently complete description of the generated sound intensity. In the present paper our attention is focused on the sound emission by fluctuations. In a bubbly flow various fluctuations may occur. Buist [1], see also Buist [4], considers fluctuations of the local void fraction caused by fluctuations of the pressure inside the bubbles. He concludes that the sound intensity produced by these can be estimated to be 90 dB, which is considerably smaller than the observed sound intensity of cavitating flows which is as mentioned earlier, 150 dB.

Another type of fluctuations mentioned by Buist [1], but not taken further into consideration, is caused by relative motion between bubbles and liquid. Relative motion in a bubbly flow occurs whenever there are accelerative forces. In vertical pipe flow, for example, this is buoyancy. In the case of the flow of a bubbly fluid along a hydrofoil it is the pressure gradient associated with the curvature of the foil. This works on both the gas and on the liquid, but because of the different inertia, both phases obtain different velocities under the same pressure gradient. The relative motion causes fluctuating velocities in the bubbly part of the fluid and therefore also of the interface between this and pure liquid outside the bubbly layer. The fluctuations of the interface, in turn, lead to sound emission in the pure liquid. Our aim is to calculate the intensity of this. We don't take a complete hydrofoil but consider instead an infinitely long wavy wall of wavelength λ only. The wavy wall is a well known concept in aerodynamics because, within the context of linear theory, the flow along a thin hydrofoil can be obtained from the analysis of one wavelength by Fourier synthesis.

The flow configuration is sketched in Fig. 2: along a wavy wall the height of which is in a two-dimensional x, y frame given as the real part of

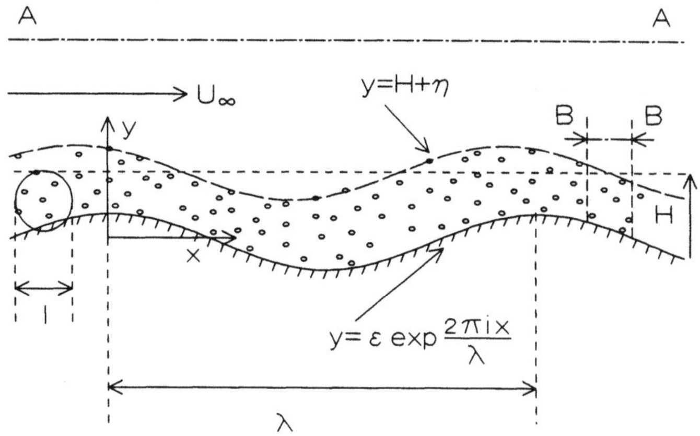

Fig. 2. Flow of a bubble/liquid mixture along a wavy wall. Between the wall and the height $y = H + \eta$ is a mixture of air bubbles and water. For $y \geq H + \eta$, there is pure water.

$$y = \varepsilon \exp\left(\frac{2\pi \mathrm{i} x}{\lambda}\right), \tag{1.1}$$

is a flow in part consisting of a bubbly liquid, and pure liquid in the remainder. The average interface is at $y = H$, the local interface at $y = H + \eta(x, t)$. The undisturbed velocity is U_∞ in both parts, independent of x, y and the time t. In the bubbly flow the macroscale is the wavelength λ, the microscale is the mean distance between bubbles. When the number density of the bubbles is indicated n, this is proportional to $n^{-1/3}$. In order that averages be meaningful, we must require

$$n^{-1/3} \ll \lambda \,. \tag{1.2}$$

The averaging is done over a region with linear dimension l, say, small with respect to λ but large with respect to $n^{-1/3}$, as symbolically indicated in Fig. 2. The part of the displacement η which is due to the mean flow (on the scale l) and which we denote $\eta_1(x)$ is steady. The part which is due to the fluctuations is denoted $\eta_2(x, t)$ and interests us here.

Flow of a bubbly liquid along a wavy wall occupying the upper half plane $y \geq 0$, has been analysed by Agostino et al. [5] some years ago. They considered flows in which relative motion between phases can be neglected and calculated for such flows distributions of void fraction, velocity and pressure. Buist [1] extended the work by Agostino et al. [5], to the situation of Fig. 2. In the next section we shall use Buist's analysis to determine the relative motion between phases in situation of Fig. 2. Then, in Section 3, we shall investigate the influence of relative motion on the fluctuations of the interface, $\eta_2(x, t)$. Finally, in Section 4, we shall relate the sound energy emitted through a plane A–A, in Fig. 2, to the fluctuations $\eta_2(x, t)$ of the interface.

2. Mean quantities including relative motion, in bubbly layer

We start with the introduction of some parameters, characterising the bubbly mixture contained, in Fig. 2, between the wavy wall and pure liquid. Let the undisturbed radius of each bubble be a_0 and $a(x, y)$ in the general case. Together with the number density n, a defines the void fraction α

$$\alpha = \frac{4}{3}\pi n a^3 . \tag{2.1}$$

Since $\alpha \sim (a/n^{-1/3})^3$, the void fraction is small when the bubble radius is small with respect to the inter bubble distance. This is always the case in cavitating flows and we may therefore in the following regard α as a small quantity, $\alpha \ll 1$. Let the (constant) liquid density be ρ_0 and let p_0 be the pressure in the undisturbed state. It is well known (see e.g. Van Wijngaarden [6]) that a bubbly suspension has a sound velocity c_m which is considerably lower than the sound velocity either in liquid or in gas, and given, for infinite wavelength, by

$$c_m^2 = \frac{\gamma p_0}{\rho_0 \alpha_0} , \tag{2.2}$$

where γ is the ratio of specific heats of the gas in the bubbles. At finite wavelength the pressure inside the bubbles is no longer equal to the pressure in the fluid adjacent to the bubble, which leads to dispersion. In the case of Fig. 2 this depends on the frequency kU_∞, where

$$k = 2\pi/\lambda , \tag{2.3}$$

as compared with the natural frequency

$$\omega_b = a_0^{-1}\{3\gamma p_0/\rho_0\}^{1/2} , \tag{2.4}$$

of the bubble.

For $\omega_b \gg kU_\infty$ we have (2.2) whereas for wavelengths k such that kU_∞ is comparable with ω_b, the modified sound speed c_m holds, given by

$$\tilde{c}_m^{-2} = c_m^{-2}\{\omega_b^2/(\omega_b^2 - k^2U_\infty^2)\} . \tag{2.5}$$

In our present application it is highly unlikely that kU_∞ would ever become close to ω_b. With $a_0 \simeq 10^{-3}$ m, $\omega_b \sim 10^4\ \mathrm{s}^{-1}$. Since we are under cavitating circumstances, $\rho_0 U_\infty^2 \sim 10^5 P_a$, which with $\rho_0 = 10^3\ \mathrm{kg/m^3}$ means $U_\infty \sim 10$ m/s. Then $kU_\infty = \omega_b$ for wavelengths of a few millimeter. With hydrofoils such small wavelengths do not occur, perhaps approximately at the very leading edge. Elsewhere λ is of the order of the chord length and in consequence $kU_\infty \ll \omega_b$. We shall therefore in the following neglect kU_∞ with respect to ω_b. Then, a Mach number M follows naturally as

$$M = U_\infty/c_m . \tag{2.6}$$

In the present application M will be low. For example, take $\alpha = 5\%$, $p_0 = 10^5 P_a$, ρ_0 as before and $\gamma = 1.4$. With, as above, $U_\infty = 10$ m/s, we obtain $M = 0.1$.

Buist [1] derived expressions for pressure, velocity and void fraction for the configuration of Fig. 2, using in the region occupied by the bubbly suspension, the averaged equations given in Biesheuvel and van Wijngaarden [3]. From the analysis in Buist [1], it follows that the pressure perturbation is the real part of

$$p_m - p_0 = -\frac{\rho_0 \varepsilon k U_\infty^2 \exp(ikx)}{(1-M^2)^{1/2}} \Bigg[\exp\{-(1-M^2)^{1/2}ky\} + \frac{\{(1-\alpha_0)(1-M^2)^{1/2} - (1-2\alpha_0)\}(\exp\{(1-M^2)^{1/2}k(y-H)\} + \exp\{-(1-M^2)^{1/2}k(y+H)\})}{2\cosh\{(1-M^2)^{1/2}kH\} + 2(1-M^2)^{1/2}\sinh\{(1-M^2)^{1/2}kH\}}\Bigg] \tag{2.7}$$

This is a rather formidable expression, which for our purpose can be simplified considerably, since we are interested in low values of both M and α. To see the contributions to $p_m - p_0$ of various effects, let us take first $M = 0$ and $H \to \infty$. Then

$$p_m - p_0 = -\rho_0 \varepsilon k U_\infty^2 \exp(ikx)\exp(-ky)\,,$$

which satisfies Laplace's equation and is $-\rho_0 U_\infty$ times the velocity associated with potential flow along the linearized wavy wall. Next take $H \to \infty$ and $M \neq 0$. Then the second term in the square brackets on the right-hand side of (2.7) vanishes and the remainder is the pressure distribution of the wavy wall flow of a compressible flow with Mach number M, to be found in many textbooks on gas dynamics, e.g. Liepmann & Roshko [7]. At finite values of H, the second term in the square brackets in (2.7) appears to be small at small M and α_0. Take for example $M = 0$ and $y = H$. Then

$$p_m - p_0 = -\rho_0 \varepsilon k U_\infty^2 \exp(ikx)\exp(-kH)\left[1 + \frac{1}{2}\,\alpha_0(1 + \exp(-2kH))\right].$$

We see that the contribution of the second term is of order α_0. In other words, the pressure distribution in the bubbly layer deviates at zero Mach number to order α_0 from the expression valid in pure liquid. This remains the case at small Mach numbers when M^2 is of order α_0, as can be readily verified.

We shall therefore base our calculation of the relative velocity between liquid and bubbles on the pressure distribution given in (2.7), without the second term in the square brackets,

$$p_m - p_0 = -\frac{\rho_0 \varepsilon k U_\infty^2}{(1-M^2)^{1/2}} \exp(ikx)\exp\{-(1-M^2)^{1/2}ky\}\,, \qquad 0 \leqslant y \leqslant H\,. \tag{2.8}$$

The velocity perturbations $u - U_\infty$ and v, in x and y direction respectively associated with this pressure perturbation, are (real parts being implied)

$$u - U_\infty = \frac{\varepsilon k U_\infty}{(1-M^2)^{1/2}} \exp(ikx)\exp\{-(1-M^2)^{1/2}ky\}\,, \tag{2.9}$$

$$v = \varepsilon k i U_\infty \exp(ikx)\exp\{-(1-M^2)^{1/2}ky\}\,. \tag{2.10}$$

Relations (2.9) and (2.8) show the well known facts that the x-component of the velocity is in phase with the wall elevation, the pressure in anti-phase.

The pressure gradient, following from (2.8) produces relative motion between bubbles and liquid. Let us first assume that this is dominated by inertia forces rather than by viscous forces. Denoting the velocity in the liquid $\mathbf{U}(u, v)$ and of the bubbles $\mathbf{V}(V_x, V_y)$, we have, D/Dt being a material derivative following the bubbles (Biesheuvel and van Wijngaarden [3]),

$$\frac{D}{Dt} m(\mathbf{V} - \mathbf{U}) = -V\nabla p_m = \rho_0 V \frac{D\mathbf{U}}{Dt} . \tag{2.11}$$

Here m is the virtual mass of a bubble, which has for spherical bubbles the value $1/2\rho_0 V$, V being the volume of one bubble. The above relation gives, for spherical bubbles, the result that the bubbles move at three times the liquid velocity. Incidentally, we note that using a material derivative following the liquid where $\mathbf{U}$ is concerned, which is sometimes advocated in the multiphase literature, would not make any difference because of the linearisation around U_∞.

The relative velocity of the bubbles with respect to the liquid has therefore the components

$$V_x - u = \frac{2\varepsilon k U_\infty}{(1 - M^2)^{1/2}} \exp(ikx) \exp\{-(1 - M^2)^{1/2} ky\} . \tag{2.12}$$

$$V_y - v = 2i\varepsilon k U_\infty \exp(ikx) \exp\{-(1 - M^2)^{1/2} ky\} . \tag{2.13}$$

Of course, frictional forces affect the relative motion. The time which it takes viscous forces to influence the motion, the so-called relaxation time, σ, say, is the product of the virtual mass m and the mobility $(12\pi\mu a)^{-1}$, or for a sphere

$$\sigma = a_0^2/18\nu , \tag{2.14}$$

where μ and ν are the dynamic and kinematic viscosities of the liquid respectively. With $a \simeq 10^{-3}$ m and water as liquid, for which $\nu = 10^{-6}\ \mathrm{m^2/s}$, σ is of the order of $10^{-1} - 10^0$. The characteristic time for changes in the velocity is λ/U_∞. The importance of viscous forces depends on whether the ratio between these times, i.e.

$$\frac{U_\infty \sigma}{\lambda} , \tag{2.15}$$

is small or large. In the first case viscosity is important, in the second case it is not. If we take, like we did earlier, $U_\infty = 10\ \mathrm{m/s}$, and if we take $\lambda = 10^{-1}$, then with $\sigma = 10^{-1}$, this parameter is 10, with $\sigma = 10^0$ it is 10^2. In practice therefore the parameter $U_\infty\sigma/\lambda$ is likely to be large, and in consequence (2.12) and (2.13) are appropriate. Nevertheless, for completeness, we give also the relative velocity in the opposite case in which (2.15) is small. In that case the pressure force represented by the right-hand side of (2.11) is not balanced, like in that equation, by the rate of change of the impulse but rather by the frictional force $12\pi\mu a(\mathbf{V} - \mathbf{U})$. This gives the relative velocities, using (2.8)

$$V_x - u = -\frac{1}{9} \frac{a_0^2}{\nu} \frac{\varepsilon k^2 U_\infty^2}{(1 - M^2)^{1/2}} \sin(kx) \exp\{-(1 - M^2)^{1/2} y\} , \tag{2.16}$$

$$V_y - v = -\frac{1}{9} \frac{a_0^2}{\nu} \varepsilon k^2 U_\infty^2 \cos(kx) \exp\{-(1 - M^2)^{1/2} y\} . \tag{2.17}$$

Note that the ratio of the magnitude of these velocities to those given in (2.12) and (2.13) is just the parameter described in (2.15), σ being given by (2.14).

3. Sound emitted by fluctuations of $\eta_2(x, t)$

We consider a small portion of a wave length, for example B–B in Fig. 2. When enlarged this looks as in Fig. 3. We see the interface η_2 under which pass bubbles with relative velocity indicated by the arrows. Far above the interface in the pure liquid is the plane A–A, the sound emission through which is of interest here. How is the sound produced? The bubbles are randomly distributed through the liquid. Their motions relative to the liquid (the magnitude of which has been calculated in Section 2), induce a velocity in each point P' in the liquid. This also is a randomly varying quantity. In particular we are interested in the velocity at a point P located in the interface. The y component v of this is related to the time derivative of η_2 by

$$\frac{\partial \eta_2}{\partial t} + U_\infty \frac{\partial \eta_2}{\partial x} = v . \tag{3.1}$$

In its turn, the motion of the interface can be related to the intensity of the sound emitted through a plane like A–A in Fig. 3. For this it is convenient to follow Buist [1] and to choose a frame moving with the free stream velocity U_∞. Such a frame has coordinates X, Y given by

$$\begin{aligned} X &= x - U_\infty t , \\ Y &= y . \end{aligned} \tag{3.2}$$

In the moving frame (3.1) reduces to

$$\frac{\partial \eta_2}{\partial t} = v . \tag{3.3}$$

Pictures similar to Fig. 3 are obtained at other locations in the wave of Fig. 2 when the arrows are tilted over an angle corresponding with the local value of $(V_y - v)/(V_x - u)$.

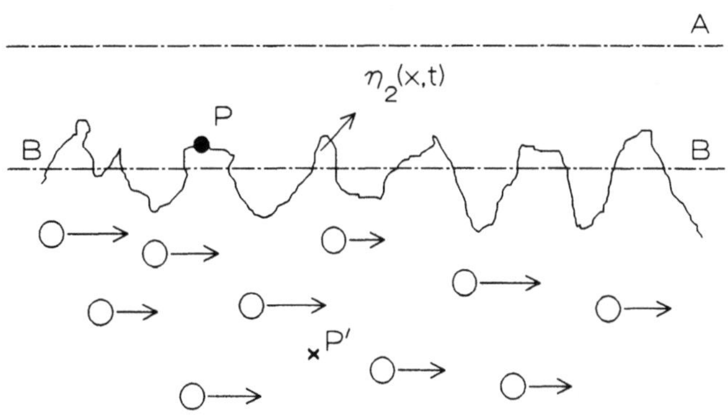

Fig. 3. Fluctuations $\eta_2(x, t)$ of the interface between bubbly fluid and pure water, caused by the motion, relative to the water, of bubbles.

In the region located in Fig. 3 above the bubbly layer, there is motion of pure liquid and when we ignore viscosity this motion is irrotational and has in consequence a velocity potential, φ, say. The y component of its gradient is the vertical velocity at the interface which, by the kinematic surface condition, must be equal to v in (3.3). Hence we have

$$\frac{\partial \eta_2}{\partial t} = \left(\frac{\partial \varphi}{\partial y}\right)_{y=H} . \tag{3.4}$$

The flow potential satisfies the wave equation

$$\frac{\partial^2 \varphi}{\partial X^2} + \frac{\partial^2 \varphi}{\partial y^2} = \frac{1}{c_l^2} \frac{\partial^2 \varphi}{\partial t^2} , \tag{3.5}$$

where c_l is the velocity of sound in liquid, 1500 m/s for the important case in which the liquid is water. The pressure disturbance $p - p_0$ in the liquid is related to φ by,

$$p - p_0 = -\rho_0 \frac{\partial \varphi}{\partial t} . \tag{3.6}$$

The energy flux, in Watt/m^2, through A–A, has the momentary value

$$I = (p - p_0)v , \tag{3.7}$$

and its mean value, indicated with the overbar, is

$$\bar{I} = \overline{(p - p_0)v} . \tag{3.8}$$

The mean can be taken with respect to time, at one location X, or with respect to X, at one time. When the process is statistically stationary these mean values are identical and both equal to the ensemble average, which is the average over all the possible configurations of a large collection of bubbles like in Fig. 3. Eventually we are interested in the spectral distribution of I, which calls for the use of Fourier transforms. Since, however, the Fourier transforms of stochastic variables as η_2 and φ do not exist recourse has to be made to either generalised functions or Fourier–Stieltjes transforms. We follow Buist [1] in employing the latter and introduce the Fourier–Stieltjes coefficient $\mathrm{d}\phi$, associated with φ by†

$$\varphi(X, y, t) = (2\pi)^{-1} \int_\omega \int_\kappa \mathrm{d}\phi(\kappa, \omega, y)\, \mathrm{e}^{\mathrm{i}\omega t}\, \mathrm{e}^{\mathrm{i}\kappa X} . \tag{3.9}$$

Likewise we define the Fourier–Stieltjes coefficient $\mathrm{d}\chi(\kappa, \omega)$ by

$$\eta_2(X, t) = (2\pi)^{-1} \int_\omega \int_\kappa \mathrm{d}\chi(\kappa, \omega)\, \mathrm{e}^{\mathrm{i}\omega t}\, \mathrm{e}^{\mathrm{i}\kappa X} . \tag{3.10}$$

Note that k is the wave number $k = 2\pi/\lambda$ of the wavy wall, whereas κ is used here as transform variable.

†The definition of the Fourier or Fourier–Stieltjes transform is chosen here in which there is a factor $(2\pi)^{-1/2}$ in front of the transform with respect to one variable and consequently also in front of the inverse integral. In that case the spectral relations used here, are in accordance with those in the well known book on stochastic phenomena by Stratonovich [8].

The inverse transformations corresponding with (3.9) and (3.10) are

$$\frac{\mathrm{d}\phi}{\mathrm{d}\omega\,\mathrm{d}\kappa} = (2\pi)^{-1} \iint \varphi(X, y, t)\,\mathrm{e}^{-\mathrm{i}\omega t}\,\mathrm{e}^{\mathrm{i}\kappa X}\,\mathrm{d}t\,\mathrm{d}X\,, \tag{3.11}$$

$$\frac{\mathrm{d}\chi}{\mathrm{d}\omega\,\mathrm{d}\kappa} = (2\pi)^{-1} \iint \eta_2(X, t)\,\mathrm{e}^{-\mathrm{i}\omega t}\,\mathrm{e}^{\mathrm{i}\kappa X}\,\mathrm{d}t\,\mathrm{d}X\,. \tag{3.12}$$

Using (3.11), (3.12) and the boundary condition (3.4), it follows from transformation of the differential equation (3.5) that

$$\mathrm{d}\phi(\omega, \kappa, y) = \left[\omega\,\frac{\exp\{-\mathrm{i}(y - H)(\omega^2/c_l^2 - \kappa^2)^{1/2}\}}{(\omega^2/c_l^2 - \kappa^2)^{1/2}}\right]\mathrm{d}\chi(\omega, \kappa)\,. \tag{3.13}$$

We are interested in the spectral density of the energy flux (3.8). According to the theory of Fourier transforms this is equal to the Fourier transform, or in the present case the Fourier–Stieltjes transform of the cross correlation

$$\iint (p - p_0)(x, t)v(x + r, t + \tau)\,\mathrm{d}x\,\mathrm{d}t\,, \tag{3.14}$$

where $(p - p_0)$ and v are taken at the level y_A of the plane A–A. In (3.14) the spatial shift r and the time shift τ are positive.[†] Taking the Fourier–Stieltjes transform of (3.14) with respect to r and τ and dividing by $(2\pi)^{-2}$ gives

$$(2\pi)^{-2} \iint (p - p_0)\exp(\mathrm{i}\omega t)\exp(\mathrm{i}\kappa x)\,\mathrm{d}x\,\mathrm{d}t \iint v(x + r, t + \tau)$$
$$\times \exp(-\mathrm{i}\omega(t + \tau))\exp\{-\mathrm{i}\kappa(x + r)\}\,\mathrm{d}r\,\mathrm{d}\tau\,. \tag{3.15}$$

Using (3.3), (3.4), (3.6), (3.11) and (3.13), this can be written as

$$\{-\rho_0\mathrm{i}\omega'\,\mathrm{d}\phi^*(\omega'\kappa')\mathrm{i}\{\omega^2/c_l^2 - \kappa^2\}^{1/2}\,\mathrm{d}\phi(\omega, \kappa)\}\,/\mathrm{d}\omega\,\mathrm{d}\kappa\,\mathrm{d}\omega'\,\mathrm{d}\kappa'\,. \tag{3.16}$$

Here, * denotes a complex conjugate. The product $\mathrm{d}\phi^*(\omega', \kappa')\,\mathrm{d}\phi(\omega, \kappa)$ is for stationary random processes zero unless $\omega' = \omega$ and $\kappa' = \kappa$, in which case

$$\frac{\mathrm{d}\Phi^*\,\mathrm{d}\Phi}{\mathrm{d}\omega\,\mathrm{d}\omega'\,\mathrm{d}\kappa\,\mathrm{d}\kappa'} = \Pi_\varphi\,. \tag{3.17}$$

Π is called the spectral density and the above derivation demonstrates that the spectral density of the emitted sound is the Fourier–Stieltjes transform of the cross correlation (3.14) and, using (3.15) and (3.16) given by

$$\Psi(\kappa, \omega) = \rho_0\omega(\omega^2/c_l^2 - \kappa^2)^{1/2}\Pi_\varphi\,.$$

Alternative forms are, when we use (3.13)

[†]Results for negative r and τ can be obtained by making use of the symmetry properties of the correlation functions. Further x or X may be used, because the difference is an unimportant shift in the spectra.

$$\Psi(\kappa, \omega) = \frac{\rho_0 \omega^3}{(\omega^2/c_l^2 - \kappa^2)^{1/2}} \Pi_\eta , \tag{3.18}$$

where

$$d\chi^* \, d\chi = \Pi_\eta \, d\omega \, d\kappa , \tag{3.19}$$

and

$$\Psi(\kappa, \omega) = \frac{\rho_0 \omega}{(\omega^2/c_l^2 - \kappa^2)^{1/2}} \Pi_v , \tag{3.20}$$

where (3.3) and (3.4) are used.

Relation (3.18) is used by Buist [1], whereas we shall in the following section employ the relation (3.20).

4. Estimate for Π_v

Referring to Fig. 4, we consider the correlation

$$v(x_1, t) v(x_1 + r, t + \tau) \tag{4.1}$$

between the y-component of the velocity induced in x_1 at time t with that induced in $x_1 + r$ at time $t + \tau$. The mean value of this over all possible x_1 and t is, with a stationary stochastic process, a function of the shift r and the time shift τ,

$$C(r, \tau) = \overline{v(x_1, t) v(x_1 + r, t + \tau)} , \tag{4.2}$$

where the mean is indicated with an overbar.

This can be taken as an ensemble average where the ensemble consists of all the possible configurations at time t and at time $t + \tau$. For each configuration bubbles which are close to the point of observation (x_1 in Fig. 4[a] and $x_1 + r$ in Fig. 4[b]) have more influence that those at a larger distance. Those at a distance larger than a multiple of the bubble radius a hardly contribute. Since the probability of finding a bubble centre in a sphere around x_1 (or $x_1 + r$ in Fig. 4[b]) of such a radius, is just the concentration α we can, in an accuracy to order α, consider each configuration to consist of only one bubble. This is located in Fig. 4[a] at a distance R_1 from x_1 and in Fig. 4[b] at a distance R_2 from $x_1 + r$. If the Fourier–Stieltjes transform of v is indicated with dB, we have, in analogy with (3.14)–(3.17)

$$\frac{dB^* \, dB}{d\omega \, d\kappa} = \Pi_v = (2\pi)^{-1} \iint C(r, \tau) \, e^{-i\omega\tau - i\kappa r} \, dr \, d\tau , \tag{4.3}$$

$C(r, \tau)$ being defined by (4.2).

We start with calculating the velocity induced in y direction in Fig. 4[a]. The sound problem formulated in Section 1 and Section 2 is two-dimensional, whereas here we have a three-dimensional problem. We employ spherical coordinates R, θ and $\tilde{\theta}$, R and θ being indicated in Fig. 4[a]. The angular coordinate $\tilde{\theta}$ is needed when we have to determine the y

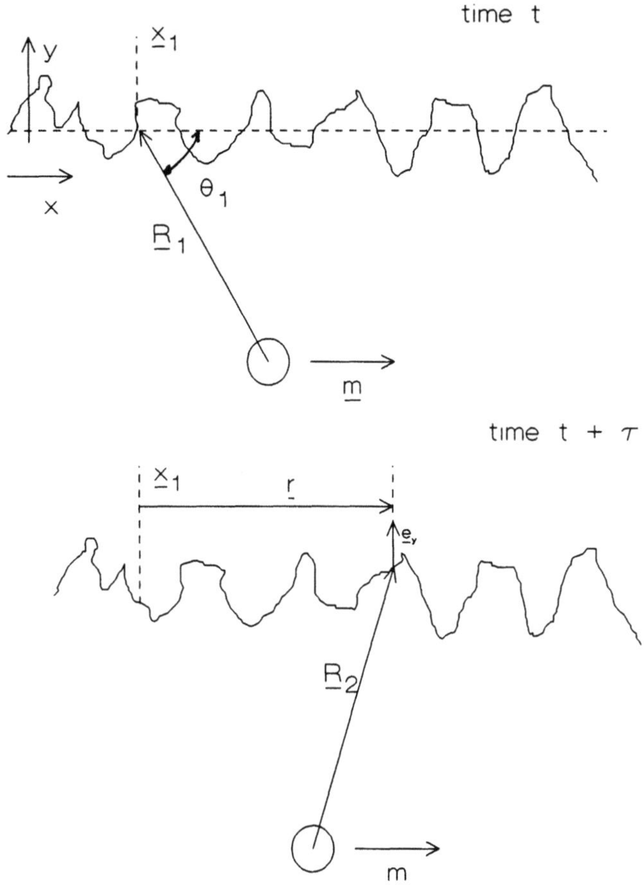

Fig. 4. (a) At time t, motion in the point $\mathbf{x}_1$ of the interface is induced by a bubble at distance $\mathbf{R}_1$ from $\mathbf{x}_1$, with dipole strength $\mathbf{m}$. (b) At time $t+\tau$ the influence is considered in a point $\mathbf{x}_1+\mathbf{r}$ of the interface separated from the bubble by a distance $\mathbf{R}_2$. The dipole strength of the bubble is $\mathbf{m}$.

component of the velocity induced in x_1. With constant bubble volume the motion of the bubble is represented with a dipole of strength **m** proportional with and in the direction of the momentary relative velocity of the bubble, **q**, say,

$$\mathbf{m} = \mathbf{q}a^3/2 \,. \tag{4.4}$$

The flow potential Ω is

$$\Omega = \frac{\mathbf{m}\cdot\mathbf{R}}{R^3} \,. \tag{4.5}$$

The velocity induced in any point at distance **R** from the centre is

$$\nabla\Omega = \frac{\mathbf{m}}{R^3} - 3\,\frac{\mathbf{m}\cdot\mathbf{R}}{R^5}\,\mathbf{R} \,. \tag{4.6}$$

The component of this in y-direction is, $\mathbf{e}_y$ being the unit vector normal to the interface

$$v(x_1, t) = \frac{\mathbf{m}\cdot\mathbf{e}_y}{R^3} - 3\,\frac{\mathbf{m}\cdot\mathbf{R}}{R^5}\,(\mathbf{R}\cdot\mathbf{e}_y) \,. \tag{4.7}$$

The dipole strength is equal to the instantaneous relative velocity of the bubble under consideration. In Section 2 we have calculated the average, over the mesoscale l. The true relative velocity fluctuates around this value. For present purposes it is sufficient to take a dipole strength $\mathbf{m}$ based on the mean velocity, given by (2.9) and (2.10).

The relative velocity is parallel to the surface, so the first term on the right-hand side of (4.7) has no component in y-direction. With the coordinates indicated in Fig. 4[a], the second term in the right-hand side of (4.7) becomes

$$v(x_1, R_1) = \frac{3mR_1^2 \cos\theta_1 \sin\theta_1 \sin\tilde{\theta}_1}{R_1^5} . \tag{4.8}$$

A similar expression can be written down for the situation sketched in Fig. 4[b] with $R_2, \theta_2, \tilde{\theta}_2$. Calculating the ensemble average amounts to multiplying the latter expression $v(x_1 + r, R_2)$ with $v(x_1, R_1)$, in (4.8), with the volume element in the half space $y \leqslant 0$ in Fig. 4[a] and 4[b] and finally with the probability of finding a bubble centre in $\mathbf{R}_1$ at time t and one in $\mathbf{R}_2$ at time $t + \tau$. To calculate the ensemble average exactly, one should know more about the fluctuations of the velocity of each bubble, which result from interactions. When these fluctuations are completely random, the configurations at time t and time $t + \tau$ are completely uncorrelated and since the average of both $v(x_1, R_1)$ and $v(x_1 + r, R_2)$ are zero, the contribution to the mean given in (4.2) from times τ, other than zero, will be small. We therefore confine ourselves to the correlation for $\tau = 0$ and consider

$$C(r) = \overline{v(x_1, t)v(x_1 + r, t)} . \tag{4.9}$$

We have seen in (4.3) how the wave number-frequency power spectrum Π_v is related to $C(r, \tau)$. The inverse of (4.3) is

$$C(r, \tau) = (2\pi)^{-1} \iint \Pi_v(\omega, \kappa)\, e^{i\omega\tau}\, e^{i\kappa r}\, d\omega\, d\kappa .$$

Similarly, putting $\tau = 0$

$$C(r) = (2\pi)^{-1} \int_\kappa \Pi_v^{|}(\kappa)\, e^{i\kappa r}\, d\kappa , \tag{4.10}$$

where

$$\Pi_v^{|}(\kappa) = \int_\omega \Pi_v(\omega, \kappa)\, d\omega . \tag{4.11}$$

We see that while in (3.20) $\Pi_v(\kappa, \omega)$ occurs at the right-hand side, the wave number spectrum associated with $C(r)$ gives us only the integral of Π_v over all frequencies. However if the main contribution to $C(r, \tau)$ comes from $\tau = 0$ as is made plausible above we put, approximately

$$C(r, \tau) = C(r)\delta(\tau)$$

and then, from (4.3)

$$\Pi_v = (2\pi)^{-1} \iint C(r)\delta(\tau)\, e^{-i\omega\tau}\, e^{i\kappa r}\, dr\, d\tau$$

$$= (2\pi)^{-1} \int C(r)\, e^{-i\kappa r}\, dr\,, \tag{4.12}$$

which is the inverse of (4.10).

In other words, the (reasonable) assumption that the mean contribution to $C(r, \tau)$ comes from the autocorrelation $C(r, 0)$ enables us to determine Π_v, which we need in order to evaluate $\Psi(\kappa, \omega)$ in (3.20).

If we take $\tau = 0$, we have to consider the situation in Fig. 5. Taking $\theta_1 = \theta_2 = \theta$ etc., we have to formulate the ensemble average (4.9). The probability of finding a bubble centre in the element of volume $R^2 \sin\theta\, dR\, d\theta\, d\tilde{\theta}$ is

$$nR^2 \sin\theta\, dR\, d\theta\, d\tilde{\theta}\,,$$

where n is the number density. Using (4.8) for $v(x_1, R)$ and a similar expression for $v(x_1 + r, R)$ we have

$$C(r) = -\frac{1}{2}\pi n \int_0^{\pi} d\theta \int_a^{\infty} \frac{9m^2(R\cos\theta - r)\cos\theta \sin^3\theta\, dR\, d\theta}{(R^2 + r^2 - 2Rr\cos\theta)^{5/2}}\,. \tag{4.13}$$

The factor $\pi/2$ in front of the integrals in (4.13) is from integration of $\sin^2\tilde{\theta}$ (resulting from $\sin\tilde{\theta}$ in the expression for $v(x_1, R)$ and another one in the expression for $v(x_1 + r, R)$) over the interval from 0 to π. Since the bubbles have a non zero radius, a, and the point x_1 must be in liquid, R runs from a to ∞. The spatial shift r is positive. Carrying out the integration over R and writing

$$s = \cos\theta \tag{4.14}$$

gives

$$C(r) = 3/2\pi m^2 n \Bigg[\int_{-1}^{1} \left\{2 - \frac{2(a - sr)}{(r^2 + a^2 - 2sar)^{1/2}} - \frac{(a - sr)(1 - s^2)r^2}{(r^2 + a^2 - 2sar)^{3/2}}\right\} \frac{s\, ds}{r^3}$$
$$- \int_{-1}^{1} \frac{s^2(1 - s^2)\, ds}{(r^2 + a^2 - 2sar)^{3/2}}\Bigg]\,. \tag{4.15}$$

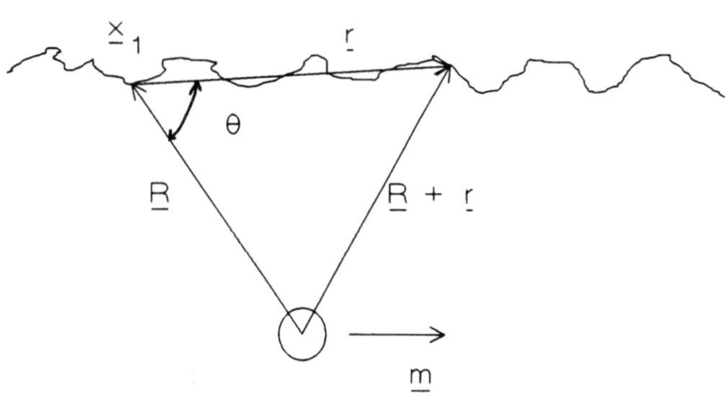

Fig. 5. Scheme for the calculation of the cross correlation between the velocity perturbation in $\mathbf{x}_1$ and $\mathbf{x}_1 + \mathbf{r}$ caused by a bubble moving relative to the fluid with dipole strength $\mathbf{m}$.

Evaluation of the integrals in (4.15), for details see the Appendix, results in

$$C(r) = 3\pi m^2 n\left(\frac{2}{3}\frac{1}{r^3} - \frac{4}{5}\frac{a^2}{r^5}\right). \tag{4.16}$$

If there is a bubble centre in $x_1 + a$, the point nearest to x_1 with which correlation can be made, is $x_1 + 2a$. Therefore r runs from $2a$ to ∞. With use of (2.1) and (4.4) we write

$$C(r) = 3/8\alpha q^2 a^3\left(\frac{1}{r^3} - \frac{6}{5}\frac{a^2}{r^5}\right), \quad r > 2a . \tag{4.17}$$

The correlation function $C(r)$ tends to zero for $r \to \infty$, as it should, but rather slowly, like r^{-3}. It is interesting to note that the decay of $C(r)$ with r is algebraic rather than the exponential behaviour which is often encountered in turbulence. The wavenumber spectrum is now obtained by taking the Fourier transform of (4.17) with respect to r,

$$\Pi_v'(\kappa) = 2(2\pi)^{-1}\int_{2a}^{\infty} C(r)\cos(\kappa r)\,\mathrm{d}r . \tag{4.18}$$

Introducing (4.17) into (4.18) and evaluating the integrals gives

$$\Pi_v'(\kappa) = \frac{3/4\alpha a^3 q^2}{2\pi}\{0.11a^{-2} + O(a^2\kappa^2)a^{-2}\} . \tag{4.19}$$

The value of Π_v' can be determined to any accuracy in (κa). We need only the first term between brackets of (4.19), for the following reason. As follows from (3.20), for given ω, κ runs from 0 to ω/c_l. Hence the maximum value of κa is $\omega a/c_l$. With $\omega \sim 10^3$, $a \sim 10^{-3}$ and $c_l = 1500$ m/s this is small enough to neglect terms in $(\kappa a)^2$.

5. Estimate for radiated sound intensity produced by relative motion

Rather than looking at the wavenumber–frequency spectral density, we consider the total emitted sound intensity $\int \psi\,\mathrm{d}\omega\,\mathrm{d}\kappa$. We have found in the previous section that to the second order in (κa), Π_v' does not depend on κ.

Hence, carrying out the integration over κ, we have from (3.19)

$$\int \psi(\omega, \kappa)\,\mathrm{d}\omega\,\mathrm{d}\kappa = \bar{\omega}\rho_0\int_0^{\omega/c_l}\frac{\Pi_v'(\kappa)\,\mathrm{d}\kappa}{(\omega^2/c_l^2 - \kappa^2)^{1/2}} .$$

Here $\bar{\omega}$ is some representative frequency, for example at the centre of gravity of the spectral density distribution. Taking now for $\Pi_v'(\kappa)$ the zeroth order term on the right-hand side of (4.19) we obtain

$$\frac{1}{\bar{\omega}}\int \psi(\omega, \kappa)\,\mathrm{d}\omega\,\mathrm{d}\kappa = (2.1 \times 10^{-2})\rho_0\alpha a q^2 , \tag{4.20}$$

as an estimate of the sound intensity emitted in a unit element along the frequency axis. The magnitude of this, of course, depends on many parameters. As an example, take $\rho_0 = 10^3$ kg/m^3, $\alpha = 10^{-2}$, $a = 10^{-3}$ m. The value of q is of the order $2\varepsilon k U_\infty$, when we base ourselves on the case where inertia forces dominate over viscous forces in the relative motion. The value of U_∞ will be close to 10 m/s under cavitating circumstances. For $\varepsilon k = 2\pi\varepsilon/\lambda$ we take $2\pi \times 0.1 = 0.2\pi$. With these values the right-hand side of (4.20) is 83×10^{-4}. With $\varepsilon k = 2\pi \times 10^{-2}$, and the other values unchanged, this is 0.83×10^{-4}.

If we express this in dB, defined as

$$10 \log_{10}(\text{intensity}) + 120 \,,$$

we obtain 79 dB in the second case and 99 dB in the first case. The intensity is here in watt sec per unit length in x direction. This can be compared with the estimate of 90 dB given in Buist [1] due to fluctuations in void fraction due to pressure fluctuations. It turns out that relative motion gives contributions of equal magnitude.

Conclusion

In this paper it has been investigated what the sound emission is from a bubble-liquid layer into pure water, when attention is focused on relative motion between bubbles and liquid. It turns out that the intensity of the emitted sound is comparable to that produced by void fraction fluctuations of other nature such as investigated in the thesis of Buist [1]. Finally it should be noted that the present investigation is of interest not only in the context of cavitation sound in which it is considered here. It can be applied also in studies of flow noise whenever relative motion is important. A possible further extension could be the relative motion due to turbulence. Such a study would be complementary to Crighton & Ffowcs Williams [9], where the sound emission of a turbulent bubbly fluid is described caused by monopole radiation.

Appendix

The first and the third integral in the square brackets in (4.13) give together

$$\frac{a}{r} \int_{-1}^{1} \frac{(1 - s^2)s \, ds}{\{r^2 + a^2 - 2sar\}^{3/2}} \,,$$

whence

$$C(r) = -\frac{3}{2} \pi m^2 n \left[\frac{a}{r} \int_{-1}^{1} \frac{(1 - s^2)s \, ds}{\{r^2 + a^2 - 2sar\}^{3/2}} - 2 \int_{-1}^{1} \frac{s \, ds}{r^3} + \frac{2}{r^3} \int_{-1}^{1} \frac{(a - sr)s \, ds}{\{r^2 + a^2 - 2sar\}^{1/2}} \right].$$

Now define

$$I_{ij} = \int_{-1}^{1} \frac{s^i \, ds}{\{r^2 + a^2 - 2sar\}^{j/2}} \,,$$

then

$$C(r) = -\frac{3}{2} \pi m^2 n \left[\frac{a}{r} (I_{13} - I_{33}) + \frac{2a}{r^3} I_{11} - \frac{2}{r^2} J_{21} \right].$$

Evaluation of the integrals gives

$$I_{13} = \frac{2a}{r^2(r^2 - a^2)} \; ; \qquad I_{33} = \frac{2}{a(r^2 - a^2)} - \frac{2}{ar^2} - \frac{4a}{5r^4} \; ;$$

$$I_{11} = \frac{2a}{3r^2} \; ; \qquad I_{21} = \frac{2}{3r} + \frac{4a^2}{15r^3} \; .$$

Introducing these values in the expression for $C(r)$ results in (4.14) in the main text.

References

1. J. Buist, On the origin and acoustical behaviour of cloud cavitation. Thesis, University of Twente (1991).
2. R. Omta, Oscillations of a cloud of bubbles of small and not so small amplitude. *J. Acoust. Soc. Am.* 82 (1987) 1018–1033.
3. J. Buist, Acoustics of travelling bubbles cavitation. In: G.E.A. Meier and P.A. Thompson (eds), *Proc. IUTAM Symp. on Adiabatic Waves in Liquid-Vapor Systems.* Springer (1990) pp. 351–360.
4. A. Biesheuvel and L. van Wijngaarden, Two-phase flow equations for a dilute dispersion of gas bubbles in liquid. *J. Fluid Mech.* 148 (1984) 304–318.
5. L. d'Agostino and C.E. Brennen, Linearized dynamics of two-dimensional bubbly and cavitating flows over slender surfaces. *J. Fluid Mechanics* 192 (1988) 485–509.
6. L. van Wijngaarden, One-dimensional flows of liquids containing small gas bubbles. *Ann. Rev. Fluid Mech.* 4 (1972) 369–396.
7. H.W. Liepmann and A. Roshko, *An Introduction in Gas Dynamics.* Wiley & Sons (1958).
8. R.L. Stratonovich, *Topics in the Theory of Random Noise*, Vol. I. Gordon & Breach (1963).
9. D.G. Crighton and J.E. Ffowcs Williams, Sound generation by turbulent two-phase flow. *J. Fluid Mech.* 36 (1969) 585–603.

Journal of Engineering Mathematics **26**: 211–230, 1992.
H.K. Kuiken and S.W. Rienstra (eds), Problems in Applied, Industrial and Engineering Mathematics.

Oscillation of a floating body in a viscous fluid

Ronald W. YEUNG and P. ANANTHAKRISHNAN

Department of Naval Architecture and Offshore Engineering, University of California at Berkeley, CA 94720, USA

Abstract. The nonlinear viscous-flow problem associated with the heaving motion of a two-dimensional floating cylinder is considered. It is formulated as an initial-boundary-value problem in primitive variables and solved using a finite-difference method based on boundary-fitted coordinates. A fractional-step procedure is used to advance the solution in time. As a case study, results are obtained for a rectangular cylinder oscillating at a Reynolds number of 10^3. The nonlinear viscous forces are compared with those of linear potential theory. An assessment on the importance of viscous and nonlinear effects is made. The solution technique is sufficiently robust that extensions to consider other single and coupled modes of motion are possible.

1. Introduction

Many investigations have been carried out in the past to understand the fluid-dynamics processes associated with oscillating bodies in a free surface. From a practical viewpoint, such investigations are significant since the results are used for estimating wave loads experienced by offshore structures and for predicting motion response of ocean vehicles. Classical first-order potential-theory results are known to be reasonably accurate for the case of small-amplitude motion in the intermediate range of wave frequencies (see Wehausen [27]). Second-order analyses (e.g. Potash [20], Sclavounos [21]) indicate that nonlinear effects are important in the high-frequency regime.

A more direct approach to the fully nonlinear potential-flow problem is by time domain analysis. Several variants of a mixed Eulerian–Lagrangian boundary-integral method, originated by Longuet–Higgins and Cokelet [18], have been developed (e.g. Vinje and Brevig [25], Baker et al. [2], Cooker et al. [9]) for obtaining nonlinear inviscid solutions to wave-body interaction problems. This mixed formulation can also be implemented using finite-difference methods (see e.g. Telste [24], Yeung and Wu [28]). Telste [24], who considered the heave oscillation of a surface-piercing cylinder, was able to demonstrate the importance of nonlinear effects at high frequency and of large-amplitudes of oscillation.

Despite some known, important effects of viscosity, such as in damping of roll motions, in loads induced by vortex shedding, etc., relatively few cases of viscous-flow analyses have been carried out in this area. Using an extended version of the Marker-and-cell (MAC) method originated by Harlow and Welch [12], Nichols and Hirt [14] computed the hydrodynamic forces on a surface-piercing cylinder heaving with small amplitudes. More elaborate versions of the MAC method have been continually developed. For example, Miyata et al. [19] solved a wave-diffraction problem using a version called TUMMAC. The computed results in [19], however, showed large deviations in horizontal exciting forces from those of experimental results.

The primary difficulty encountered in tackling viscous-flow problems is that the Navier–Stokes equations are nonlinear with the unknown variables pressure and velocity coupled together. The use of vorticity and stream-function formulation would allow one to decouple the pressure and velocity fields. However, the procedure can neither be easily extended to the three-dimensional case, nor be used to accommodate stress-type boundary conditions in any straightforward manner.

Viscous-flow problems with free boundaries have further complications. The problem associated with contriving a proper open-boundary condition is nontrivial (see Yeung [29]). Waves of large amplitude often lead to skewed or extreme geometry, imposing rather stringent demand on domain-discretization techniques. Proper discretization is imperative, otherwise boundary conditions cannot be accurately implemented and detail structures such as vorticity generation cannot be effectively resolved. Finally, no completely satisfactory model has yet been devised to model the movement of the contact line (intersection of the body and free surface).

Recently, the authors have completed the development of a numerical method based on boundary-fitted coordinates for solving fully nonlinear viscous free-surface flow problems (see Ananthakrishnan [1]). The solution of the Navier–Stokes equations in primitive variables is based on the fractional-step method originally formulated by Chorin [7], Temam [23] and others. Briefly, an intermediate velocity field (also referred to as the 'auxiliary field') is first computed by neglecting the pressure terms. The pressure field is obtained by solving a Poisson equation. The correctional effects of pressure are then incorporated into the auxiliary field to determine the divergence-free velocity field. Our fractional-step formulation is implemented using a finite-difference method based on boundary-fitted coordinates. A variational formulation involving the concept of a reference space (see Steinberg and Roache [22], Yeung and Ananthakrishnan [31]) is used to generate grids. Previously, using this grid-generation technique in conjunction with the mixed Lagrangian–Eulerian formulation, we were able to obtain highly nonlinear *inviscid-flow* solutions, such as shallow-water breaking waves [30] and wave diffraction over submerged obstacles [32].

In the present work, we use the above finite-difference procedure to solve the nonlinear viscous-flow problem associated with heaving motion of a floating body. Solutions are obtained for a range of physical parameters, including the possibility of large amplitude of body oscillation.

2. Mathematical formulation

Notations and definitions of the problem being studied are described in Fig. 1. A fixed (inertial) coordinate system with the x-axis along the mean water level and the y-axis

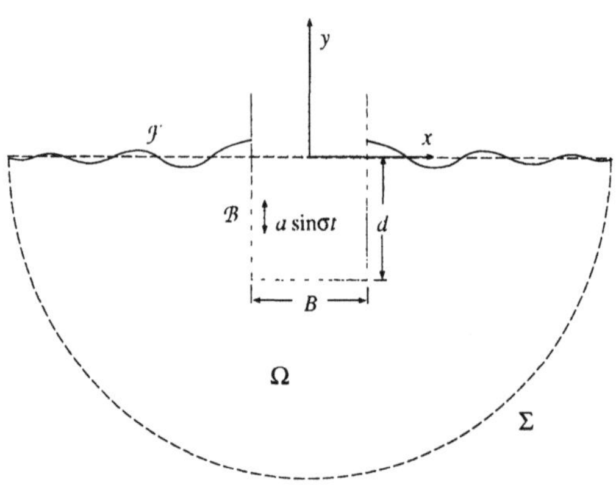

Fig. 1. Cylinder heaving in a viscous fluid, problem definition and notations (in dimensional form).

pointing upwards is chosen. The equilibrium draft of the rectangular body is denoted by d and the beam by B. The amplitude and frequency (in rad/sec) of oscillation are given by a and σ, respectively. The free surface is denoted by $\mathscr{F}$, the body contour by $\mathscr{B}$, and an open boundary by Σ. The kinematic viscosity coefficient of the fluid is denoted by ν, its density by ρ, and the gravitational acceleration by g. The governing equations will be nondimensionalized with respect to B, ρ, and σ. Accordingly, time is scaled by $1/\sigma$, length by B, velocity by σB, pressure by $\rho B^2\sigma^2$, and force by $\rho B^3\sigma^2$. We will henceforth adopt this nondimensionalized notation. Where convenient, subscripts followed by a comma (,) will be used to denote partial differentiation with respect to the subscripted variables.

2.1. *Field equations*

The field equations governing the viscous wave problem are the Navier–Stokes equations:

$$\nabla \cdot \mathbf{u} = 0 \tag{1}$$

and

$$\frac{\partial \mathbf{u}}{\partial t} + (\mathbf{u} \cdot \nabla)\mathbf{u} = -\nabla\left(p + \frac{y}{F_\sigma^2}\right) + \frac{1}{R_\sigma}\nabla^2\mathbf{u}\,, \tag{2}$$

which are derived using the conservation laws of mass and momentum, with the fluid being assumed incompressible, homogeneous, and Newtonian. The unknown variables $\mathbf{u} = \mathbf{u}(\mathbf{x}, t) = (u, v)$ and $p(\mathbf{x}, t)$ are the velocity and pressure fields, respectively. As is commonly known, the second term on the left-hand side is associated with convective effects of the fluid and the last term on the right-hand side with diffusion effects of viscosity. In the above equation, R_σ and F_σ are defined by

$$R_\sigma = \frac{\sigma B^2}{\nu}\,, \tag{3}$$

$$F_\sigma = \sigma\sqrt{\frac{B}{g}}\,, \tag{4}$$

which will henceforth be referred to as the Reynolds number and as the frequency parameter, respectively.

2.2. *Boundary conditions*

On $\mathscr{F}$, in the absence of surface tension, continuity of stress vector components along the normal and tangential directions provides the following dynamic conditions:

$$n_i T_{ij} n_j = 0\,, \tag{5}$$

$$\tau_i T_{ij} n_j = 0\,, \tag{6}$$

where

$$T_{ij} = -p\delta_{ij} + \frac{1}{R_\sigma}\left\{\frac{\partial u_i}{\partial x_j} + \frac{\partial u_j}{\partial x_i}\right\} \tag{7}$$

is the stress tensor in the indicial notation. (n_1, n_2) and (τ_1, τ_2) in equations (5) and (6) denote the components of unit normal- and tangential-vectors, respectively.

The unknown elevation of the free surface can be tracked in time by integrating the following Lagrangian description of $\mathscr{F}$

$$\frac{D}{Dt}\mathbf{x} = \mathbf{u}\,, \tag{8}$$

where D/Dt denotes the material derivative. For non-breaking waves, the following Eulerian description can also be used to advance the free-surface elevation, $y = Y(x, t)$:

$$\frac{\partial Y}{\partial t} = v - u\,\frac{\partial Y}{\partial x}\,. \tag{9}$$

The body is forced to heave (vertically) in the form of

$$y_b(t) = y_b(0) + \frac{a}{B}\sin(t)\,, \quad t \geq 0\,, \tag{10}$$

where $y_b(t)$ represents the y-coordinate value of any point fixed to the rigid body. The prescribed motion can be arbitrary, but sinusoidal motion is chosen here merely for illustration. On the body surface $\mathscr{B}$, the appropriate no-flux and no-slip conditions are imposed:

$$u = 0\,, \tag{11}$$

$$v = \frac{a}{B}\cos(t)\,. \tag{12}$$

Approximate conditions are used at the intersection of the body and the free surface to model its movement. Several experimental and theoretical investigations have been devoted solely to examine the precise mechanisms occurring near the contact line (see e.g., Dussan [10], Huh and Mason [15], Koplik et al. [17]), but the findings are still inconclusive. Hence, as an approximation, we find it plausible to assume that the fluid slips freely at the contact line; in other words,

$$\frac{\partial v}{\partial x} = 0 \quad \text{at } \mathscr{B} \cap \mathscr{F} \tag{13}$$

is assumed for computing the y-component of the velocity. The horizontal component of the velocity at $\mathscr{B} \cap \mathscr{F}$ is given by the no-flux condition (11).

Approximate conditions are also used, in the present work, for the closure of the problem at the open boundary Σ. Accordingly, at sufficiently large distance from the body we assume that

$$p = p_{\text{stat}}\,, \tag{14}$$

where p_{stat} denotes the pressure field of the quiescent fluid (see Grosenbaugh and Yeung [11] for a similar treatment of an inviscid flow problem). An auxiliary velocity at Σ is determined by first-order spatial extrapolation. A decomposition relation, which is a consequence of the fractional-step procedure, is used to determine the unknown velocity at Σ. Specific details of the fractional-step procedure are given later.

The above open-boundary condition can result in wave reflections when the radiating waves reach Σ. The steady-state, inviscid-fluid dispersion relation of deep-water gravity waves, i.e.

$$\lambda = 2\pi / F_\sigma^2 , \tag{15}$$

where λ is the nondimensional wavelength, can be used to position the open boundary sufficiently far so that wave reflections occurring at Σ during the course of the simulation are negligibly small. However, transient waves of larger wavelength may reach the boundary Σ sooner.

The field equations together with the given boundary conditions are solved as an *initial-value* problem. The initial data are taken to be those of the static (quiescent) fluid case. The body motion is started impulsively at $t = 0^+$.

2.3. *Force calculation*

At each instant of discrete time (after solving for the unknowns, namely $\mathbf{u}$, p, and the free-surface elevation, using the above field and boundary equations), the vertical component of the stress vector can be integrated along $\mathcal{B}$ to determine the nondimensional heave force F exerted by the body:

$$F = -\int_{\mathcal{B}} T_{2j} n_j \, \mathrm{d}s = -\int_{\mathcal{B}} \left\{ \frac{1}{R_\sigma} \left(\frac{\partial u}{\partial y} + \frac{\partial v}{\partial x} \right) n_1 + \left(-p + \frac{2}{R_\sigma} \frac{\partial v}{\partial y} \right) n_2 \right\} \mathrm{d}s , \tag{16}$$

where $\mathrm{d}s$ is the differential (arc-length). One can also compute the contributions of each term in the above equation in order to estimate viscous- and pressure-term contributions (bearing in mind, however, that pressure is affected by viscosity also). Thus, the viscous shear-stress component F_s of the heave force can be written as

$$F_s = -\frac{1}{R_\sigma} \int_{\mathcal{B}} \left(\frac{\partial u}{\partial y} + \frac{\partial v}{\partial x} \right) n_1 \, \mathrm{d}s, \tag{17}$$

and the normal viscous-stress component F_n as

$$F_n = -\frac{2}{R_\sigma} \int_{\mathcal{B}} \frac{\partial v}{\partial y} n_2 \, \mathrm{d}s . \tag{18}$$

Subtracting the static-pressure component $(-y/F_\sigma^2)$ from the total pressure p, we can compute the dynamic-pressure force F_p as

$$F_p = \int_{\mathcal{B}} \left(p + \frac{y}{F_\sigma^2} \right) n_2 \, \mathrm{d}s . \tag{19}$$

Once the velocity field is known, the vorticity field

$$\omega(\mathbf{x}, t) = \frac{\partial v}{\partial x} - \frac{\partial u}{\partial y} \tag{20}$$

can also be computed.

3. Method of numerical solution

As shown in Fig. 2, the physical space $(x, y; t)$ is mapped to a computational space $(\xi, \eta; T)$ so that the governing equations can be solved in a uniform rectangular mesh. The mapping introduces the following transformation relations

$$\partial_{,x} = \frac{1}{J_1} \{y_{,\eta}\partial_{,\xi} - y_{,\xi}\partial_{,\eta}\}\,, \tag{21}$$

$$\partial_{,y} = \frac{1}{J_1} \{-x_{,\eta}\partial_{,\xi} + x_{,\xi}\partial_{,\eta}\}\,, \tag{22}$$

$$\partial_{,t} = \partial_{,T} - \frac{x_{,T}}{J_1} \{y_{,\eta}\partial_{,\xi} - y_{,\xi}\partial_{,\eta}\} - \frac{y_{,T}}{J_1} \{-x_{,\eta}\partial_{,\xi} + x_{,\xi}\partial_{,\eta}\}\,, \tag{23}$$

where

$$J_1 = x_{,\xi}y_{,\eta} - x_{,\eta}y_{,\xi} \tag{24}$$

is the Jacobian of the transformation of $(x, y; t)$ to $(\xi, \eta; T)$. Similarly, one can show, as an example, that the Laplace operator in $(x, y; t)$ space when transformed to $(\xi, \eta; T)$ space becomes

$$\partial_{,xx} + \partial_{,yy} = A\partial_{,\xi\xi} - 2B\partial_{,\xi\eta} + C\partial_{,\eta\eta} + E\partial_{,\xi} + F\partial_{,\eta}\,, \tag{25}$$

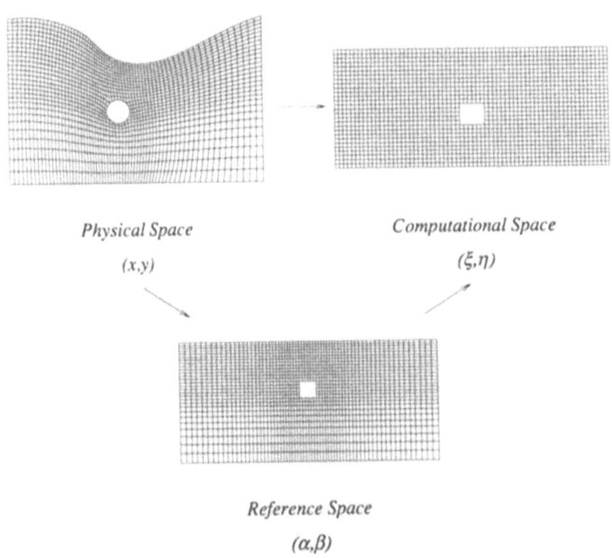

Fig. 2. Grid generation using reference space.

where

$$A = \frac{x_{,\eta}^2 + y_{,\eta}^2}{J_1^2}, \qquad B = \frac{x_{,\xi}x_{,\eta} + y_{,\xi}y_{,\eta}}{J_1^2}, \qquad C = \frac{x_{,\xi}^2 + y_{,\xi}^2}{J_1^2},$$

$$E = \xi_{,xx} + \xi_{,yy}, \qquad F = \eta_{,xx} + \eta_{,yy}.$$

These relations will be used to transform the flow equations, given in Section 2, to the computational space.

3.1. Boundary-fitted coordinates

A grid-generation procedure developed earlier (see Yeung and Ananthakrishnan [31]) has also been used in the present work to implement the coordinate mapping. This is based on a variational formulation (see Brackbill and Saltzman [6]) in conjunction with the notion of a reference space introduced by Steinberg and Roache [22]. The crux of the procedure is that a geometrically-similar intermediate space (α, β) is first discretized, and its properties are then transferred to the physical mesh. This is illustrated in Fig. 2, which is taken from Yeung and Vaidhyanathan [32]. The intermediate space is geometrically simpler than the physical space (see Fig. 2) so that it can be discretized easily, for example, by using simple algebraic relations. The intermediate space is also discretized in such a way that its grid properties such as cell-area distribution and coordinate spacings are also the ones desired in the physical space. For instance, the present problem requires finer resolution of the region near the body and the free surface so that viscosity effects can be accurately resolved. The desired grid properties, that are first achieved in the intermediate-space, are then transferable to the physical mesh by *minimizing* the following functionals:

$$I_s^*(\xi, \eta) = \iint_\Omega \mathrm{d}x\,\mathrm{d}y\, |\nabla_{x,y}\xi|^2 + |\nabla_{x,y}\eta|^2 - \iint_\Omega \mathrm{d}\alpha\,\mathrm{d}\beta\, |\nabla_{\alpha,\beta}\xi|^2 + |\nabla_{\alpha,\beta}\eta|^2, \tag{26}$$

$$I_c^*(\xi, \eta) = \iint_\Omega \mathrm{d}x\,\mathrm{d}y\, J_2, \tag{27}$$

$$I_o^*(\xi, \eta) = \iint_\Omega \mathrm{d}x\,\mathrm{d}y\, \{\nabla_{x,y}\xi \cdot \nabla_{x,y}\eta\}^2 J_1^3, \tag{28}$$

where the subscripts s, c, and o on the left-hand side represent the grid properties, viz. smoothness, cell-area, and orthogonality, respectively. Equation (26) is a measure of the difference, in coordinate spacings, between the physical and the reference spaces. The term J_2 in equation (27) denotes the Jacobian of the mapping of the physical space onto the reference space ($J_2 = x_{,\alpha}y_{,\beta} - x_{,\beta}y_{,\alpha}$). Note that equation (28), the orthogonality functional, corresponds to the direct mapping of $(x, y; t)$ to $(\xi, \eta; T)$ given originally by Brackbill and Saltzman [6]. We do not use the reference space to specify the angular properties of the coordinates. This is because it is preferable to have orthogonal or nearly orthogonal grids for accurate implementation of Neumann-type boundary conditions.

The field equations corresponding to the functionals (26) to (28) are given below:

Smoothness

$$\theta_1 x_{,\xi\xi} - 2\kappa_1 x_{,\xi\eta} + \sigma_1 x_{,\eta\eta} = -\frac{J_1}{J_3^2}\{(\beta_{,\eta}\chi - \alpha_{,\eta}\mu)x_{,\xi} + (-\beta_{,\xi}\chi + \alpha_{,\xi}\mu)x_{,\eta}\}\,, \tag{29}$$

$$\theta_1 y_{,\xi\xi} - 2\kappa_1 y_{,\xi\eta} + \sigma_1 y_{,\eta\eta} = -\frac{J_1}{J_3^2}\{(\beta_{,\eta}\chi - \alpha_{,\eta}\mu)y_{,\xi} + (-\beta_{,\xi}\chi + \alpha_{,\xi}\mu)y_{,\eta}\}\,, \tag{30}$$

where

$$\chi = \theta_3\alpha_{,\xi\xi} - 2\kappa_3\alpha_{,\xi\eta} + \sigma_3\alpha_{,\eta\eta}\,, \qquad \mu = \theta_3\beta_{,\xi\xi} - 2\kappa_3\beta_{,\xi\eta} + \sigma_3\beta_{,\eta\eta}\,.$$

The coefficients in the above equations are given by

$$\theta_1 = x_{,\eta}^2 + y_{,\eta}^2\,, \qquad \kappa_1 = x_{,\xi}x_{,\eta} + y_{,\xi}y_{,\eta}\,, \qquad \sigma_1 = x_{,\xi}^2 + y_{,\xi}^2\,,$$

$$\theta_3 = \alpha_{,\eta}^2 + \beta_{,\eta}^2\,, \qquad \kappa_3 = \alpha_{,\xi}\alpha_{,\eta} + \beta_{,\xi}\beta_{,\eta}\,, \qquad \sigma_3 = \alpha_{,\xi}^2 + \beta_{,\xi}^2\,,$$

and J_3 $(= \alpha_{,\xi}\beta_{,\eta} - \alpha_{,\eta}\beta_{,\xi})$ is the Jacobian of the mapping of the reference space to the computational space.

Cell-area distribution

$$\frac{2}{J_3^2}\{b_{v1}x_{,\xi\xi} + b_{v2}x_{,\xi\eta} + b_{v3}x_{,\eta\eta} + a_{v1}y_{,\xi\xi} + a_{v2}y_{,\xi\eta} + a_{v3}y_{,\eta\eta}\} = -J_1^2\frac{\partial}{\partial x}\left(\frac{1}{J_3^2}\right), \tag{31}$$

$$\frac{2}{J_3^2}\{a_{v1}x_{,\xi\xi} + a_{v2}x_{,\xi\eta} + a_{v3}x_{,\eta\eta} + c_{v1}y_{,\xi\xi} + c_{v2}y_{,\xi\eta} + c_{v3}y_{,\eta\eta}\} = -J_1^2\frac{\partial}{\partial y}\left(\frac{1}{J_3^2}\right), \tag{32}$$

where

$$a_{v1} = -x_{,\eta}y_{,\eta}\,, \qquad b_{v1} = y_{,\eta}^2\,, \qquad c_{v1} = x_{,\eta}^2\,,$$

$$a_{v2} = x_{,\xi}y_{,\eta} + x_{,\eta}y_{,\xi}\,, \qquad b_{v2} = -2y_{,\xi}y_{,\eta}\,, \qquad c_{v2} = -2x_{,\xi}x_{,\eta}\,,$$

$$a_{v3} = -x_{,\xi}y_{,\xi}\,, \qquad b_{v3} = y_{,\xi}^2\,, \qquad c_{v3} = x_{,\xi}^2\,.$$

Orthogonality

$$b_{o1}x_{,\xi\xi} + b_{o2}x_{,\xi\eta} + b_{o3}x_{,\eta\eta} + a_{o1}y_{,\xi\xi} + a_{o2}y_{,\xi\eta} + a_{o3}y_{,\eta\eta} = 0\,, \tag{33}$$

$$a_{o1}x_{,\xi\xi} + a_{o2}x_{,\xi\eta} + a_{o3}x_{,\eta\eta} + c_{o1}y_{,\xi\xi} + c_{o2}y_{,\xi\eta} + c_{o3}y_{,\eta\eta} = 0\,, \tag{34}$$

where

$$a_{o1} = x_{,\eta}y_{,\eta}\,, \qquad b_{o1} = x_{,\eta}^2\,, \qquad c_{o1} = y_{,\eta}^2\,,$$

$$a_{o2} = x_{,\xi}y_{,\eta} + x_{,\eta}y_{,\xi}\,, \qquad b_{o2} = 2(2x_{,\xi}x_{,\eta} + y_{,\xi}y_{,\eta})\,, \qquad c_{o2} = 2(2y_{,\xi}y_{,\eta} + x_{,\xi}x_{,\eta})\,,$$

$$a_{o3} = x_{,\xi}y_{,\xi}\,, \qquad b_{o3} = x_{,\xi}^2\,, \qquad c_{o3} = y_{,\xi}^2\,.$$

A linear combination of the above field equations taken as

$$\mathscr{L} = \lambda_s \mathscr{E}_s + \lambda_c \mathscr{E}_c + \lambda_o \mathscr{E}_o , \tag{35}$$

where the λs and $\mathscr{E}$s denote respectively, user-specified weighting values and field equations, are solved. The λ values are to be chosen as per the desired levels (or relative importance) of smoothness, cell-area variations and orthogonality, required in a particular problem. At each time step, the field equations for the grid are solved subject to the location of boundary points.

3.2. *Fractional-step formulation*

The mapping relations (21) to (25) are used to transform the governing equations to the $(\xi, \eta; T)$ space. This is a straightforward workout in calculus but resulting in lengthy expressions. We will therefore, for the purpose of exposition here, present explicitly only the terms in the transformed equations that are essential for the description of the solution method. Remaining terms will be denoted simply by using symbols. Also, for brevity, we retain the notations used in Section 2 to denote differential operators; e.g., the gradient operator ∇ now represents

$$\nabla = (\partial_{,x}, \partial_{,y}) = \left(\frac{1}{J_1} \{ y_{,\eta}\partial_{,\xi} - y_{,\xi}\partial_{,\eta} \}, \frac{1}{J_1} \{ -x_{,\eta}\partial_{,\xi} + x_{,\xi}\partial_{,\eta} \} \right) . \tag{36}$$

The auxiliary velocity field $\mathbf{u}^{\text{aux}}$ is first determined by using the momentum equations without the pressure term [8]:

$$(1-q)\mathbf{u}^{\text{aux}} = \delta T \{ \mathbf{u}^n - \mathscr{N}^n + \mathscr{D}^n \} , \tag{37}$$

where the superscript n denote the current instant of discrete time at which solutions are known. $\mathscr{N}^n$ and $\mathscr{D}^n$ are the known terms corresponding to the discretized form of the coordinate-transformed convection and diffusion terms, respectively. The term q on the left-hand side represents the possibility of implicit differencing. In the case of explicit differencing (in time), $q = 0$.

The discretized form of the full *transformed* momentum equation can be written as

$$(1-q)\mathbf{u}^{n+1} = \delta T \left\{ \mathbf{u}^n - \mathscr{N}^n + \mathscr{D}^n + \nabla\left(p + \frac{y}{F_\sigma^2} \right)^{n+1} \right\} . \tag{38}$$

A comparison of equations (37) and (38) leads to

$$\mathbf{u}^{\text{aux}} = \mathbf{u}^{n+1} + (1-q)^{-1} \delta T \, \nabla\left(p + \frac{y}{F_\sigma^2} \right)^{n+1} , \tag{39}$$

where

$$\nabla \cdot \mathbf{u}^{n+1} = 0 . \tag{40}$$

In other words, $\mathbf{u}^{\text{aux}}$ can be expressed as a direct sum of $\mathbf{u}^{n+1}$ and pressure fields. Equation

(39) can be conveniently rewritten as

$$\mathbf{u}^{\mathrm{aux}} = \mathbf{u}^{n+1} + \nabla\varphi^{n+1} , \tag{41}$$

where

$$\varphi^{n+1} \equiv (1-q)^{-1}\,\delta T\left(p + \frac{y}{F_\sigma^2}\right)^{n+1} , \tag{42}$$

which should not be mistaken as a velocity potential. Note that equation (42) gives a one-to-one relation between φ and $p + y/F_\sigma^2$ which is dependent on the scheme (i.e., dependent on q).

Once the auxiliary field has been evaluated using equation (37), equation (41) can be used to determine $\nabla\varphi^{n+1}$ and $\mathbf{u}^{n+1}$. The decomposition can be carried out either sequentially or iteratively. Following Kim and Moin [16], we use the following sequential procedure. The Poisson equation for φ

$$\nabla^2\varphi = \nabla\cdot\mathbf{u}^{\mathrm{aux}} , \tag{43}$$

obtained by taking the divergence of equation (41), is solved first for φ (and hence p through equation (42)). The correctional effects are then added to the auxiliary field through equation (41) to obtain $\mathbf{u}^{n+1}$:

$$\mathbf{u}^{n+1} = \mathbf{u}^{\mathrm{aux}} - \nabla\varphi^{n+1} . \tag{44}$$

We will elaborate on the boundary conditions needed for the evaluation of $\mathbf{u}^{\mathrm{aux}}$, φ etc. in the next subsection, after providing a glimpse of some related theoretical aspects in the following paragraph. Details can be found in the cited references.

In problems where the normal component of velocity is zero on the boundary, as in flows bounded by rigid boundaries, one can show using Gauss theorem that $\mathbf{u}^{n+1}$ and $\nabla\varphi$ fields are orthogonal to each other (see Chorin [7]). The orthogonality of the pressure-gradient and the divergence-free-velocity fields, as stated by Bell et al. [5], "effectively eliminates pressure from the system while enforcing 1.2 [the equation of continuity]; in fact, specifying pressure boundary conditions overdetermines the system". Consequently, the decomposition (41) can be expresses as

$$\mathbf{u}^{n+1} = \mathscr{P}\mathbf{u}^{\mathrm{aux}} \tag{45}$$

$$\nabla\varphi^{n+1} = \mathscr{Q}\mathbf{u}^{\mathrm{aux}} = (\mathscr{I} - \mathscr{P})\mathbf{u}^{\mathrm{aux}} , \tag{46}$$

where $\mathscr{P}$ and $\mathscr{Q}$, not explicitly given here, are orthogonal operators which project $\mathbf{u}^{\mathrm{aux}}$ onto the divergence-free $\mathbf{u}^{n+1}$ and $\nabla\varphi$ fields, respectively. The term $\mathscr{I}$ in equation (46) denotes the unit operator. In the case of free-surface flows, however, the divergence-free velocity and pressure-gradient fields are not orthogonal. Hence, operating the *free-surface* Navier–Stokes equations with the projection operator $\mathscr{P}$ would not eliminate the pressure term. The projected pressure term, nevertheless, is determined by the free-surface pressure condition (see Beale [4]). These observations indicate that the velocity boundary conditions on the body and the stress conditions on the free surface, as given in Section 2, are sufficient for the solution of the pressure and velocity fields.

3.3. Implementation of boundary conditions

In the present work, physical conditions (given in Section 2) and the decomposition relation (41) are taken into account for obtaining numerical boundary conditions for $\mathbf{u}^{\text{aux}}$, $(\nabla\varphi)^{n+1}$, and $\mathbf{u}^{n+1}$. Uniform validity of the equation (41) in space is important for obtaining accurate solutions in a consistent manner (see Kim and Moin [16]).

On $\mathcal{B}$, we use

$$\mathbf{u}^{\text{aux}} \cdot \mathbf{n} = \mathbf{u}^{n+1} \cdot \mathbf{n}\,, \tag{47}$$

$$\mathbf{u}^{\text{aux}} \cdot \boldsymbol{\tau} \approx \mathbf{u}^{n+1} \cdot \boldsymbol{\tau} + \frac{\partial}{\partial \boldsymbol{\tau}} (\varphi^{n})\,, \tag{48}$$

(where $\mathbf{n}$ and $\boldsymbol{\tau}$ denote unit vectors in the normal and tangential directions, respectively) to specify the values of $\mathbf{u}^{\text{aux}}$ on $\mathcal{B}$. Equation (47), in view of equation (41), provides the following homogeneous Neumann condition for φ^{n+1},

$$\frac{\partial}{\partial \mathbf{n}} (\varphi^{n+1}) = 0\,. \tag{49}$$

Note that $\mathbf{u}^{n+1}$ is given by the no-slip conditions (11) and (12). It is determined by the free-slip equations (11) and (13) at the contact points $\mathcal{B} \cap \mathcal{F}$. The location of the body is determined by the forced oscillation, i.e. by equation (10).

On the free surface $\mathcal{F}$, we use an approximation of equation (41) to specify the auxiliary velocity value:

$$\mathbf{u}^{\text{aux}} \approx \mathbf{u}^{n} + \nabla\varphi^{n}\,. \tag{50}$$

A two-step predictor-corrector type iterative procedure is applied to the free-surface stress equations for determining $\mathbf{u}^{n+1}$ and φ^{n+1} (see Ananthakrishnan [1] for details). Basically, we begin with the known value of φ from previous instant of time and use it as the Dirichlet condition for the solution of the Poisson equation. Equation (41) is then used to predict $\mathbf{u}^{n+1}$ in the fluid domain. Stress relations, rewritten as Neumann-type relations for u and v in (ξ, η, T) space, are then used to get the first estimate of $\mathbf{u}^{n+1}$ on $\mathcal{F}$. The normal-stress condition with the predicted velocity thus provides the Dirichlet condition for φ at the corrector stage. The remaining steps in the above predictor stage are again repeated in a similar manner for correcting $\mathbf{u}^{n+1}$ and for updating φ^{n+1}. Concurrently, the kinematic condition (8) or (9) is also integrated using the predictor-corrector method to advance the free surface. The above two-step procedures thus require the solution of Poisson equation for φ twice at each instant of time.

At the open boundary Σ, we obtain a Dirichlet condition for φ from the approximation (14), using the scheme-dependent relation (42) between p and φ. The auxiliary-field $\mathbf{u}^{\text{aux}}$ is determined by first-order spatial extrapolation. $\mathbf{u}^{n+1}$ is then computed using the decomposition relation (41), i.e. as

$$\mathbf{u}^{n+1} = \mathbf{u}^{\text{aux}} - \nabla\varphi^{n+1}\,. \tag{51}$$

3.4. Numerical schemes

Several schemes to discretize the convection and diffusion terms in the *transformed* momentum equation – e.g. first-order upwind, second-order Adams–Bashforth, forward-time central-space, implicit Crank–Nicholson, etc. – have been considered by Ananthakrishnan [1] in order to evaluate the comparative merits of these schemes applied to viscous-wave problems. In the present work, we use one of the many sets of procedures that have been studied. This procedure consists of using an upwind differencing for the treatment of *coordinate transformed* convection terms and central-differencing for discretizing the diffusion terms; both of these schemes are explicit, and hence $q = 0$ in equation (42).

The Poisson equation (43) is centrally-differenced and solved by a Gaussian elimination based on LU-decomposition; the code requires the storage of (and operations in) only the non-zero band of the coefficient matrix. The grid equations, also centrally differenced, are solved iteratively using mixed over-under relaxation method.

4. Results and discussion

In this section, as a case study, results obtained for a simple section over a range of F_σ and a/B values are presented. A mesh size of (121×41), with coordinates clustered near the body and the free surface, is used for the discretization of the physical space. The value of R_σ is set to be 10^3 for all cases, which is appropriate for the above mesh size. A time-step size of 0.02, which satisfy stability criteria posed by *linear* convection and diffusion problems on the numerical schemes (see Hirsch [13]), is used in all cases. Velocity scales associated with both body velocity and wave celerity are taken into account in satisfying the Courant condition. For grid generation, the physical space with free surface replaced by flat horizontal surface is used as the reference space. This allows discretizing the reference space by simple algebraic interpolation. We remark that this is not a restriction on the grid-generation method. For example, for breaking-wave type problems, grids generated at previous instants of time can effectively be used as the reference mesh (see Yeung and Vaidhyanathan [32]).

4.1. Flow structure

First, in order to illustrate the intricate details of the vorticity/vortex generation around the oscillating body, velocity-vector and vorticity-contour plots corresponding to a typical case of large-amplitude of oscillation ($a/B = 0.2$, $d/B = 0.5$, $F_\sigma = 2.0$) are shown in Figs 3a–3d. These figures correspond to instants of discrete time that are roughly one-quarter period apart. Figure 3a corresponds to $t = 0.999\hat{T}$ (where $\hat{T}$ here denotes the period of oscillation) when the body is ascending. As can be observed, the flow is primarily into the void generated by the upward movement. Note the incipient formation of wake eddies near the sharp edges. Fully developed wake vortices can be seen at a later time during the periodic motion, see e.g. at $t = 1.247\hat{T}$ (Fig. 3b). A small asymmetry in the flow field about the centerline ($x = 0$) can be noticed in the vorticity-contour plot. In this plot, generation of free-surface vorticity, which is rather weak because of small surface curvature, can also be seen. The entire flow picture changes as the body is descending (see Fig. 3c, corresponding to $t = 1.499\hat{T}$). The wake eddies that have been formed earlier (during the ascent) undergo

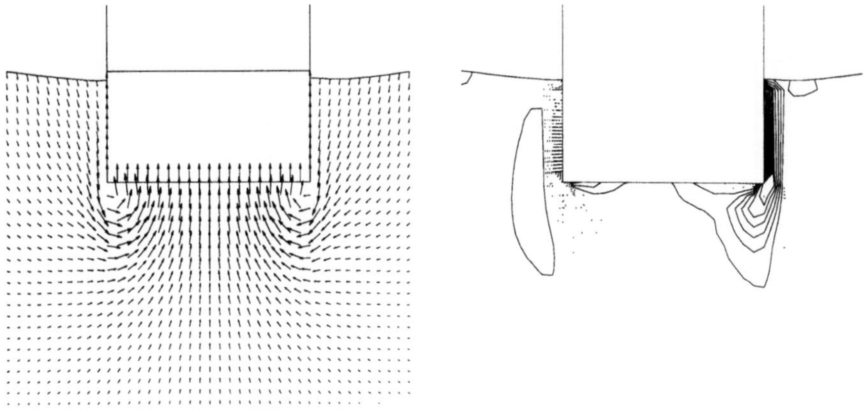

Fig. 3a

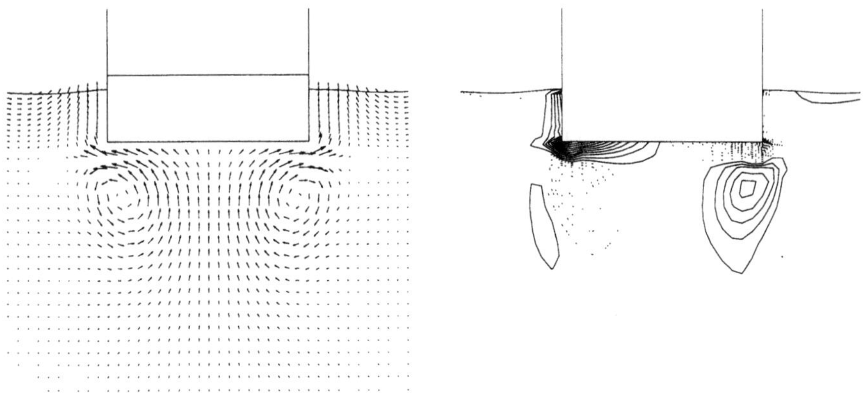

Fig. 3b

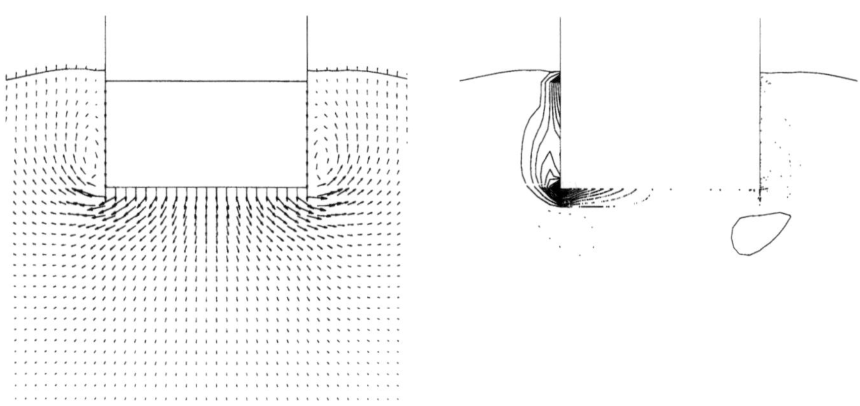

Fig. 3c

Fig. 3. Velocity-vector and vorticity contour plots for $R_\sigma = 10^3$, $F_\sigma = 2.0$, $a/B = 0.2$, $d/B = 0.5$, at (a) $t = 0.999\hat{T}$, (b) $t = 1.247\hat{T}$, (c) $t = 1.499\hat{T}$, and (d) $t = 1.814\hat{T}$. $\hat{T}$ is the period of oscillation. In the velocity-vector plots, the horizontal line in the body denotes the calm water level. The solid lines in the vorticity-contour plots denote negative (clockwise) vorticity and dotted lines positive (counterclockwise) vorticity.

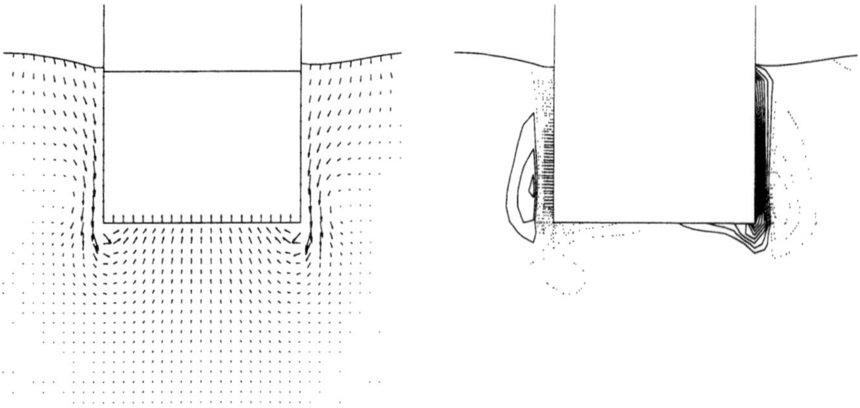

Fig. 3d

Fig. 3 (cont.).

diffusion while being convected downwards. The vorticity field is now dominant in the shear layer generated along the sides near the sharp edges. One can notice the formation of vortices in the side shear layers. Finally, in Fig. 3d, which corresponds to $t = 1.814\hat{T}$, the outward translation of the vortices formed earlier (in Fig. 3c) can be clearly seen. The primary flow at this time is again (as in Fig. 3a) into the bottom void as the body is ascending. In all of the above vorticity-contour plots, partial shedding of vorticity into the fluid can be observed.

4.2. Force calculations

First, heave-force components for a typical case ($d/B = 1.0$, $F_\sigma = 2.0$ and $a/B = 0.3$) are shown in Fig. 4. These shear-stress, viscous normal-stress, and dynamic-pressure components

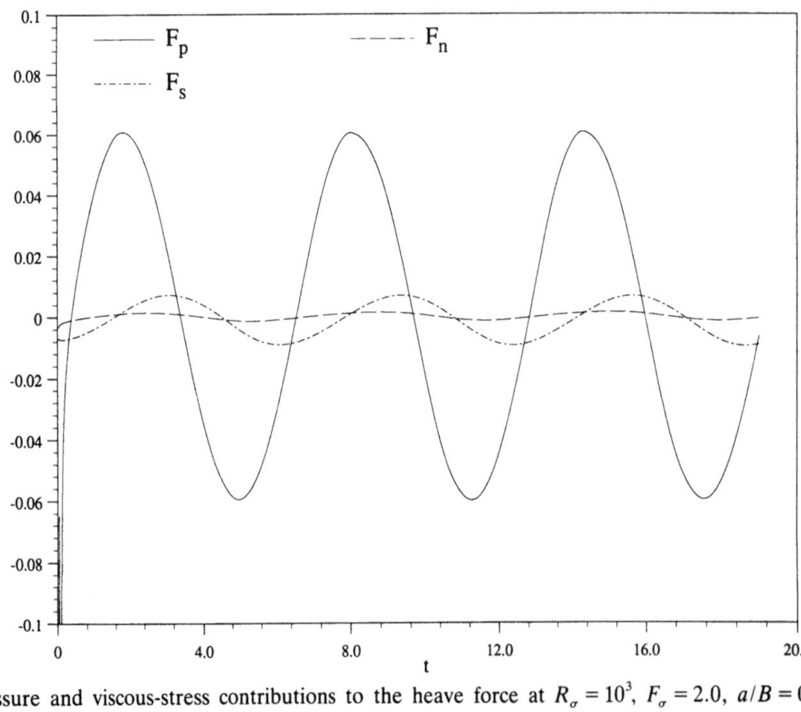

Fig. 4. Pressure and viscous-stress contributions to the heave force at $R_\sigma = 10^3$, $F_\sigma = 2.0$, $a/B = 0.3$, $d/B = 1.0$.

of the heave force are computed using equations (17) to (19). As can be observed in the figure, the pressure component is much larger than those of the viscous stress components. This is a well known fact for flows past bluff bodies with sharp edges. In the present case, the shear-stress contribution is about 10%, while that of the normal viscous stress is only about 1%, of the pressure force. The shear stress contribution will be even smaller at higher Reynolds-number flows. It should also be pointed out that the calculation of the shear stress requires a fine spatial resolution of the boundary layers. Because of the smallness of the viscous stress contribution, we will henceforth retain only the pressure term in the force computations.

In order to determine the influence of the amplitude and frequency of oscillation in the nonlinear heave forces, the following cases are studied:

$F_\sigma = 2.0$, Case (i): $a/B = 0.1$, Case (ii): $a/B = 0.2$;

$F_\sigma = 1.5$, Case (iii): $a/B = 0.1$, Case (iv): $a/B = 0.2$;

$F_\sigma = 1.0$, Case (v): $a/B = 0.1$, Case (vi): $a/B = 0.2$.

The draft-to-beam ratio d/B is set to be 0.5. Cases (i) & (ii) are presented in Fig. 5, (iii) & (iv) in Fig. 6, and (v) & (vi) in Fig. 7. The heave forces are normalized with respect to the amplitude of oscillation; i.e. presented as $F/(a/B)$, where F now is only the pressure component given by equation (19). The nonlinear viscous results are compared with those of linear, potential-flow, *frequency-domain*, steady-state results. The linear results are obtained using the solution method developed by Yeung in [33].

For purpose of reference, a *sine* curve $y(t) = 0.2 \sin t$, is drawn to indicate the position of the body. The value 0.2 is chosen merely for clarity.

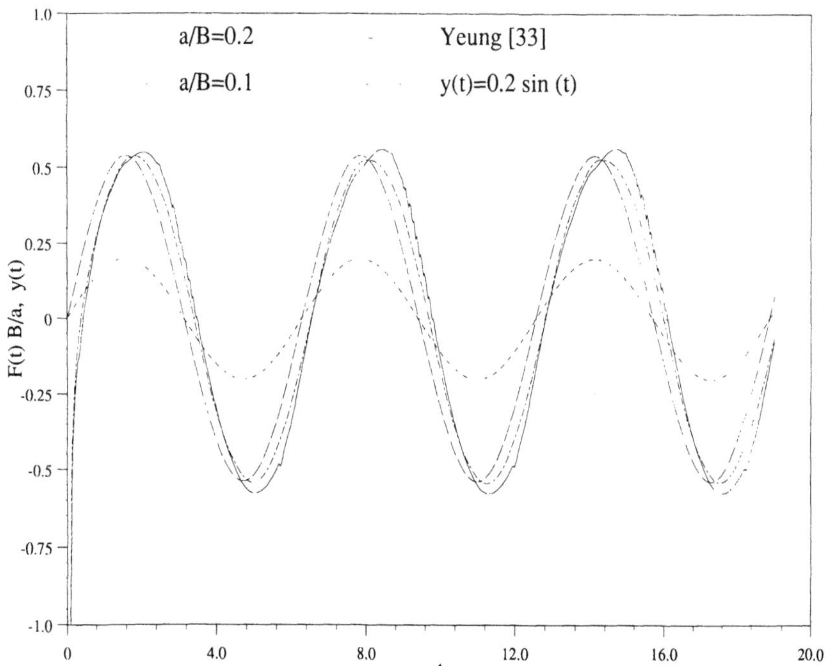

Fig. 5. Normalized viscous heave forces for $F_\sigma = 2.0$, $R_\sigma = 10^3$, $d/B = 0.5$ at amplitudes $a/B = 0.2$ and $a/B = 0.1$. For comparison, linear potential results (Yeung [33]) is given. Sine curve $y(t) = 0.2 \sin t$ indicates body position.

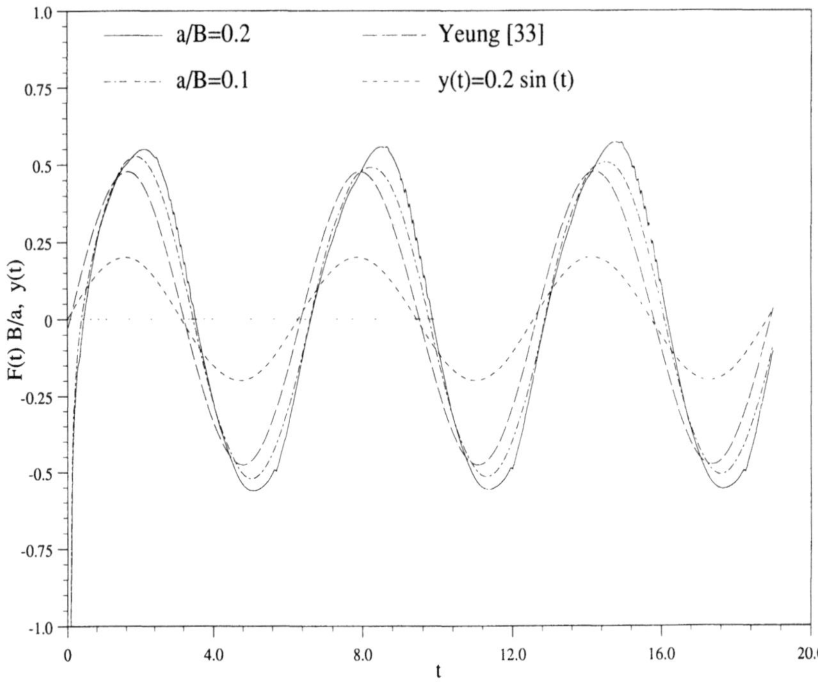

Fig. 6. Normalized viscous heave forces for $F_\sigma = 1.5$, $R_\sigma = 10^3$, $d/B = 0.5$ at amplitudes $a/B = 0.2$ and $a/B = 0.1$. For comparison, linear potential results (Yeung [33]) is given. Sine curve $y(t) = 0.2 \sin t$ indicates body position.

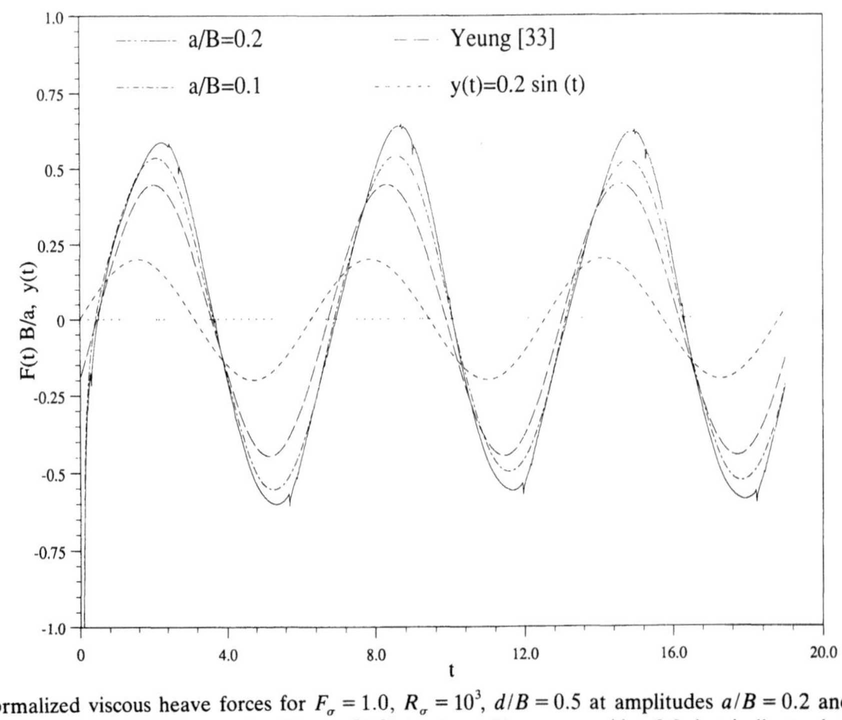

Fig. 7. Normalized viscous heave forces for $F_\sigma = 1.0$, $R_\sigma = 10^3$, $d/B = 0.5$ at amplitudes $a/B = 0.2$ and $a/B = 0.1$. For comparison, linear potential results (Yeung [33]) is given. Sine curve $y(t) = 0.2 \sin t$ indicates body position.

In all of the nonlinear viscous-flow calculations, negatively large values are found at the start of the simulation. Recall that the body motion is stated impulsively at $t=0^+$. For such impulsively started motion in a viscous fluid, it has been shown that the stress on the body surface is square-root singular in t at $t=0^+$ (see Bar-Lev and Yang [3]). However, this negatively large value seems to perpetuate only for a short duration of time.

Since the force curves are normalized with respect to the amplitude of oscillation, deviations in force curves of different a/B values would indicate nonlinear effects. Deviations of the computed results with respect to that of the linear, potential-flow theory is a measure of both nonlinear and viscosity effects. The phase shift of the force with respect to the body motion is attributed to the damping component (in phase with velocity) of the hydrodynamic forces.

Based on the results given in Figs 5–7, the following observations can be made:

- Inspection of results reveals that steady-state is reached within about two periods of oscillation.
- The negative phase shift of the linear potential force is small at $F_\sigma = 2.0$, since wave damping is almost negligible in this relatively high frequency regime (Fig. 5). However, it can be seen that the wave-damping component is quite large at the lower frequency $F_\sigma = 1.0$ of oscillation (Fig. 7).
- The phase shift is more negative (compared to linear potential results) in the case of nonlinear viscous solutions. It can also be observed, in general, that the phase shift increases with the increase in the amplitude of oscillation. This indicates the effect of viscosity on damping.
- In the high-frequency regime (e.g. Fig. 5), the amplitude of the computed viscous forces is close to that of the linear, potential-theory results when the amplitude of oscillation is small. This is consistent with the analytical results of Yeung and Wu [34]. It can also be observed that the negative peak of the force curve is larger in magnitude than that at the positive peak, a known nonlinear effect. The computed viscous forces deviate substantially from those of potential flow at low frequency (see Fig. 7).
- In general, the normalized heave forces are larger when the amplitude of oscillation is large. This increase is more dramatic at low frequency (Fig. 7).
- At large amplitude of oscillation, a rather uniform spiky behavior can be observed during the second quarter of the periodic motion (see Figs 5 and 6). During this interval, as shown in Figs 3b and 3c, eddies formed in the wake are convected and diffused, and the vortices at the sides near the sharp edges are also developing. It is believed that the spikes in the force results are caused by the drastic pressure changes accompanying the generation of these vortices at the edges. However, no such spiky pattern is observed at the smaller amplitude of oscillation.

Finally, to better elucidate the structure of the vorticity fields, we show in Fig. 8 a color plot of the vorticity contours corresponding to two different motion amplitudes ($a/B = 0.1$, 0.2). Both plots in this figure correspond to $d/B = 0.5$, $F_\sigma = 2.0$, and $t = 1.247\hat{T}$. Note that the contours are given in the computational space. The actual vortical structure in the physical space will be slightly distorted from these plots because of coordinate mapping. It can be seen that the intensity of the vorticity field is larger in the case of larger amplitude of oscillation. Vorticity generation at the free surface, because of its curvature, can also be seen in the case of large-amplitude motion.

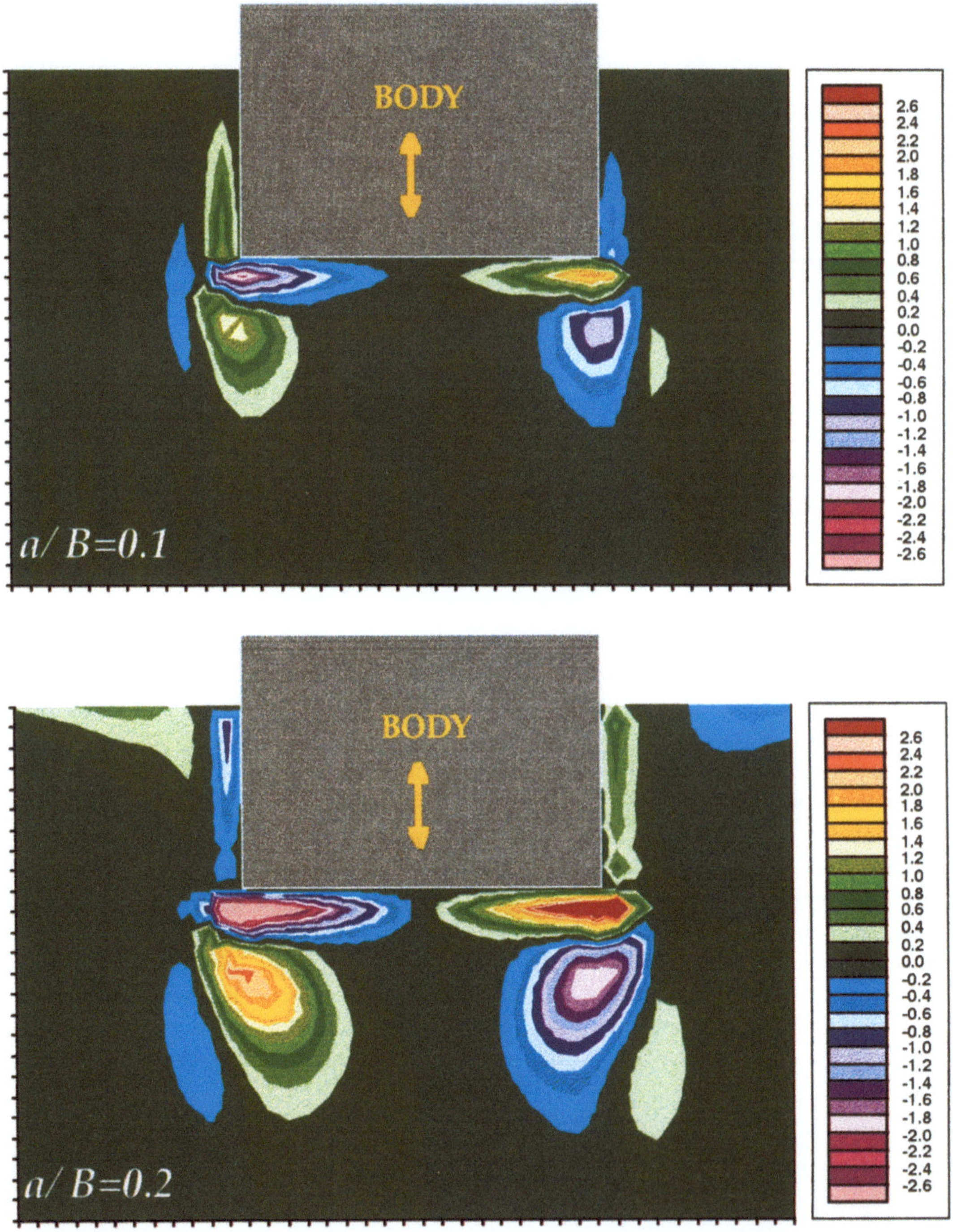

Fig. 8. Vorticity structure generated around cylinder, effects of amplitude of motion (shown in computational space); $F_\sigma = 2.0$, $R_\sigma = 10^3$, $d/B = 0.5$, and $a/B = 0.1$, $a/B = 0.2$. Positive contours (green and red spectrum) correspond to counterclockwise vorticity and negative contours (blue and pink spectrum) to clockwise vorticity.

5. Conclusions

A new method for the accurate solution of nonlinear, wave-body interaction problems in a viscous fluid has been developed. As has been shown in this paper, the present method is robust and is especially capable of coping with large amplitude of body oscillation. Although only heave motion is considered here, other modes of oscillation, such as sway and roll, or their combination, can be easily handled.

For the heave-motion study, we have provided some typical results corresponding to a fairly wide range of frequency and amplitude of heave motion. At high frequency, small-amplitude forces agree well with those of linear potential theory in magnitude. The difference in phase between the viscous and inviscid cases is due to viscosity effects. Nonlinear and quadratic-damping effects are clearly evident for the case of large amplitude of oscillation. At low frequency, it is shown that heave forces are strongly dependent on the amplitude of oscillation.

We were not able to provide results at extremely low frequencies, which would have required a very large numerical domain in order to avoid the ill effects of open-boundary reflections. Similar difficulty is also encountered in laboratory experiments because of tank-size limitations (see e.g. Vugts [26]). The scatter in the experimental data at low frequencies are customarily attributed to spurious effects of finite tank sizes. On the contrary, the results of Yeung and Wu [34], which were based on the linearized viscous-flow equations, show that viscosity effects could be important in the low-frequency regime in the laboratory scale. The present nonlinear work also seems to point towards that possibility. A more elaborate treatment of the open boundary is being developed. Once accomplished, we hope to resolve completely the precise role played by viscosity in the low-frequency regime.

Acknowledgement

Research reported in this paper has been supported primarily by the Office of Naval Research, under grant N00014-91-J1614 and N00014-91-J1155. We gratefully acknowledge such support as well as a Research & Development grant from Cray Research Inc., funded through the University of California at Berkeley.

References

1. P. Ananthakrishnan, *Surface waves generated by a translating two-dimensional body: effects of viscosity*. Ph.D. thesis, Department of Naval Architecture and Offshore Engineering, University of California at Berkeley, USA (1991).
2. G.B. Baker, D.I. Meiron and S.A. Orszag, Generalized vortex method for free-surface flow problems. *Journal of Fluid Mechanics* 123 (1982) 477–501.
3. M. Bar-Lev and H.T. Yang, Initial flow field over an impulsively started circular cylinder. *Journal of Fluid Mechanics* 72 (1975) 625–647.
4. J.T. Beale, Large-time regularity of viscous surface waves. *Archive for Rational Mechanics and Analysis* 84(4) (1984) 307–352.
5. J.B. Bell, P. Collela and H.M. Glaz, A second-order projection method for the incompressible Navier–Stokes equations. *Journal of Computational Physics* 85 (1989) 257–283.
6. J.U. Brackbill and J.S. Saltzman, Adaptive zoning for singular problems in two dimensions. *Journal of Computational Physics* 46 (1982) 342–368.

7. A.J. Chorin, Numerical solution of incompressible flow problems. *Studies in Numerical Analysis* 2 (1968) 64–71.
8. A.J. Chorin, Numerical solution of the Navier–Stokes equations. *Mathematics of Computations* 22 (1968) 745–762.
9. M.J. Cooker, D.H. Peregrine, C. Vidal and J.W. Dold, The interaction between a solitary wave and a submerged semicircular cylinder. *Journal of Fluid Mechanics* 215 (1990) 1–22.
10. E.B. Dussan V., The moving contact line: the slip boundary condition. *Journal of Fluid Mechanics* 77 (1976) 665–684.
11. M.A. Grosenbaugh and R.W. Yeung, Nonlinear free-surface flow at a two-dimensional bow. *Journal of Fluid Mechanics* 209 (1989) 57–75.
12. F.H. Harlow and J.E. Welch, Numerical calculation of time dependent viscous incompressible flow of fluid with free surface. *Physics of Fluids* 8 (1965) 2182–2183.
13. C. Hirsch, *Numerical Computation of Internal and External Flows I & II.* A Wiley-Interscience Publication (1988).
14. B.D. Nichols and C.W. Hirt, Methods for calculating multi-dimensional, transient, free surface flows past bodies. In *Proceedings*, First International Conference on Numerical Ship Hydrodynamics, Gaithersburg, Maryland (1975) 253–277.
15. C. Huh and S.G. Mason, The steady movement of a liquid meniscus in a capillary tube. *Journal of Fluid Mechanics* 81 (1977) 401–419.
16. J. Kim and P. Moin, Application of a fractional-step method to incompressible Navier–Stokes equations. *Journal of Computational Physics* 59 (1985) 308–323.
17. J. Koplik, J.R. Banavar and J.F. Willemsen, Molecular dynamics of fluid flow at solid surfaces. *Physics of Fluids A* 1(5) (1989) 781–794.
18. M.S. Longuet-Higgins and E.D. Cokelet, The deformation of steep surface waves on water: I. A numerical method of computation. *Proceedings, Royal Society of London* A350 (1976) 1–26.
19. H. Miyata, H. Kajitani, M. Zhu and T. Kawano, Nonlinear forces caused by breaking waves. In *Proceedings*, Sixteenth Symposium on Naval Hydrodynamics, Berkeley, California (1986) 514–536.
20. R.L. Potash, Second-order theory of oscillating cylinders. *Journal of Ship Research* 15(4) (1971) 295–324.
21. P.D. Sclavounos, Radiation and diffraction of second-order surface waves by floating bodies. *Journal of Fluid Mechanics* 196 (1988) 65–91.
22. S. Steinberg and P.J. Roache, Variational grid generation. *Numerical Methods in Partial Differential Equations* 2 (1986) 71–96.
23. R. Temam, *Navier–Stokes Equations – Theory and Numerical Analysis.* North-Holland Publishing Company (1979).
24. J.G. Telste, Calculation of fluid motion resulting from large-amplitude forced heave motion of a two-dimensional cylinder in a free surface. In: *Proceedings*, Fourth International Conference on Numerical Ship Hydrodynamics, Washington D.C. (1985).
25. T. Vinje and P. Brevig, Numerical simulation of breaking waves. *Advances in Water Resources* 4 (1981) 77–82.
26. J.H. Vugts, *The hydrodynamic coefficients for swaying, heaving, and rolling cylinders in a free surface.* Report No. 194, Laboratorium voor Scheepsbouwkunde, Technische Hogeschool, Delft, The Netherlands (1968).
27. J.V. Wehausen, The motion of floating bodies. *Annual Review of Fluid Mechanics* 3 (1971) 237–268.
28. R.W. Yeung and C-F. Wu, Nonlinear wave-body motion in a closed domain. *Computers and Fluids* 17 (1989) 351–370.
29. R.W. Yeung, Numerical methods in free-surface flows. *Annual Review of Fluid Mechanics* 14 (1982) 395–442.
30. R.W. Yeung and P. Ananthakrishnan, Solution of nonlinear water-wave and wave-body interaction problems using a new boundary-fitted coordinates method. In *Proceedings*, Fourth International Workshop on Water Waves and Floating Bodies, Øystese, Norway (1989) 269–274.
31. R.W. Yeung and P. Ananthakrishnan, Numerical grid generation for water-wave problems using reference space, *to be published* (1992).
32. R.W. Yeung and M. Vaidhyanathan, Nonlinear interaction of water waves with submerged obstacles, to appear in *International Journal on Numerical Methods in Fluids* (1992).
33. R.W. Yeung, A hybrid integral-equation method for the time-harmonic free-surface flows. In *Proceedings*, First International Conference on Numerical Ship Hydrodynamics, Gaithersburg, Maryland (1975) 581–608.
34. R.W. Yeung and C-F. Wu, Viscosity effects on the radiation hydrodynamics of horizontal cylinders. *Journal of Offshore Mechanics and Arctic Engineering* 113 (1991) 334–343.

Journal of Engineering Mathematics **26**: 231–265, 1992.
H.K. Kuiken and S.W. Rienstra (eds), Problems in Applied, Industrial and Engineering Mathematics.

Electromagnetic response of composite superconducting wires

L.J.M. van de KLUNDERT, E.M.J. NIESSEN* and P.J. ZANDBERGEN[1]

*Faculty of Applied Physics, [1]Faculty of Applied Mathematics, University of Twente, P.O. Box 217, 7500 AE Enschede, The Netherlands (*author for correspondence)*

Key words: superconductors, magnet stability, ac losses

Abstract. In this article a review is given of analytical and numerical calculations on the electromagnetic properties of composite superconducting wires. The review is based on the research performed at the University of Twente during the last ten years. Due attention is given to related results in the literature.

The basic elements of the description are the Maxwell equations supplemented with a set of constitutive equations, relating the electric field **E** and the current density **j** in the composite. The problem is non-linear due to the non-linear **E**–**j** relation describing the superconducting filaments.

The basic analytical and numerical tools for analyzing engineering problems are presented. Furthermore a synopsis is given of characteristic types of numerical results. Some comparisons between analytical and numerical results are also given.

Dedication

The first author suddenly passed away, when this paper was nearly finished. We therefore dedicate this article to the memory of Professor Louis van de Klundert, who was one of the world's leading researchers in the industrial application of superconductivity.

1. Introduction

1.1. Application of large current superconductivity

The practical use of superconductors in large electromagnets has steadily grown since the discovery of the so-called 'hard' type II superconductors, like NbTi and $NbSn_3$, in the late fifties. Already in the early sixties bubble chamber magnets with more than $10\,m^3$ free volume were constructed and used continuously for many years. Also other magnet systems like 4–6 km rings of accelerator dipoles, TEVATRON at Fermilab and HERA at DESY, have been constructed successfully and are operational. Design studies and prototypes testing for Accelerator rings of 27 km and 10 T (LHC, CERN) and 80 km at 6 T (SSC, Dallas) are underway. The first superconducting tokamak T-7 was constructed in the Kurchatov Institute in Moscow and nowadays several much larger systems, TORE SUPRA in Cadarache and T-15 in Moscow, are in operation. Still larger tokamak systems, like NET and ITER, are now under design and are intended to demonstrate the possibility of energy production by controlled fusion in the next decade.

The application of superconductors in levitated trains both for levitation and propulsion with speeds up to 512 km/hr have been demonstrated more than 10 years ago. Still under design and construction are large components for energy production and transport systems at power frequencies, 50–60 Hz. Here we can mention generators, transformers and transport lines. The advantage of using superconductors in quasi DC systems, the large ones mentioned above and small ones, e.g. MRI and laboratory magnets, is the reduction of the power consumption by at least 6 orders of magnitude compared to conventional ones. In AC systems the large reduction in weight and size and better controllability seems to be of major importance, although the reduction in power loss is still considerable.

The average current density in superconducting systems ranges from 50–500 A/mm^2 in external fields up to 12 T. Field rates in AC systems, and in DC systems under fault conditions, may locally be in the range from 10–1000 T/s, either parallel or perpendicular to the local conductor direction. Consequently, larger demands exist on the mechanical stability of superconducting systems than in conventional ones. Moreover, under these conditions the superconducting state is often required to be maintained or the transition to the normal state (quench) has to be performed in a predictable way, such as not to damage the system.

In this paper a review will be given of the electrodynamics of superconductors based primarily on the research performed at the University of Twente during the last ten years. Of course relevant results as reported elsewhere in the literature are taken into account. The review is restricted to the analytical and numerical investigation of twisted wires. Three topics are of main interest: the current carrying capacity, the dissipation and the electromagnetically induced quench. For simplicity all calculations have been performed under isothermal conditions.

1.2. Superconductors, wires and cables

Superconducting wires generally consist of many (10^2–10^6) filaments of superconducting material embedded in a normal conducting matrix: Cu, CuNi or Al. For reasons of loss power reduction and electromagnetic stability the wire is – after drawing it down to almost the desired diameter – twisted in the last reduction step with a twist length L_p ranging from 10 to 100 times the wire radius R.

If η_{sc} is the volume fraction of superconducting material and N_f the number of filaments, the filament radius R_f is given by

$$R_f = R\sqrt{\eta_{sc}/N_f}\,. \tag{1.2.1}$$

Typical values are 0.1–50 μm for R_f, 0.1–0.5 mm for R and 0.1–0.5 for η_{sc}.

Figure 1.1 gives cross sections of some wires. It may be seen that the filament area may or may not include the centre part of the wire. The outer shell of the wire that does not contain filaments may be thick or thin. If we assume that the ring shaped filamentary zone extends from r_1 till r_2, $0 \leq r_1 < r_2 \leq R$, so occupying the fraction $\eta_f = (r_2^2 - r_1^2)/R^2$ of the total cross section, the local fraction of superconductor in the filamentary zone equals

$$\eta = \eta_{sc}/\eta_f\,. \tag{1.2.2}$$

This value of η plays an important role in the next section. In many examples in this paper, however, we will assume that the filaments are uniformly distributed over the cross section of the wire, unless a specific configuration is mentioned.

Figure 1.2 shows the cross section of superconducting cables used for accelerator dipoles and tokamaks. The contact points between the individual wires, usually called strands, will allow for closed current paths and consequently for extra loss power induced by external

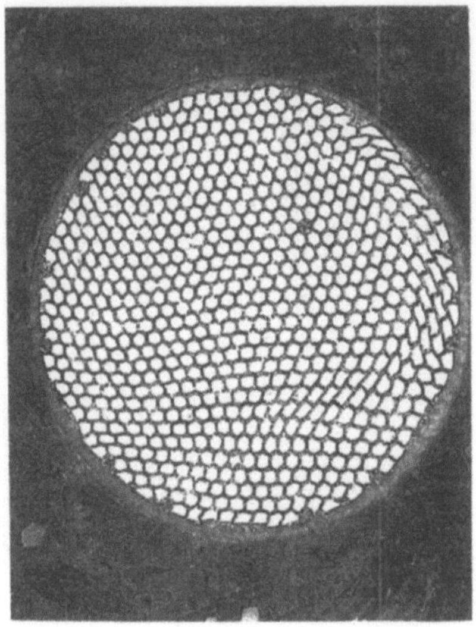

(a)

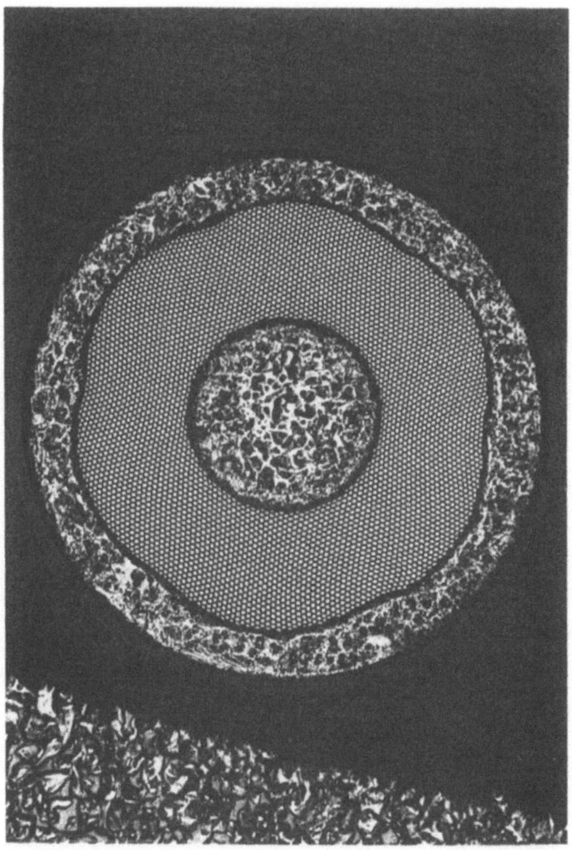

(b)

Fig. 1.1. Cross sections of some multifilamentary superconducting composite wires. The filament area may (a) or may not (b) include the centre part of the wire.

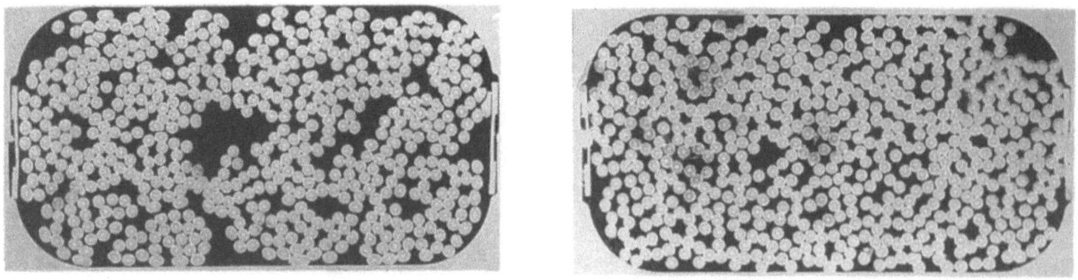

(a) (b)

Fig. 1.2. Cross section of superconducting cables used for accelerator dipoles and tokamacs: *T*-type (a), *L*-type (b).

field changes. The *L*-type gives a much more regular distribution of the voids than the *T*-type (see Fig. 1.2).

The evaluation of the electrodynamic response in a twisted multifilamentary wire can be achieved by solving the Maxwell equations supplemented with an appropriate set of constitutive equations and boundary conditions. In earlier works also simplified potential theory has been used, whereas network approaches still find applications.

2. The mathematical formulation of the problem

2.1. The Maxwell equations

In general the Maxwell equations read:

$$\nabla \times \mathcal{E} = -\partial_t \mathcal{B} , \qquad \nabla \times \mathcal{H} = j + \partial_t \mathcal{D} , \tag{2.1.1}$$

where $\mathcal{E}$ is the electric field, $\mathcal{B}$ the magnetic field, $\mathcal{H}$ the magnetic field strength, j the current density and $\mathcal{D}$ the electric displacement [1]. Using Carr's continuum model [2] the behaviour of composite superconductors will be described in terms of averaged quantities $\mathbf{E}$, $\mathbf{B}$, $\mathbf{j}$ and the so-called magnetization $\mathbf{M}$. The Maxwell equations in a conducting non-magnetic composite and in slowly time varying fields are then given by:

$$\nabla \times \mathbf{E} = -\partial_t \mathbf{B} , \tag{2.1.2}$$

$$\nabla \times (\mathbf{B} - \mu_0 \mathbf{M}) = \mu_0 \mathbf{j} . \tag{2.1.3}$$

Since $\mathbf{M}$ results from macroscopic currents (see Section 2.5) in the filaments and varies much weaker in space and time than $\mathbf{B}$ the term $\mu_0 \mathbf{M}$ can be neglected in (2.1.3). In cylindrical coordinates we can write:

$$\begin{aligned}
&\frac{1}{r}\,\partial_\varphi E_z - \partial_z E_\varphi = -\partial_t B_r , &&\frac{1}{r}\,\partial_\varphi B_z - \partial_z B_\varphi = \mu_0 j_r , \\
&\partial_z E_r - \partial_r E_z = -\partial_t B_\varphi , &&\partial_z B_r - \partial_r B_z = \mu_0 j_\varphi , \\
&\frac{1}{r}\,(\partial_r r E_\varphi - \partial_\varphi E_r) = -\partial_t B_z , &&\frac{1}{r}\,(\partial_r r B_\varphi - \partial_\varphi B_r) = \mu_0 j_z ,
\end{aligned} \tag{2.1.4}$$

and, since $\nabla \cdot (\nabla \times A) = 0$ for any vector, it may be beneficial to replace one of the above equations by:

$$\nabla \cdot \mathbf{B} = \frac{1}{r}(\partial_r r B_r + \partial_\varphi B_\varphi) + \partial_z B_z = 0 ,$$
$$\nabla \cdot \mathbf{j} = \frac{1}{r}(\partial_r r j_r + \partial_\varphi j_\varphi) + \partial_z j_z = 0 . \tag{2.1.5}$$

2.2. Phenomenology of composite superconductors

Taking into account that $R_f/L_p \sim 10^{-3}$, i.e. the filaments are almost parallel to each other, it is useful to define a local coordinate system for the constitutive equations with one coordinate, $e_\parallel$, parallel to the filament direction and two, e_1 and e_2, in the plane perpendicular to $e_\parallel$. Then we have:

$$j_\parallel = \sigma_\parallel E_\parallel + \eta j_p(\mathbf{E}, \mathbf{B}, \dot{\mathbf{B}}) ,$$
$$j_1 = \sigma_{11} E_1 + \sigma_{12} E_2 , \tag{2.2.1}$$
$$j_2 = \sigma_{12} E_1 + \sigma_{22} E_2 .$$

j_p is the superconducting current density and σ_{ij} are elements of the conductivity matrix. The coefficient $\sigma_\parallel$ is given by $\sigma_\parallel = (1-\eta)\sigma_0$ where σ_0 is the bulk conductivity of the matrix material.

For arrangements of filaments with 3-, 4- or 6-fold rotational symmetry, the coarse grained values of the conductivity matrix reduce to

$$\sigma_{12} = 0 \quad \text{and} \quad \sigma_{11} = \sigma_{22} = \sigma_\perp .$$

Carr [3] states that this isotropic value $\sigma_\perp$ is given by

$$\sigma_\perp = \sigma_0(1+\eta)/(1-\eta) \quad \text{or} \quad \sigma_\perp = \sigma_0(1-\eta)/(1+\eta) , \tag{2.2.2}$$

whenever the filament material does or does not contribute to the transverse current conduction, respectively. This is mainly determined by the properties of the matrix superconductor interface. Usually non-conducting filaments are assumed in Cu matrix, whereas perfect conducting material is assumed, when a highly resistive CuNi matrix is used. Kanbara [4] and Rem [5] investigated the validity of relation (2.2.2). Recently it was shown [6] that for hexagonal filaments in a 6-fold symmetry the maximum deviation of the numerically calculated $\sigma_\perp$ from the value given above is about 2% at $\eta = 0.5$.

A generally accepted expression for the superconducting current density j_p does not exist and only simplified expressions, valid in some regimes can be given. The strong non-linear behaviour of the superconducting material under DC conditions can be approximated by either a power law

$$E_\parallel = E_0^{\mathrm{DC}}(|j_p|/j_c)^n \operatorname{sign}(j_p) \quad \text{or} \quad j_p = j_c\sqrt[n]{|E_\parallel|/E_0^{\mathrm{DC}}} \cdot \operatorname{sign}(E_\parallel) \tag{2.2.3}$$

or an exponential expression [7, 8]

$$E_\parallel = E_0^{\rm DC} \frac{j_p}{j_c} \exp[(|j_p| - j_c)/j_1] \quad \text{or} \quad j_p = j_c\left[1 + (j_1/j_c)\ln\left(E_\parallel \Big/ \left(E_0^{\rm DC} \frac{j_p}{j_c}\right)\right)\right] \operatorname{sign}(E_\parallel)\,, \tag{2.2.4}$$

whenever $|E_\parallel| < E_0^{\rm DC}$. j_c can be referred to as the critical current density. The ratio j_1/j_c, called the smoothness factor, is typically smaller than $2 \cdot 10^{-2}$.

If we assume the usual linear critical state relation

$$E_\parallel = \rho_s[j_p - j_c \operatorname{sign}(E_\parallel)] + E_0^{\rm DC}\,, \quad \text{for } |E_\parallel| > E_0^{\rm DC} \tag{2.2.5}$$

with ρ_s the flux flow resistivity, for larger values of $|E_\parallel|$, a continuous function $j_p(E_\parallel)$ exists over the entire j_p domain. The 'critical' current density j_c is a function of $B_\perp = (B_1^2 + B_2^2)^{1/2}$ and $B_\parallel$ [9] and temperature T.

Under AC conditions, i.e. both $\dot{B}_\perp$ and $\dot{B}_\parallel \neq 0$, filaments exhibit a dynamic resistance described by the following relations

$$\begin{aligned} j_p &= j_c E_\parallel / E_0^{\rm AC}\,, \qquad |E_\parallel| < E_0^{\rm AC}\,, \\ j_p &= j_c \operatorname{sign}(E_\parallel)\,, \qquad |E_\parallel| > E_0^{\rm AC}\,. \end{aligned} \tag{2.2.6}$$

For round filaments $E_0^{\rm AC} = (8/3\pi)R_f|\dot{B}_\perp|$ [10]. The influence of $B_\parallel$ on the dynamic resistance never has been investigated. Since the DC resistivity will become much smaller than the dynamic resistivity already close to j_c we will neglect the DC resistivity in all numerical AC calculations. Filaments are said to be unsaturated or saturated, whenever $|j_p|$ is smaller or equal to j_c respectively. A sharp boundary separates the unsaturated from the saturated regions in the filamentary zone.

In cylindrical coordinates the constitutive equations then read

$$\begin{aligned} j_r &= \sigma_\perp E_r\,, \\ j_\varphi &= \beta r j_s + \sigma_{\varphi\varphi} E_\varphi + \sigma_{\varphi z} E_z\,, \\ j_z &= j_s + \sigma_{\varphi z} E_\varphi + \sigma_{zz} E_z\,, \end{aligned} \tag{2.2.7}$$

with: $\beta = 2\pi/L_p$, $\operatorname{tg}\psi = \beta r$, j_s is the superconducting component of j_z:

$$\begin{aligned} j_s &= \eta j_p \cos\psi\,, \\ \sigma_{\varphi\varphi} &= \sigma_\perp \cos^2\psi + \sigma_\parallel \sin^2\psi = \sigma_\perp + \Delta\sigma \sin^2\psi\,, \qquad \sigma_{\varphi z} = \Delta\sigma \sin\psi \cos\psi\,, \\ \sigma_{zz} &= \sigma_\parallel \cos^2\psi + \sigma_\perp \sin^2\psi = \sigma_\parallel - \Delta\sigma \sin^2\psi\,, \end{aligned} \tag{2.2.8}$$

where $\Delta\sigma = \sigma_\parallel - \sigma_\perp$.

The above model originates from Carr and is called the anisotropic continuum model.

A further simplification of the constitutive equations can be obtained by putting $E_0^{\rm DC} = E_0^{\rm AC} = 0$. This means that in case of AC conditions the filament radius and its resulting dynamic resistivity is neglected. This is allowed whenever the (induced) components of **E** are much larger than E_0^{AC}. Filaments thus will be saturated whenever $E_{\parallel} = E_{\varphi} \sin \psi + E_z \cos \psi \neq 0$. In the unsaturated case the constitutive equations further reduce to

$$\begin{aligned} j_r &= \sigma_{\perp} E_r , \\ j_{\varphi} &= \beta r j_s + \sigma_{\perp} E_{\varphi} , \\ j_z &= j_s + \sigma_{\perp} E_z , \qquad E_z = -\beta r E_{\varphi} . \end{aligned} \tag{2.2.9}$$

This reduction makes the Maxwell equations accessible for analytic considerations. It may be noted that $\sigma_{\parallel}$ disappears from the equations regardless of its value.

2.3. The applied field $\mathbf{B}^{A}$

Each of the components of the applied magnetic field $\mathbf{B}^{A}$ may be periodic functions of φ and z. We separate the total field $\mathbf{B}$ in an applied and an induced part $\mathbf{B} = \mathbf{B}^{A} + \mathbf{B}^{I}$. $\mathbf{B}^{A}$ then is the field generated in the volume of the conductor (wire, cable, braid) by currents outside the conductor, $\mathbf{B}^{I}$ is the field generated in all space by currents flowing inside the conductor. In all cases considered here no current passes through the conductor surface. If $\mathbf{B}^{A} = \mathbf{0}$ one speaks of a self field problem.

If we follow the path of one strand in a fully transposed cable the local applied field components will vary periodically since the direction of the strand will change compared with the general direction of the applied field. Moreover, if a multi coil system like a tokamak is considered, also other period lengths of the applied field have to be considered.

In general the applied field can be thought of as the sum of four basic components. In cylindrical coordinates the two components uniform in z are

$$\mathbf{B}_1^{A} = B_1(\sin \varphi, \cos \varphi, 0) , \qquad \mathbf{B}_2^{A} = B_2(0, 0, 1)$$

and the two periodic in z:

$$\begin{aligned} \mathbf{B}_3^{A} &= B_2\left(I_1'(pr) \sin \varphi \sin pz, \frac{I_1(pr)}{pr} \cos \varphi \sin pz, I_1(pr) \sin \varphi \cos pz \right) , \\ \mathbf{B}_4^{A} &= B_4(I_1(pr) \sin pz, 0, I_0(pr) \cos pz) . \end{aligned} \tag{2.3.1}$$

$\mathbf{B}_1^{A}$ and $\mathbf{B}_3^{A}$ have their main contribution in the y-direction, $\mathbf{B}_2^{A}$ and $\mathbf{B}_4^{A}$ in the z-direction. I_0 and I_1 are Bessel functions of the second kind. $p = 2\pi / L_z$, L_z is the period length in the z-direction. These four components all are regular solutions of $\nabla \cdot \mathbf{B}^{A} = \nabla \times \mathbf{B}^{A} = 0$ for $0 \leqslant r \leqslant R$.

The coefficients B_i, $i = 1, 4$, are functions of time. Two cases are considered here: the periodic time dependence $B_i = B_0 \, e^{-i\omega t}$ and the linear time dependence $B_0 = 0$ for $t < 0$, $B_0 = \alpha t$ for $t \geqslant 0$. In the latter case we have to deal with an in time decaying effect of the transient occurring at $t = 0$. For large times compared to specific response times of the system, $\dot{\mathbf{B}}^{I}$ will vanish and we will refer to this situation as the stationary case.

2.4. The induced field $\mathbf{B}^{I}$ *and the boundary conditions*

The shape of the induced field $\mathbf{B}^{I}$ outside the wire, resulting from the current distribution induced in the wire by the applied field $\mathbf{B}^{A}$ and the transport current I^{A} in general has a very complicated structure. When the applied magnetic field is described using some Fourier terms in the φ and z-direction, higher Fourier modes can appear in the induced field outside the wire due to:

a. The non-linear $j_\parallel(E_\parallel)$ relation, which may result in the occurrence of boundaries $r(\varphi, z, t)$, separating saturated and unsaturated regions;
b. the finite sample length, due to boundary conditions at the ends, i.e. at $z = \pm L$, where $2L$ is the length of the sample.

The induced field can be calculated from $\nabla \cdot \mathbf{B}^{\mathrm{I}} = \nabla \times \mathbf{B}^{\mathrm{I}} = 0$ under the condition $\mathbf{B}^{\mathrm{I}}(r \to \infty) = \mathbf{0}$.

For each of the applied fields $\mathbf{B}_i^{\mathrm{A}}$ given in (2.3.1) we get

$$\mathbf{B}_1^{\mathrm{I}} = \sum_m a_m(\sin m\varphi, \cos m\varphi, 0)\,,$$

$$\mathbf{B}_2^{\mathrm{I}} = \mathbf{0}\,,$$

$$\mathbf{B}_3^{\mathrm{I}} = \sum_n \sum_m b_{nm}(K_m'(npr) \sin m\varphi \sin npz\,, \quad K_m(npr) \sin m\varphi \sin npz/npr,\ K_m(npr) \sin m\varphi \cos npz)\,, \tag{2.4.1}$$

$$\mathbf{B}_4^{\mathrm{I}} = \sum_n c_n(K_1(npr) \sin npz, 0, -K_0(npr) \cos npz)\,.$$

Here a_m, b_{nm} and c_n are functions of t which still must be calculated; K_0 and K_1 are Bessel functions of the second kind.

The induced field from the transport current I^{A} is given by:

$$B_\varphi^{\mathrm{I}} = \mu_0 I^{\mathrm{A}}/2\pi r\,. \tag{2.4.2}$$

In general the (double) summations have to be taken over an infinite number of terms. Notice that the exact form of the applied field is given, equation (2.3.1), but only the shape of the induced field outside the wire is given, equation (2.4.1). The currents and field components inside the wire must now be calculated using that the values of the coefficients in (2.4.1) follow from the continuity of $\mathbf{B}^{\mathrm{I}}$ at $r = R$. In numerical calculations it is sufficient to impose the boundary conditions in an integral way [5] (see Section 4.2.2). Other boundary conditions, to be fulfilled when determining the solution of the Maxwell equations, are $E_r(R) = 0$ and $B_r^{\mathrm{I}}(0) = B_\varphi^{\mathrm{I}}(0) = 0$.

It may be noted that the response to a linear combination of B_i^{A} given in (2.3.1) is not necessarily a linear combination of B_i^{I} given in (2.4.1). Such complex problems, however, have not been studied so far.

2.5. The magnetization and dissipated power density

In a virgin filament, changes of the external field will induce screening currents which hold up the penetration of the external field into the filament. Flux penetration occurs because the current density is limited by the critical current density. The penetration field is the magnetic field amplitude beyond which the interior of the superconductor can no longer by shielded from the external magnetic field. As a consequence, for magnetic field changes larger than the penetration field, the total amount of superconducting material is positively or negatively saturated. The flux penetration process is not reversible, i.e. the magnetization shows a hysteresis behaviour. Subjecting a superconductor to an alternating magnetic field therefore involves dissipation. So inside the filament a closed current loop appears having an effective magnetic moment. The averaged **M** in a multifilamentory superconductor results from adding all the small magnetic moments of current distributions inside the filaments.

We already saw this in Carr's continuum model where for averaged quantities we found equation (2.1.3). In this equation only **M** is reminiscent of the variations in current density within the filaments.

The induced screening currents will penetrate the whole filament if the field change amounts [11]

$$\begin{aligned} B_{p\|} &= \mu_0 j_{c\varphi} R_f\,, \quad \text{in parallel field change,} \\ B_{p\perp} &= \frac{2}{\pi}\,\mu_0 j_{cz} R_f\,, \quad \text{in perpendicular field change.} \end{aligned} \tag{2.5.1}$$

$j_{c\varphi}$ and j_{cz} may have different values due to the production process of the wire. A field change of $2B_p$ is needed to bring the filament from one state of saturation into the opposite one. In an array of widely spaced filaments with 6-fold symmetry and a perpendicular field change larger than $2B_p$ a constant magnetization is obtained

$$\begin{aligned} M_\perp &= -\frac{4}{3\pi}\,\eta j_{cz} R_f\,\mathrm{sign}(\dot{B}_\perp)\,, \quad \text{if } \dot{B}_\| = 0\,, \\ M_\| &= -\frac{1}{3}\,\eta j_{c\varphi} R_f\,\mathrm{sign}(\dot{B}_\|)\,, \quad \text{if } \dot{B}_\perp = 0\,. \end{aligned} \tag{2.5.2}$$

If the filament carries a transport current equivalent to an average current density j_p the magnetization is in a good approximation reduced by a factor $1-(j_p/j_c)^2$ [5]. The local loss power density p is given by

$$\begin{aligned} p &= \mathbf{j}\cdot\mathbf{E} - \mathbf{M}\cdot\dot{\mathbf{B}} \\ &= \sigma_\perp(E_r^2 + E_\perp^2) + \sigma_\| E_\|^2 + \eta j_p E_\| - M_\perp \dot{B}_\perp - M_\| \dot{B}_\|\,, \end{aligned} \tag{2.5.3}$$

where either $\dot{B}_\perp$ or $\dot{B}_\| = 0$.

Values for $M_\perp$ and $B_{p\perp}$ for hollow filaments and filaments of square cross section have been reported by [12] and [10], respectively.

Notice that in the Maxwell equations (2.1.4) the magnetization **M** is neglected while in the loss calculations, equation (2.5.3), the loss term due to the magnetization is present. The reason for neglecting **M** in equation (2.1.4) is twofold:

1) $|\nabla \times \mu_0 \mathbf{M}| \ll |\nabla \times \mathbf{B}|$,
2) $\nabla \times \mathbf{M} = \mathbf{0}$ because **M** is constant if $B > 2B_p$.

3. Analytical solutions

The solution of the Maxwell equations can be found analytically only if a considerable reduction in the complexity of the problem can be obtained. Four examples will be given below. In the first one the conductor, e.g. a hollow wire of twisted filaments, is represented by a surface current, with linear properties, only. It should be remarked that this model is also valid for a hollow cable of twisted strands. In the second and third example the solution for uniform fields perpendicular and parallel to the wire are given. In the last example the solution in one turn of a cylindrical coil placed in a uniform field perpendicular to the coil axis is presented.

The non-linear character of the problem, resulting from the non-linear $j_{\|}$-component of the constitutive equation, can be removed by assuming that the whole interior of the wire is unsaturated. In order to account for the boundary condition $E_r(R)=0$ or $B_\varphi(R)=B_\varphi^0$ in rotational symmetric problems a surface current $\mathbf{J}$, which also behaves linearly, is introduced. $\mathbf{J}$ flows in the filament direction and has components J_z and J_φ, with $J_\varphi=\beta RJ_z$. If $E_r(R^-)\neq 0$ the condition $E_r(R^+)=0$ is fulfilled if

$$\partial_z J_z + R^{-1}\partial_\varphi J_\varphi = \sigma_r E_r(R^-)\,, \tag{3.1}$$

which relation follows from conservation of current at the surface of the wire. $E_r(R^-)$ and $E_r(R^+)$ denote the limiting values for $r\to R$, if $r<R$ or $r>R$ respectively. The continuity of $\mathbf{B}^{\mathrm{I}}$ at $r=R$ is maintained, but for B_φ^{I} and B_z^{I} the following relations replace the continuity requirement in absence of the surface current $\mathbf{J}$:

$$B_\varphi^{\mathrm{I}}(R^+) - B_\varphi^{\mathrm{I}}(R^-) = \mu_0 J_z\,, \qquad B_z^{\mathrm{I}}(R^-) - B_z^{\mathrm{I}}(R^+) = \mu_0 J_\varphi\,. \tag{3.2}$$

These discontinuity relations follow from the Stokes versions of the Maxwell equations. The analytical approximations should be compared to numerical solutions taking into account the spatial extension of the shielding currents at the surface and the non-linear **E–j** relation. Numerical solutions will be presented in Section 4.

3.1. *A hollow cylinder in a periodically applied field*

If we consider a hollow cylinder of infinitesimal thickness and surface conductivity σ perpendicular to the filament direction, we can write for the most general applied field $\mathbf{B}^{\mathrm{A}}$ with a fixed frequency ω [10]:

$$\mathbf{B}^{\mathrm{A}} = \sum_n \sum_m a_{nm}^{\mathrm{A}}\, \mathrm{e}^{\mathrm{i}(m\varphi+npz-\omega t)}(I_m'(npr), \mathrm{i}I_m(npr)/npr, \mathrm{i}I_m(npr))\,.$$

For $r<R$ the induced field $\mathbf{B}^{\mathrm{I}}$ can be written

$$\mathbf{B}^{\mathrm{I}} = \sum_n \sum_m a_{nm}\, \mathrm{e}^{\mathrm{i}(m\varphi+npz-\omega t)}(I_m'(npr), \mathrm{i}I_m(npr)/npr, \mathrm{i}I_m(npr))$$

and for $r>R$, since B_r is continuous

$$\mathbf{B}^{\mathrm{I}} = \sum_n \sum_m a_{nm}\, \mathrm{e}^{\mathrm{i}(m\varphi+npz-\omega t)} I_m'(npR)/K_m'(npR)$$
$$(K_m'(npr), \mathrm{i}K_m(npr)/npr, \mathrm{i}K_m(npr))\,.$$

If we apply equation (3.2) we find

$$\mu_0(J_\varphi, J_z) = \sum_n \sum_m \mathrm{i}a_{nm}\, \mathrm{e}^{\mathrm{i}(m\varphi + npz - \omega t)}/npR \cdot K_m'(npR)(-1, m/npr).$$

If we now write the constitutive equations in the form

$$\sigma E_\varphi = \cos^2\psi J_\varphi - \sin\psi\cos\psi J_z\,,$$

$$\sigma E_z = \sin^2\psi J_z - \sin\psi\cos\psi J_\varphi\,,$$

we find from $R^{-1}\partial_\varphi E_z - \partial_z E_\varphi = -\partial_t B_r$

$$a_{nm} = \frac{-\mathrm{i}\mu_0\omega\sigma R(npr)^2 K_m'(npR) I_m'(npr) a_{nm}^{\mathrm{A}}}{(m\sin\psi + npR\cos\psi)^2 + \mathrm{i}\omega\mu_0\sigma R(npR)^2 K_m'(npR) I_m'(npR)}$$

and for the Ohmic loss P per unit cylinder surface

$$P = \sum_{n,m} P_{nm} = \sum_n \sum_m \frac{1}{8\sigma}\left(\frac{m\sin\psi + npR\cos\psi}{\mu_0(npR)^2 K_m'(npR)}\right)^2 |a_{nm}|^2\,.$$

In Fig. 3.1 $P_{4,1}/P_0$ is shown for various values of $\omega\tau = \omega\mu_0\sigma R/(2\sin^2\psi)$, P_0 is the loss in a uniform steadily increasing perpendicular field $\bar{B}_x = \alpha t$:

$$P_0 = \frac{\sigma}{8\pi^2}(L_p\dot{B}_x^{\mathrm{A}})^2(1 + \beta^2R^2) = \frac{\sigma}{2}\,\dot{B}_x^{\mathrm{A}2}(R^2 + \beta^{-2}).$$

The limit of $P_{4,1}/P_0$ for $L_z/L_p \to \infty$ equals 1/2 because of the time averaging.

3.2. *Twisted wire in uniform perpendicular field*

In analogy with the previous treatment the case of a perpendicular applied field can be given. For reasons of brevity only the case of z-invariant sinusoidal time dependence of the applied field will be presented. For more details see [13], [14] and [15].

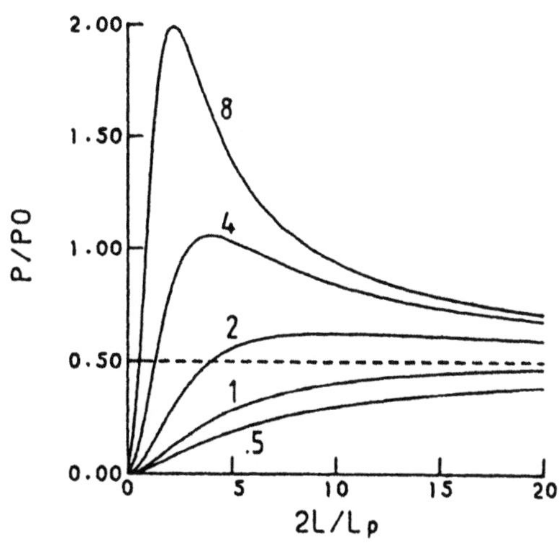

Fig. 3.1. $P_{4,1}/P_0$ for various values of $\omega\tau$ [10].

The Maxwell equations for this special case read

$$\begin{aligned}
&\partial_\varphi E_z = \mathrm{i}\omega r B_r\,,\\
&-\partial_r E_z = \mathrm{i}\omega B_\varphi\,,\\
&\partial_r r E_\varphi - \partial_\varphi E_r = \mathrm{i}\omega r B_z\,,\\
&\partial_\varphi B_z = \mu_0 \sigma_r r E_r\,,\\
&-\partial_r B_z = \beta r \mu_0 j_s + \mu_0 \sigma_\perp E_\varphi\,,\\
&\partial_r r B_\varphi - \partial_\varphi B_r = r(\mu_0 j_s + \mu_0 \sigma_\perp E_z)\,.
\end{aligned} \tag{3.2.1}$$

Assuming

$$\begin{aligned}
&B_r = g \sin\varphi\, \mathrm{e}^{-\mathrm{i}\omega t}\,,\\
&E_r = \frac{\mathrm{i}\omega}{\beta} h \sin\varphi\, \mathrm{e}^{-\mathrm{i}\omega t}\,,\\
&\mu_0 j_s = \mathrm{i}\omega\tau k \cos\varphi\, \mathrm{e}^{-\mathrm{i}\omega t}\,,
\end{aligned}$$

with $\tau = \mu_0\sigma_\perp/\beta^2$, we get for the surface current

$$\mu_0 J_z = -\mathrm{i}\omega\tau h(R) \cos\varphi\, \mathrm{e}^{-\mathrm{i}\omega t}$$

and $J_\varphi = \beta R J_z$, and the remaining components of E and B

$$\begin{aligned}
&E_\varphi = \frac{\mathrm{i}\omega}{\beta} g \cos\varphi\, \mathrm{e}^{-\mathrm{i}\omega t}\,,\\
&B_z = -\mathrm{i}\omega\tau\beta r h \cos\varphi\, \mathrm{e}^{-\mathrm{i}\omega t}\,,\\
&B_\varphi = \partial_r(rg) \cos\varphi\, \mathrm{e}^{-\mathrm{i}\omega t}\,.
\end{aligned} \tag{3.2.2}$$

Putting

$$g = A \sum_{n=0} a_n u^{2n}\,, \qquad h = A \sum_{n=0} b_n u^{2n}\,, \quad \text{with } u = \beta r \tag{3.2.3}$$

recurrence relations for a_n and b_n can be derived.

$$\begin{aligned}
&a_{-1} = b_{-1} = 0\,, \qquad a_0 = b_0 = 1\,,\\
&4n(n+1)(1-\mathrm{i}\omega\tau)a_n = \mathrm{i}\omega\tau[((2n+1)^2 - 1 + \mathrm{i}\omega\tau)b_{n-1} - (2n+1)a_{n-1}]\,,\\
&4n(n+1)(1-\mathrm{i}\omega\tau)b_n = \mathrm{i}\omega\tau[\mathrm{i}\omega\tau(2n+1)b_{n-1} - a_{n-1}]\,.
\end{aligned}$$

Notice that the convergence interval of the power series is bounded. From the boundary conditions A can be determined to be

$$A = 2B_0/[A + (1 - i\omega\tau(1 + \beta^2 R^2))A]\,, \tag{3.2.4}$$

and consequently the power loss per unit length

$$S = \frac{\pi R}{2\mu_0} E_z B_\varphi^* \,.$$

Here B_φ^* is the complex conjugate of B_φ. If we write

$$S = \frac{\pi R^2}{2\mu_0} i\omega B_0^2(\mu' + i\mu'') \tag{3.2.5}$$

the values of μ' and μ'' can be plotted as shown in Fig. 3.2 [15].

For small β the $\mu''(\mu')$ curve has an ellipse as a limiting envelope whereas for $\beta \to \infty$, where the filaments degenerate in rings, the relation for a solid copper wire

$$\mu = 1 + J_2(kr)/J_0(kr)\,, \tag{3.2.6}$$

where $k^2 = i\omega\mu_0\sigma_\perp$ and J_0 and J_2 are Bessel functions of the first kind, must be obtained. This limit cannot be verified using these series expansions for convergence reasons.

It may be noted that for small ω the power loss is [15]

$$P = \frac{\omega B_0^2}{\mu_0} \frac{\omega\tau'}{1 + (\omega\tau')^2}\,, \quad \text{with } \tau' = \frac{\mu_0\sigma_\perp}{2}\left[\left(\frac{L_p}{2\pi}\right)^2 + R^2\right] \tag{3.2.7}$$

rather than the widely used expression for τ' where the term R^2 is omitted.

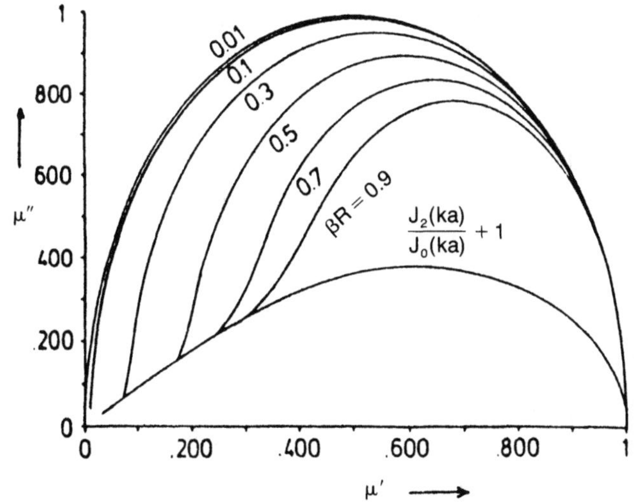

Fig. 3.2. Representation of the Poynting vector in terms of a complex permeability for a uniform perpendicular applied field [15].

3.3. Twisted wire in parallel field

The shape of the applied field is chosen to be rotational symmetric. Then the Maxwell equations read

$$\begin{aligned}
&-\partial_z E_\varphi = -\partial_t B_r ,\\
&\partial_z E_r - \partial_r E_z = -\partial_t B_\varphi ,\\
&\partial_r (rE_\varphi) = -r\partial_t B_z ,\\
&-\partial_z B_\varphi = \mu_0 \sigma_r E_r ,\\
&\partial_z B_r - \partial_r B_z = \beta r \mu_0 j_s + \mu_0 \sigma_\perp E_\varphi ,\\
&\partial_r (rB_\varphi) = r(\mu_0 j_s + \mu_0 \sigma_\perp E_z) .
\end{aligned} \tag{3.3.1}$$

If we assume that the problem is also z-invariant, we immediately get $E_r = B_r = 0$ irrespective of the time dependence of $\mathbf{B}^A$. Moreover, it can be shown that $E_z + \beta r E_\varphi = 0$ implies $B_\varphi - \beta r B_z = 0$ [16]. $E_r = 0$ implies that no current exchange in the r-direction occurs, irrespective of the value of σ_r [17, 18].

Two cases can be considered with respect to the time dependence:

$$B_z^A = 0 \quad t < 0, \qquad B_z^A = \alpha t \quad t \geq 0 \quad \text{or} \quad B_z^A = B_0 \, e^{-i\omega t} .$$

For E_φ a partial differential equation can be obtained,

$$\partial_r \frac{1}{r}(1 + \beta^2 r^2)\partial_r r E_\varphi = (1 + \beta^2 r^2)\mu_0 \sigma_\perp \dot{E}_\varphi . \tag{3.3.2}$$

In principle the equation can be solved applying a separation of variables technique. This has so far not been done. In the stationary case, i.e. after the decay of the transient effect, the solution reads

$$\begin{aligned}
&E_\varphi = -\alpha \ln(1 + \beta^2 r^2)/2\beta^2 r ,\\
&B_\varphi = \beta r \alpha t/(1 + \beta^2 r^2) ,\\
&B_z = \alpha t/(1 + \beta^2 r^2) ,
\end{aligned} \tag{3.3.3}$$

and

$$\mu_0 j_s = 2\beta \alpha t/(1 + \beta^2 r^2)^2 , \tag{3.3.4}$$

From (3.3.2) it can be seen, that if $\sigma_\perp = 0$ $(\tau = 0)$ saturation occurs first at $r = 0$ and

$$B_z^A = \alpha t = \mu_0 \eta j_c L_p / 4\pi . \tag{3.3.5}$$

Thus saturation in the inner region occurs after a change in B_z^A independent of the rate of change of B_z^A.

In case a sinusoidal time dependence of B_z^A is assumed $B_z^A = B_0\,e^{-i\omega t}$ it is useful to put

$$
\begin{aligned}
&E_\varphi = i\omega g\,, \qquad E_z = -i\omega\beta r g\,, \\
&B_\varphi = \beta\partial_r(rg)\,, \qquad B_z = \frac{1}{r}\,\partial_r(rg)\,, \qquad \mu_0 j_s = k\,.
\end{aligned}
\tag{3.3.6}
$$

Writing

$$
g = A\sum_{n=0} a_n u^{2n+1}\,, \qquad k = A\sum_{n=0} b_n u^{2n}\,, \quad \text{with } u = \beta r
$$

the following recurrence relations for a_n and b_n exist:

$$
\begin{aligned}
&a_{-1} = b_{-1} = 0\,, \quad a_0 = 1\,, \\
&4(n+1)(n+2)a_{n+1} = -[4(n+1)^2 a_n + i\omega\tau(a_n + a_{n-1})]\,, \\
&b_n = \beta^2[(2n+1)^2 a_n + i\omega\tau a_{n-1}]\,, \\
&\tau = \mu_0\sigma_\perp/\beta^2\,.
\end{aligned}
$$

The constant A follows from the boundary condition for $B_\varphi^A = 0$

$$
A = B_0/\beta(1+\beta^2R^2)\sum_{n=0}(2n+2)a_n(\beta R)^{2n}\,. \tag{3.3.7}
$$

The loss power per unit of length can be determined from the real part of the Poynting vector

$$
S = \frac{\pi R}{\mu_0}\,E \times B^*\,. \tag{3.3.8}
$$

If we write

$$
S = \frac{\pi R^2}{2\mu_0}\,i\omega B_0^2(\mu' + i\mu'')
$$

the real and imaginary part of $\mu = \mu' + \mu''$ can be plotted for $\omega\tau$ and βR as a parameter. Figure 3.3 shows these curves [15]. For $\beta R = 0$, i.e. no twist, the curve for a pure metallic bar is found whereas for $\beta R \to \infty$, i.e. the filaments degenerate in rings, the semicircle of a paramagnetic material must be obtained.

3.4. Field in the plane of a turn of a cylindrical coil

In important practical applications wires are used in ring-shaped configurations, e.g. the *D*-coils in the Next European Torus fusion reactor [19]. In this subsection the properties of a

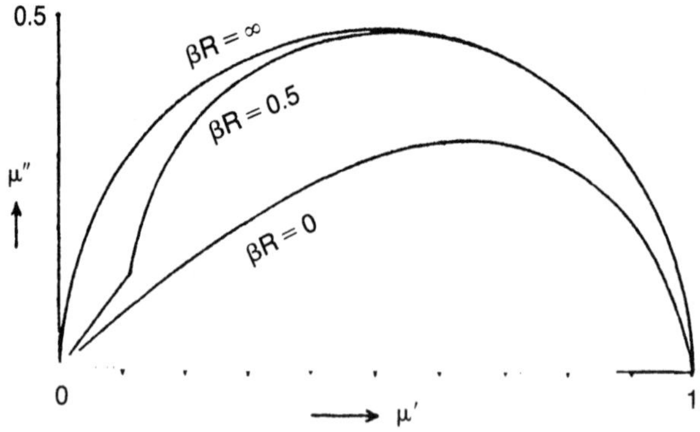

Fig. 3.3. Representation of the Poynting vector in terms of a complex permeability for a uniform parallel applied field [15].

D-coil in an AC magnetic field are studied, using an ideal torus configuration for the wire, which represents one turn of the coil. The radius of the wire is R and the mean radius of the torus is $R_0 \gg R$ (see Fig. 3.4).

The applied time dependent magnetic field is considered to be uniform and parallel to the plane of the torus with constant time derivative $\dot{B}^A$. This field, chosen in the z-direction ($\dot{B}_z^A$) will be perpendicular to the wire for coordinates $(x, z) = (0, \pm R_0)$ and parallel for $(x, z) = (\pm R_0, 0)$. For other coordinates (x, z) on the wire the field is partly perpendicular and partly parallel.

We want to perform the calculations in the natural r, φ, θ system where θ represents the angle on the torus with the positive x-axis and r and φ are the cylindrical coordinates perpendicular to θ. The configuration and definitions are outlined in Fig. 3.4. The (r, φ, θ) system is a positive oriented orthogonal coordinate system. The coordinate transformation for the vector **x** with x, y and z coordinates to the new coordinate system reads:

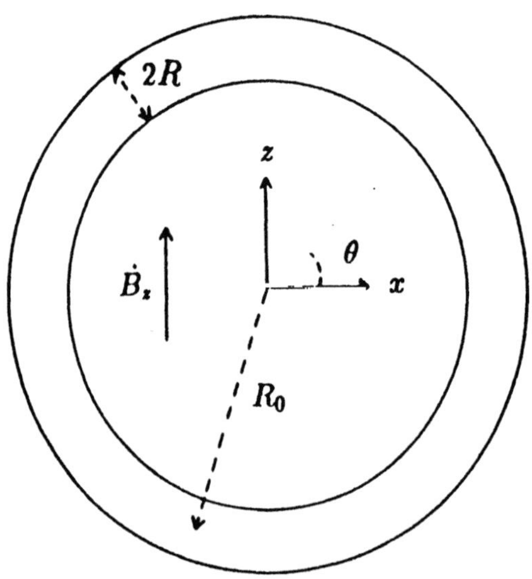

Fig. 3.4. Circular configuration for the wire with radius R in a torus shape with mean radius R_0. Here $R \ll R_0$ [20].

$$(x, y, z) = ((R_0 + r\cos\varphi)\cos\theta, r\sin\varphi, (R_0 + r\cos\varphi)\sin\theta)\,, \tag{3.4.1}$$

which gives for the elementary lengths:

$$h_r = 1\,, \qquad h_\varphi = r\,, \qquad h_\theta = R_0 + r\cos\varphi\,. \tag{3.4.2}$$

In this new coordinate system the Maxwell equations in case of a stationary problem read:

$$(\nabla\times\mathbf{E})_r = \frac{1}{r}\,\partial_\varphi E_\theta - \frac{1}{R_0 + r\cos\varphi}\,[\partial_\theta E_\varphi + \sin\varphi E_\theta] = -\dot{B}_r^{\mathrm{A}}\,, \tag{3.4.3}$$

$$(\nabla\times\mathbf{E})_\varphi = -\partial_r E_\theta - \frac{1}{R_0 + r\cos\varphi}\,[-\partial_\theta E_r + \cos\varphi E_\theta] = -\dot{B}_\varphi^{\mathrm{A}}\,, \tag{3.4.4}$$

$$(\nabla\times\mathbf{E})_\theta = \frac{1}{r}\,[\partial_r(rE_\varphi) - \partial_\varphi E_r] = -\dot{B}_\theta^{\mathrm{A}}\,. \tag{3.4.5}$$

The applied magnetic field $\dot{B}_z^{\mathrm{A}}$ can be written in its r, φ, θ components: $\dot{\mathbf{B}}^{\mathrm{A}} = \dot{B}_z^{\mathrm{A}}(\cos\varphi\sin\theta, -\sin\varphi\sin\theta, \cos\theta)$. Conservation of the bulk current density $\mathbf{j}$ in the interior of the wire is written as:

$$\nabla\cdot\mathbf{j} = \frac{1}{r}\,[\partial_r(rj_r) + \partial_\varphi j_\varphi] + \frac{1}{R_0 + r\cos\varphi}\,[\cos\varphi j_r - \sin\varphi j_\varphi + \partial_\theta j_\theta] = 0\,. \tag{3.4.6}$$

At the surface of the wire $r = R$ we consider a surface current with components J_φ and J_θ in the φ and θ direction respectively. From the fact that no current flows out of the wire at the boundary $r = R$, conservation of current is written as:

$$-(R_0 + R\cos\varphi)Rj_r(r = R) + (R_0 + R\cos\varphi)\partial_\varphi J_\varphi - J_\varphi R\sin\varphi + R\partial_\theta j_\theta = 0\,. \tag{3.4.7}$$

The differential equations cannot be solved directly so we use a power series in r because $R \ll R_0$. We show the calculation of E_φ in detail. Substituting $E_\parallel = 0$ so $E_\theta = -\beta r E_\varphi$ in equation (3.4.3) we obtain:

$$2\pi\partial_\varphi E_\varphi + \frac{L_p}{R_0}\,\partial_\theta E_\varphi = L_p\dot{B}_r^{\mathrm{A}} + \frac{r}{R_0}\,[L_p\cos\varphi\dot{B}_r^{\mathrm{A}} + 2\pi\sin\varphi E_\varphi - 2\pi\cos\varphi\partial_\varphi E_\varphi]\,. \tag{3.4.8}$$

The homogeneous solution of (3.4.8) is zero [20]. For finding the particular solution we write E_φ as a power series in r: $E_\varphi = E_\varphi^{(0)} + rE_\varphi^{(1)} + r^2E_\varphi^{(2)} + \cdots$. This gives:

$$2\pi\partial_\varphi E_\varphi^{(0)} + \frac{L_p}{R_0}\,\partial_\theta E_\varphi^{(0)} = L_p\dot{B}_r^{\mathrm{A}}$$

and

$$2\pi\partial_\varphi E_\varphi^{(1)} + \frac{L_p}{R_0}\,\partial_\theta E_\varphi^{(1)} = \frac{L_p}{R_0}\,\cos\varphi\dot{B}_r^{\mathrm{A}} + \frac{2\pi}{R_0}\,\sin\varphi E_\varphi^{(0)} - \frac{2\pi}{R_0}\,\cos\varphi\partial_\varphi E_\varphi^{(0)}$$

for $E_\varphi^{(0)}$ and $E_\varphi^{(1)}$ respectively. The solution is:

$$E_{\varphi}^{(0)} = \frac{L_p \dot{B}_z^{\mathrm{A}}}{2\pi(1-k^2)} [\sin\varphi \sin\theta + k\cos\varphi\cos\theta]$$

and

$$E_{\varphi}^{(1)} = -\frac{\dot{B}_z^{\mathrm{A}}}{2}\cos\theta - \frac{\dot{B}_z^{\mathrm{A}} k(1+2k^2)}{(1-k^2)(4-k^2)}\sin 2\varphi \sin\theta$$

$$- \frac{\dot{B}_z^{\mathrm{A}} k^2(5+k^2)}{2(1-k^2)(4-k^2)}\cos 2\varphi\cos\theta$$

with

$$k = \frac{L_p}{2\pi R_0} .$$

The higher order Fourier terms in φ appear because h_θ contains the term $r\cos\varphi$, see equation (3.4.2). For the other electric field terms and the currents the power series approach is equivalent.

Due to the shielding currents the maximal transport current at low losses (I_{max}) is for small k approximated by:

$$I_{\mathrm{max}} = \pi R^2 \left[\eta j_c - 2\sigma_\perp |\dot{B}_z^{\mathrm{A}}| \frac{L_p}{2\pi} \cdot \frac{1}{k^2} \right] . \tag{3.4.9}$$

The maximum value of $|\dot{B}_z^{\mathrm{A}}|$ for which the centre of the wire is unsaturated is given by:

$$|\dot{B}_z^{\mathrm{A}}| < \frac{\pi\eta j_c k^2}{\sigma_\perp L_p} \quad \text{(for small } k\text{)} . \tag{3.4.10}$$

Notice that for this value of $|\dot{B}_z^{\mathrm{A}}|$, I_{max} is zero. Furthermore, the coupling losses per unit volume are given by:

$$\frac{P}{\mathrm{Vol}} = \frac{\sigma_\perp \dot{B}_z^{\mathrm{A}2}}{2}\left(\frac{L_p}{2\pi}\right)^2 + \frac{\sigma_\perp \dot{B}_z^{\mathrm{A}2}}{4} \cdot \frac{a^2}{k^2} \quad \text{(for small } k\text{)} , \tag{3.4.11}$$

where the first and second term are due to the perpendicular and parallel component of the applied field $\dot{B}_z^{\mathrm{A}}$ respectively. The first term is half of the value arising in case the field is perpendicular over the whole length of the wire.

Another commonly used approach for calculating the field and currents in a toroidal configuration is as follows: consider a straight wire and apply a spatially periodic time dependent magnetic field. The torus problem can be approximated by a straight wire in a spatially dependent magnetic field for small values of k [20].

4. Numerical solutions

4.1. Introduction

In the previous section, where analytical solutions were calculated we had to idealize the problem significantly in order to be able to perform the calculations. In this section we investigate aspects concerning numerical solutions. The reason for performing numerical calculations is threefold:

(1) to find bounds for the validity of the analytical approximations (for design purposes),
(2) to compare non-trivial analytical solutions with numerical data (checks on the solution methods),
(3) to be able to obtain results where no analytical calculations are possible, mainly due to the non-linear **E**–**j** relation. The most important cases that cannot be treated analytically are:

    (1) $R_f \neq 0$, this case will not be considered in this article, see [10],
    (2) time dependency/transient phenomena including saturation,
    (3) $I^{\mathrm{A}} \neq 0$,
    (4) calculation of the exact form of boundaries between positive/negative/unsaturated regions.

These items will be considered in this section. We treat the basic ideas of the numerical model and present some specific numerical results. The comparison between numerical and analytical results is performed for parallel applied magnetic fields.

Even the largest supercomputers are not able to cope with general 4 dimensional space-time problems due to large cpu times and high storage requirements. Therefore in general we have to reduce the dimensions of the problem using symmetry arguments like rotational symmetry (φ-invariance) or z-invariance. Consequence is that some terms in the Maxwell equations are not present anymore. The number of unknowns and the cpu time are reduced considerably.

4.2. *Numerical model*

Three important aspects can be determined concerning the numerical model:

*) Grid using the method of grid staggering,
*) Boundary conditions,
*) Non-linear **E**–**j** relation.

We now investigate every aspect in more detail.

4.2.1. *Grid staggering*

When considering a convenient grid, use is made of the so-called staggered space grid. Maxwell's equations in integral form read:

$$\begin{aligned} \int_{\partial s} \mathbf{B} \cdot \mathrm{d}\mathbf{l} &= \iint_S \mu_0 \mathbf{j} \cdot \mathbf{e}_n \, \mathrm{d}S \,, \\ \int_{\partial s} \mathbf{E} \cdot \mathrm{d}\mathbf{l} &= \iint_S -\dot{\mathbf{B}} \cdot \mathbf{e}_n \, \mathrm{d}S \,. \end{aligned} \tag{4.2.1.1}$$

These equations are discretized second-order accurate in time and space. For the time discretization mostly the three point backward method is used and for the space discretization the midpoint rule is used. These discretizations can be implemented very conveniently on a staggered space grid. As an example we consider a two dimensional $r-\varphi$ grid given in Fig. 4.1. Recognize that $\nabla\cdot\mathbf{B}$ and $\nabla\times\mathbf{E}$ are calculated at the same position (solid curve in Fig. 4.1), as well as $\nabla\times\mathbf{B}$ and $\nabla\cdot\mathbf{j}$ (dashed curve). The formulae for discretizing the equations using the solid basic cell as considered in Fig. 4.1 read:

$$\nabla\cdot\mathbf{B}=0: \quad (i+1)\,\Delta\varphi\; B_r(i+1,j)-(i-1)\,\Delta\varphi\; B_r(i-1,j) + B_\varphi(i,j+1)-B_\varphi(i,j-1)=0\,,$$

$$(\nabla\times\mathbf{E})_z=-\dot{B}_z: \quad (i+1)\,\Delta\varphi\; E_\varphi(i+1,j,k)-(i-1)\,\Delta\varphi\; E_\varphi(i-1,j,k)+E_r(i,j-1,k) - E_r(i,j+1,k)=-2i\,\Delta r\,\Delta\varphi\;\dot{B}_z(i,j,k)\,,$$

with

$$\dot{B}_z(i,j,k)=(3B_z(i,j,k)-4B_z(i,j,k-1)+B_z(i,j,k-2))/(2\,\Delta t)\,.$$

The indices i, j, k are related to the r, φ, t coordinates, respectively. The discretization of the equations using the dashed basic cell are similar to the above given discretizations.

The basic molecules in three dimensions for both equations are given in Fig. 4.2,

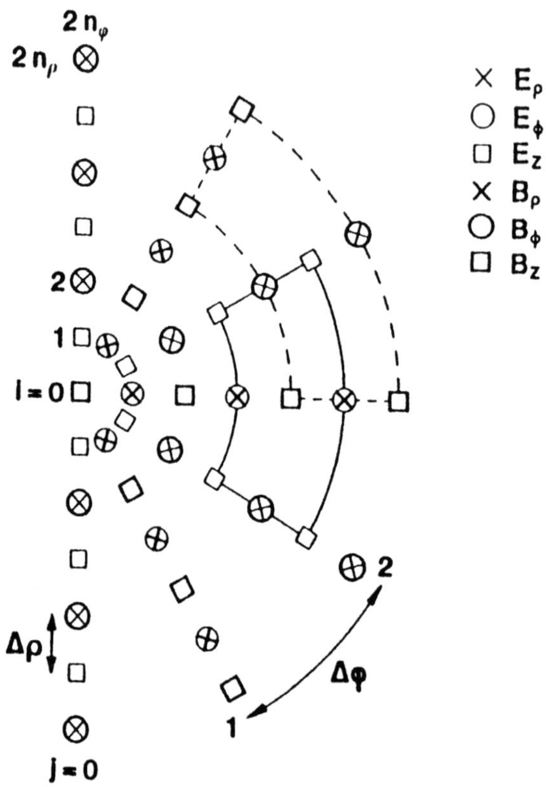

Fig. 4.1. A small version of the grid, used for the discretization of the electromagnetic field inside the wire ($n_\rho=2$, $n_\phi=3$) [5].

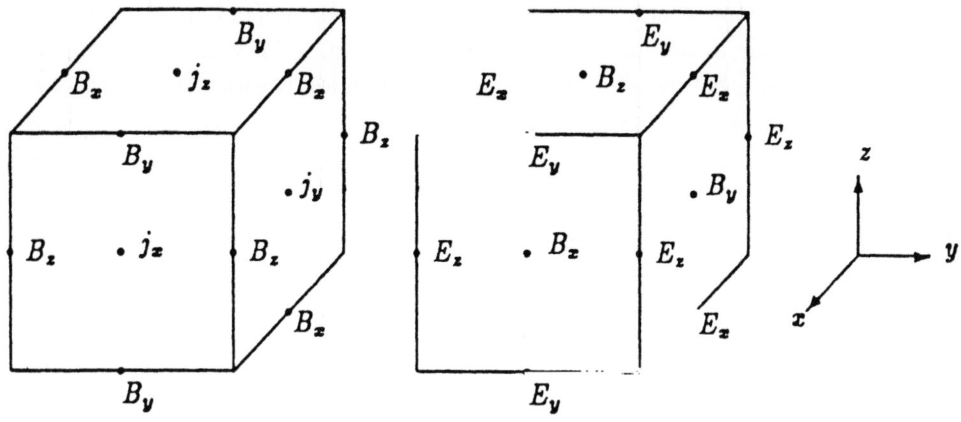

Fig. 4.2. The two basic molecules for the 3-D staggered grid [10].

respectively [10]. These two molecules can be combined to a staggered 3D grid (Fig. 4.3). Notice that the staggering of these two molecules is imposed by the coupling of the two sets of Maxwell's equations. This grid, useful for isotropic and weak anisotropic media, can also be used for superconducting media. For practical superconducting wires with twistlength very large compared to the radius, the superconducting constitutive equation is merely a relation between the axial component of the electric field and the current density. One has to make an interpolation of the axial component of the electric field for calculating the superconducting current density in the other direction. This interpolation is a weak point of the numerical method because it influences the iteration process, which will be described later. Normally this interpolation can be circumvented for a 2D grid, by adapting the grid. The numerical scheme is second order consistent at those points where **E** and **B** are sufficiently smooth. At the free boundaries, however, the consistency is only first order in the discretization steps, because then j_φ and j_z are not continuously differentiable functions of **E**. Furthermore, there is a limitation on the time step because the initial guess (due to the iteration process, treated in Section 4.2.3) may be too inaccurate for large time steps [5].

4.2.2. Boundary conditions

The boundary condition for the currents is $j_r(R) = 0$ which can easily be implemented on the grid.

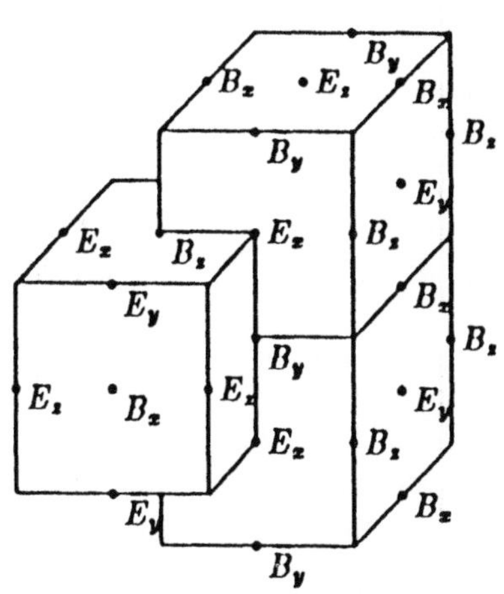

Fig. 4.3. The three dimensional staggered grid for the computation of the electromagnetic fields [10].

In general, for solving Maxwell's equations, the magnetic field in all space must be considered: so not only in the interior of the wire but also in the vacuum surrounding it or even in the centre for a hollow wire. The boundary condition is formulated for $r \to \infty$ where the induced magnetic field must vanish. However, we are only interested in the solution in the interior of the wire and need boundary conditions on $r = R$. They can be found considering that outside the wire the magnetic field satisfies $\Delta \mathbf{B} = 0$ because no currents are present. The magnetic field components outside the wire can be written using double summations in φ and z over an infinite number of terms which then are matched to the magnetic field at the surface of the wire itself. In this way the boundary condition for $r \to \infty$ is translated into a boundary condition at $r = R$. Result is that we only have to discretize the equations in the interior of the wire with correct boundary conditions at $r = R$ and $r = 0$.

This translation of the boundary condition will now be explained in detail for a z-invariant $r - \varphi$ grid (Section 4.2.2.1) and a φ-invariant $r - z$ grid (Section 4.2.2.2) where only one summation in the Laplacian solution remains.

4.2.2.1. Boundary condition at $r = R$ for the magnetic field for a z-invariant problem

For a uniform applied magnetic field in the y-direction perpendicular to an infinitely long circular wire (so $B_z(R) = 0$) we can match B_r and B_φ at $r = R$ with the $r - \varphi$ dependent Laplacian solution outside the wire ($r \geq R$) (see Section 2.4):

$$
\begin{aligned}
B_r(r, \varphi) &= B_y^{\mathrm{A}}(t) \sin \varphi + \sum_{n=1}^{\infty} a_n \left(\frac{R}{r}\right)^{n+1} \sin n\varphi \,, \\
B_\varphi(r, \varphi) &= B_y^{\mathrm{A}}(t) \cos \varphi + \sum_{n=0}^{\infty} - a_n \left(\frac{R}{r}\right)^{n+1} \cos n\varphi \,.
\end{aligned}
\qquad (4.2.2.1.1)
$$

Here use is made of the symmetry arguments $A_r(r, \varphi - \frac{1}{2}\pi) = -A_r(r, \frac{1}{2}\pi - \varphi)$ and $A_\varphi(r, \varphi - \frac{1}{2}\pi) = A_\varphi(r, \frac{1}{2}\pi - \varphi)$ with A_r, A_φ the r and φ components of any vector field $\mathbf{A}$. The terms a_n can be eliminated using the orthogonality of $\sin n\varphi$ and $\cos n\varphi$. This provides a relation between B_r and B_φ, which at the wire surface $r = R$ gives:

$$
\int_0^\pi B_\varphi(R)\, \mathrm{d}\varphi = \frac{\mu_0 I^{\mathrm{A}}(t)}{2R} \quad (\text{for } n = 0) \,,
$$

$$
\int_0^\pi B_\varphi(R, \varphi) \cos n\varphi \, \mathrm{d}\varphi + \int_0^\pi B_r(R, \varphi) \sin n\varphi \, \mathrm{d}\varphi = \pi B_y^{\mathrm{A}}(t) \delta_{1,n} \quad (\text{for } n \geq 1) \,.
$$

Here $\delta_{i,j}$ is the Kronecker delta. For the numerical grid refer to Fig. 4.1. With $\Delta\varphi = \pi / N_\varphi$ for any $(N_\varphi + 1)$ equally spaced B_φ points on the wire surface ($r = R$), this relation can be used N_φ times in its discretized form:

$$
\sum_{j=1}^{N_\varphi + 1} B_\varphi(R, \varphi_j) \cos n\varphi_j + \sum_{j=1}^{N_\varphi} B_r'(R, \varphi_j) \sin n\varphi_j = \Delta\varphi \quad (n = 1, N_\varphi)
$$

and the symbol B_r' is introduced, because in our grid B_r is not defined at the surface, but at positions half a cell dimension inside the wire. To calculate B_r' it is approximated by a linear interpolation of the value $B_r(\varphi_j)$ at a grid point near the surface and $B_r''(\varphi_j)$ at an imaginary point half a cel dimension outside the wire. Using the Maxwell equation $\nabla \cdot \mathbf{B} = 0$ at the wire surface, the boundary equations for B_φ are found. The $(N_\varphi + 1)^{\text{th}}$ equation for $B_\varphi(R, N_\varphi + 1)$ is

$$\sum_{j=1}^{N_\varphi+1} B_\varphi(R, j) = \frac{\mu_0 I^{\mathrm{A}}(t)}{2R\,\Delta\varphi}\,.$$

4.2.2.2. Boundary conditions at $r = R$ for the magnetic field for a φ-invariant problem
Consider an infinitely long circular wire in a periodic applied magnetic field with periodicity length L_z with only r and z components. This φ-invariant problem will be discussed in more detail in Section 4.3.2B. We match B_r and B_z at $r = R$ with the $r - z$ dependent Laplacian solution outside the wire ($r \geqslant R$) where use is made of the symmetry argument $\partial_z B_z = 0$ at $z = 0$:

$$\begin{aligned} B_z(r, z) &= B^{\mathrm{A}}(t) I_0(pr) \cos pz + \sum_{n=1}^{\infty} - a_n K_0(npr) \cos npz\,, \\ B_r(r, z) &= B^{\mathrm{A}}(t) I_1(pr) \sin pz + \sum_{n=1}^{\infty} a_n K_1(npr) \sin npz\,. \end{aligned} \tag{4.2.2.2.1}$$

Elimination of a_n gives, substituting $r = R$:

$$K_1(npR) \int_0^{L_z} B_z(R) \cos npz\,\mathrm{d}z + K_0(npR) \int_0^{L_z} B_r(R) \sin npz\,\mathrm{d}z = \frac{L_z}{2} B^{\mathrm{A}}(t) \delta_{1,n} \frac{L_z}{2\pi R}$$

using the Wronskian [21]: $I_\nu(z)K_{\nu+1}(z) + I_{\nu+1}(z)K_\nu(z) = 1/z$. Using the symmetry condition we see that $\int_0^{L_z} = 4 \int_0^{L_z/4}$, so only the interval $0 \leqslant z \leqslant L_z/4$ must be calculated. The numerical implementation is similar to the $r - \varphi$ grid (Section 4.2.2.1).

4.2.3. Non-linear **E–j** *relation*
Due to the non-linearity in the **E–j** relation the numerical solution cannot be found directly but the problem is linearized and iterated.

The iteration process [5, 10] is fully based on the non-linear relation between the parallel component of the electric field $E_\parallel$ and the superconducting current density j_p. The iteration process is as follows: every grid point is labelled. A grid point is labelled to be positively saturated if the previous or initial value of $E_\parallel > 0$, negatively saturated if $E_\parallel < 0$ and unsaturated otherwise. We use the constitutive equations related to the labelling and solve the set of equations. With the solution we check the predicted $E_\parallel$. If the prediction is not equal to the solution we change the label, but for convergence reasons one should not allow an element to change from a negatively saturated element into a positively saturated one or vice versa, but only allow saturated elements to turn into unsaturated ones or vice versa. There is no mathematical proof that this iteration scheme converges but in practice it works very well. Because $E_\parallel$ is calculated using E_φ and E_z it is best to calculate both electric field components at the same position.

4.3. Numerical results

In this section we present numerical results. The most general grid is 4 dimensional in space-time but due to symmetry arguments the time dependency can be included or not and the space grid can be 1, 2 or 3 dimensional. Subsection 4.3.1 deals with DC and stationary problems and in subsection 4.3.2 non-stationary problems are considered. In each subsection the dimension of the space grid is used to classify the results. Furthermore we compare some numerical and analytical results, mainly in case of parallel applied magnetic fields.

When not specifically mentioned, we consider a circular multifilamentary twisted wire.

4.3.1. *DC and stationary problems*

The dimension of the space grid is used to classify the results:

(A) 2-dimensional space grid

An infinitely long multifilamentary wire carrying a DC transport current in a constant perpendicular applied magnetic field [22]. The set of equations is: $\nabla \times \mathbf{B} = \mu_0 \mathbf{j}$; $\nabla \times \mathbf{E} = \mathbf{0}$. The problem is z-invariant so we need a 2D grid in the $r - \varphi$ coordinates. We have to translate the boundary conditions for $\mathbf{B}$ at $r \to \infty$ to a condition for $\mathbf{B}$ at $r = R$ as was already explained (Section 4.2.2.1). The non-linear **E–j** relation (2.2.7) is used, with $j_c(B_\perp)$ given by the Kim relation [23] and $E_0^{\mathrm{DC}} = 0$. Zero, one or two boundaries separating saturated regions can appear. The mesh contains about 10×10 grid points on the $r - \varphi$ surface given by $0 \leqslant r \leqslant R$ and $0 \leqslant \varphi \leqslant \pi$. In Fig. 4.4 the saturated and non saturated regions are given in case 2 boundaries appear. In Figs 4.5 and 4.6 the $E_z - I_z$ relation for the whole wire is given at constant applied field $B^{\mathrm{A}} = 2T$ and at constant twistpitch $L_p/R_w = 20$ respectively. As can be seen, there is a very large increase in E_z as saturation occurs.

Another DC problem is described by Boschman [24] where he calculates the solution for superconducting films connected through matrix material. J_c of every film is a periodic function of x: $J_{c,1}(x) = J_0(1 - 2|x|/L)$ and $J_{c,2}(x) = J_0 - J_{c,1}(x)$ with total critical current J_0. For this z-invariant problem we need a 2D xy grid. For 2 values of I^{A} the saturation pattern is shown in Fig. 4.7. This current transfer through the matrix material between the filaments causes Ohmic losses.

In this class of problems we can also consider the stationary solution (i.e. $\dot{\mathbf{B}}^{\mathrm{I}} = \mathbf{0}$) for an infinitely long wire subject to a perpendicular AC magnetic field, constantly increasing in time. The only difference with the already treated problem is the modified

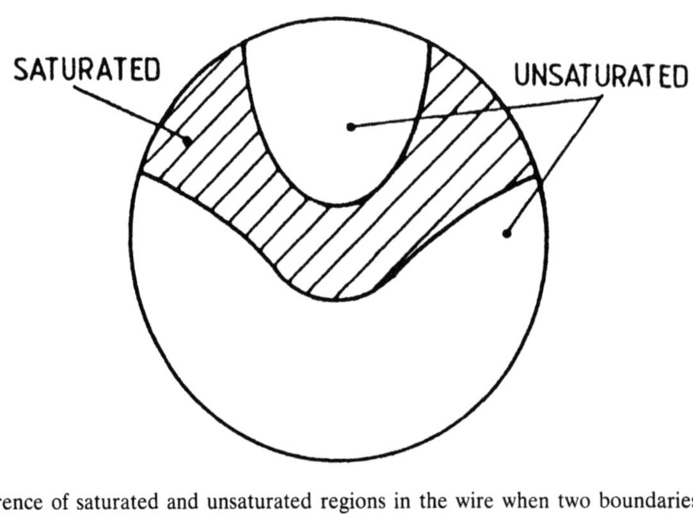

Fig. 4.4. Occurrence of saturated and unsaturated regions in the wire when two boundaries are present [22].

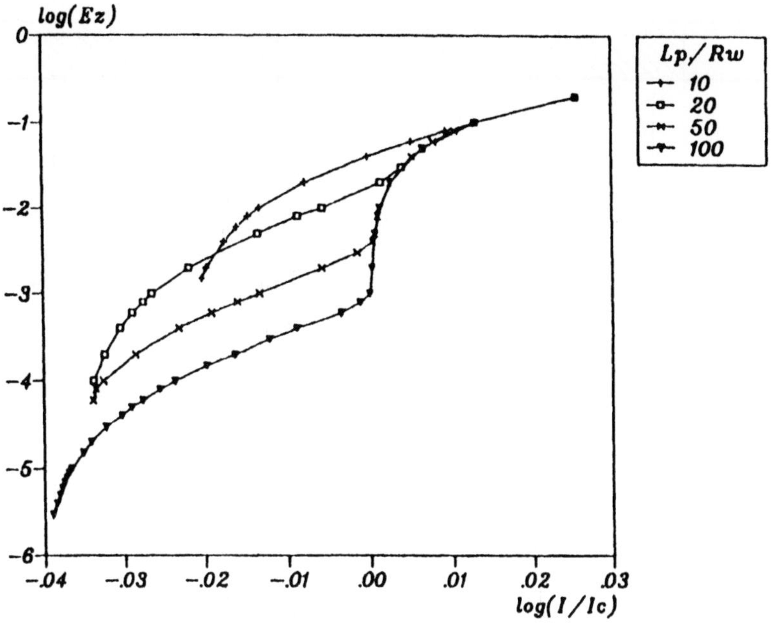

Fig. 4.5. Electric field as a function of the scaled current, for various values of L_p/R_w at an applied field of 2 tesla [22].

equation: $\nabla \times \mathbf{E} = -\dot{\mathbf{B}}^A$. The saturation boundaries for zero applied current are given in Fig. 4.8 [10]. Region 3 is saturated by coupling currents and cannot contribute to the transport current. Only region 1 can carry transport current which means that the maximum transport current I_{max} for a perpendicular field rate is reduced as described by [10]:

$$I_{max} = I_c\left[1 - \frac{\sigma_\perp R|\dot{\mathbf{B}}_\perp|}{\eta j_c} \cdot \frac{1}{\beta^2 R^2}\right]. \tag{4.3.1.1}$$

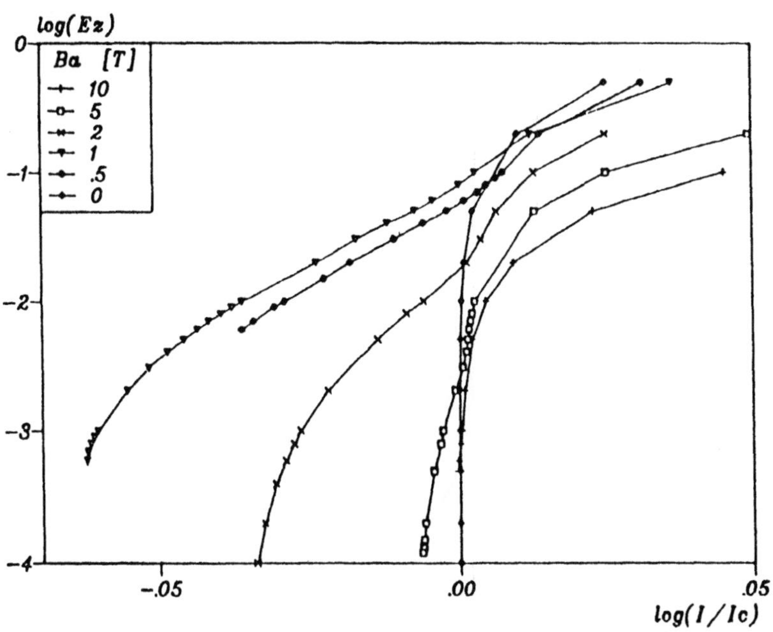

Fig. 4.6. Electric field as a function of the scaled current, for various values of the applied field, and $L_p/R_w = 20$ [22].

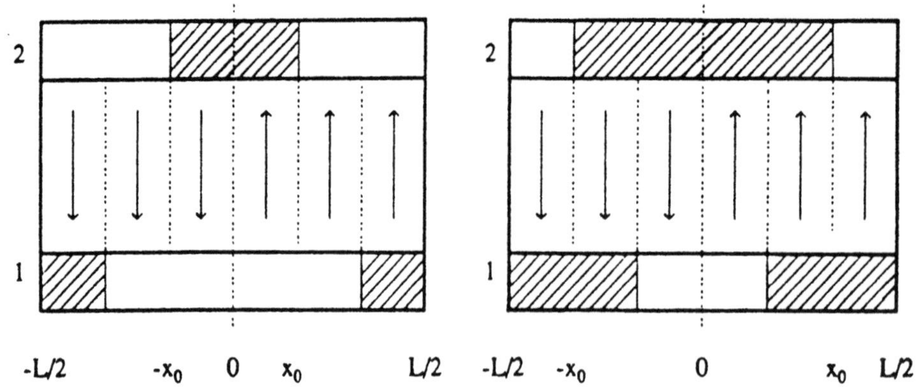

Fig. 4.7. Saturated (shaded) and unsaturated regions of the superconducting surface layers. The arrows indicate the direction of the current transfer between the layers [24].

The $r-\varphi$ grid size is about 18×42 in the $r-\varphi$ surface given by $0 \leq r \leq R$ and $0 \leq \varphi \leq 2\pi$.

(B) 3-dimensional space grid

An infinitely long multifilamentary wire carrying a DC transport current with no applied field. The wire properties are not constant, but a periodic function of the axial coordinate z. For the i-th superconducting filament we consider e.g.:

$$j_{c_i}(z) = j_c\left[1 + \alpha_i \sin\left(\frac{2\pi z}{L_z} + \varphi_i\right)\right], \tag{4.3.1.2}$$

with $0 \leq \alpha_i \leq 1$, L_z the periodicity length and φ_i some phase. This problem cannot be calculated analytically using Fourier expansions because of the non-linearity in the **E**–**j** relation.

The set of equations is: $\nabla \times \mathbf{E} = \mathbf{0}$ and $\nabla \cdot \mathbf{j} = 0$. A 3D x, y, z-grid is used taking into account one period length L_z. These calculations are performed investigating the influence of non-uniformity of wire properties on n values as described by equation (2.2.3).

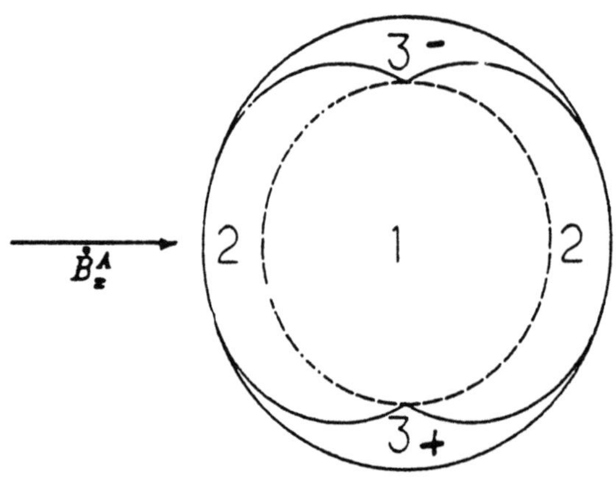

Fig. 4.8. The three different regions for the current distribution of a wire in an applied field $\dot{B}_x^A$. Regions 1 and 2 are unsaturated, region 3 is saturated [10].

4.3.2. Non-stationary problems

Here also the dimension of the space grid is used to classify the results:

(A) 2-dimensional space-time grid

An infinitely long circular wire in an AC magnetic field parallel to the z-axis. The set of equations is: $\nabla \times \mathbf{B} = \mu_0 \mathbf{j}$; $\nabla \times \mathbf{E} = -\dot{\mathbf{B}}$. Furthermore the non-linear **E–j** relation (2.2.7) is used. For this $\varphi - z$-invariant problem the boundary conditions are:

$$E_\varphi(r=0, t) = 0\,, \qquad B_\varphi(r=0, t) = 0\,, \qquad B_\varphi(r=R, t) = \frac{\mu_0 I^A(t)}{2\pi R}\,, \quad \text{and}$$

$$B_z(r=R, t) = B_z^A(t)\,.$$

The initial conditions are $\mathbf{E} = \mathbf{B} = \mathbf{0}$ for $t \leq 0$. Using Maxwell's equations we directly obtain: $E_r = B_r = 0$ and so we have to solve 5 unknowns: E_φ, E_z, j_s, B_φ, B_z using the remaining 4 Maxwell's equations and constitutive equation (2.2.7). Notice that in the unsaturated regions where $E_\parallel = 0$ the relation $\dot{B}_\varphi = \beta r \dot{B}_z$ holds. Starting from the virgin state, we see that $B_\varphi = \beta r B_z$ until saturation occurs.

Concerning the constitutive equations E_φ, E_z and j_s should be calculated at the same position. The magnetic field components should be calculated in between. Because 3 of the 4 boundary conditions are given in terms of the magnetic field, it is most convenient to calculate the magnetic field components at $r = 0$ and $r = R$. Then we only have to calculate E_φ $(r = 0)$ using an extrapolation of the internal E_φ values. The grid we arrive at is given in Fig. 4.9.

In the previous section an analytical approximation for the maximum amplitude not saturating the interior of the wire and the loss power were given under the assumption that all return current flows in a surface shell. Numerical approaches to the problem show that some peculiar and unexpected effects are, however, present. Figure 4.10 shows the time development of the saturated regions when a z-invariant $B_z^A = \sin(\omega t)$ is applied at $t = 0$, furthermore $I^A = 0$. When $\omega t < \pi/2$ a negative saturated region starts in the outer region of the wire. At $\omega t = \pi/2$ a positive saturated region starts at $r = R$ but the negative saturated region moves inward and disappears at $\omega t = 3\pi/2$. This pattern repeats each period. In Fig. 4.11 a small time independent transport current is present and, as can be seen, the positive saturated regions penetrate much further than the negative ones. This means that a parallel applied field in an unexpected way reduces

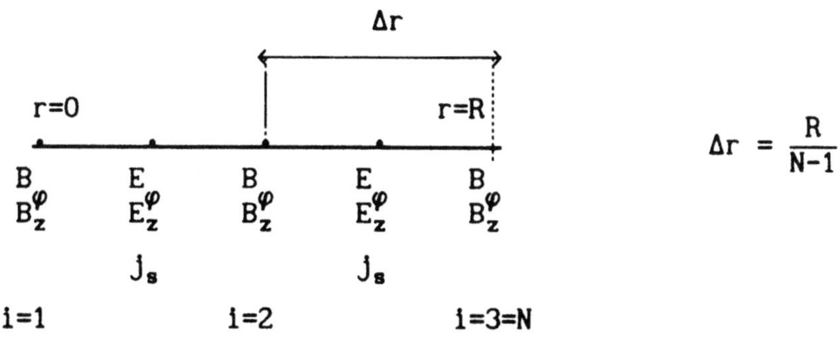

Fig. 4.9. Grid for case 4.3.2A.

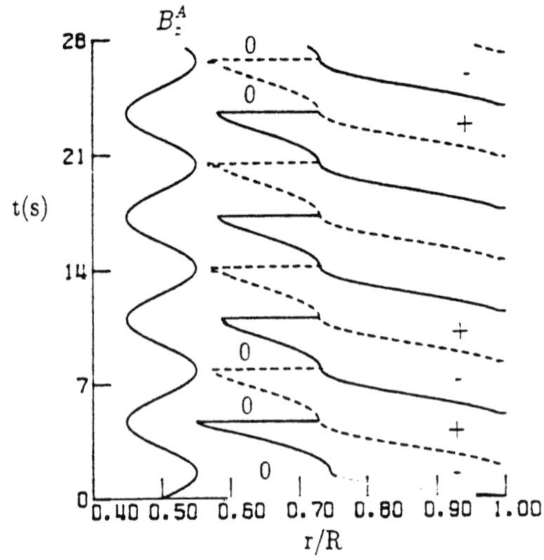

Fig. 4.10. Numerical solution of the saturation of the filaments for a uniform B_z^A with sinusoidal time dependence and no transport current.

the current carrying capacity of the wire since at the outside a double layer of positive and negative saturated current exists. When we apply a sinusoidal magnetic field:

$$B_z^A(t) = B_0 \sin \omega t = \frac{B_0}{2i} (e^{i\omega t} - e^{-i\omega t}) \tag{4.3.2.1}$$

the response of the electric field is (see Section 3.3):

$$E_\varphi(t) = -\frac{\omega}{2} (g\, e^{-i\omega t} + g^* e^{i\omega t}) . \tag{4.3.2.2}$$

Notice that both B_z^A and E_φ are real functions of time. Furthermore there is no applied

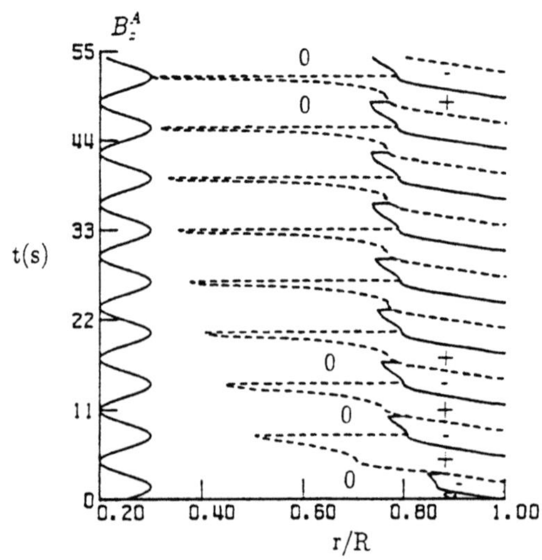

Fig. 4.11. Numerical solution of the saturation of the filaments for a uniform B_z^A with sinusoidal time dependence with a transport current of about 0.2 of the critical current.

current so $B_\varphi(R) = 0$. Calculating the mean loss per unit length ($\bar{P}$) using Poynting's theorem we get:

$$\bar{P} = -\frac{\omega}{2\pi}\int_0^{2\pi}\int_0^{2\pi/\omega}\frac{1}{\mu_0}(E_\varphi(R)B_z^{A*}(R) - E_z(R)B_\varphi^*(R))R\,\mathrm{d}\phi\,\mathrm{d}t\left[\frac{W}{m}\right]. \tag{4.3.2.3}$$

Using the analytical approach from Section 3.3 we obtain:

$$\bar{P} = -\frac{2\pi R}{\mu_0}\,\omega\left(\frac{\mathrm{i}g(R) - \mathrm{i}g^*(R)}{4}\right)B_0 = -\frac{\pi R\omega B_0}{2\mu_0}(\mathrm{i}g(R) - \mathrm{i}g^*(R)),$$
$$\bar{P} = \frac{\pi R\omega B_0}{\mu_0}\cdot \mathrm{Im}(g(R)), \tag{4.3.2.4}$$

with Im the imaginary part of a complex number.

Only the imaginary part of $g(R)$ appears in the loss term. This is easy to see because for real values of $g(R)$ the phase difference between B_z^A and $E_\varphi(R)$ is $\pi/2$, resulting in no loss because of the orthogonality between the sine and cosine function. $\bar{P}$ can also be found by calculating the dissipation directly from **E** and **j**:

$$\bar{P} = \frac{\omega}{2\pi}\int_0^{2\pi/\omega}\int_0^R\int_0^{2\pi}\mathbf{E}\cdot\mathbf{j}r\,\mathrm{d}r\,\mathrm{d}\varphi\,\mathrm{d}t, \tag{4.3.2.5}$$

with

$$\mathbf{E}\cdot\mathbf{j} = \sigma_\| E_\|^2 + \sigma_\perp E_\perp^2 + E_\| \eta j_p\,. \tag{4.3.2.6}$$

Numerically we can calculate either the contour integral (equation (4.3.2.3)) or the surface integral (equation (4.3.2.5)). Despite the higher work load, the surface integral is calculated because:

(1) Using the contour integral we need $E_\varphi(R)$ which should be calculated *extrapolating* from interior values. Using equation (4.3.2.4) we see that the loss is very sensitive to the phase difference between $B_z(R)$ and $E_\varphi(R)$. Calculating $E_\varphi(R)$ by extrapolation, this phase difference is calculated very inaccurately.
(2) The time integration of the surface integral is a summation of only positive terms which is not the case for the time integration of the contour integral.

For this type of problem (sinusoidally applied uniform B_z field, no transport current) we now compare numerical and analytical results. In Fig. 4.12 the scaled loss $2\pi\bar{P}/(\omega^2B_0^2)$ is given as function of the amplitude of the applied field B_0. The complete parameter setting is: $B_z^A = B_0 \sin t$; $I^A = 0$; $R = 10^{-3}$ m; $L_p = 0.1$ m; $R_f = 0$; $j_c = 10^9$ A/m^2; $\eta = 0.5$ and $\sigma = 10^9$ $(\Omega\mathrm{m})^{-1}$. Grid parameters: $\Delta r = R/50$ and $\Delta t = 2\pi/200$. Also the analytically calculated loss term (see equation (4.3.2.4)) using the linearized set of equations (see Section 3.3) is indicated in the figure.

For small values of B_0 there is perfect agreement between the numerical and analytical results because on the numerical grid no saturation occurs, resulting in a linear problem. The only loss term that contributes to the total loss is $\sigma_\perp E_\perp^2$. The deviation which starts at $B_0 = B_1$ can be explained as follows: for $B_0 > B_1$ outer grid cells saturate. The loss terms $\sigma_{\|} E_{\|}^2$ and $E_{\|} \eta j_p$ then also contribute to the total loss term. When saturation occurs the reason for deviation between numerical and analytical result is twofold:

(1) in general: the saturated shielding current is dissipative while in the analytical approximation the surface current is considered to be non-dissipative;
(2) for this specific situation: at the surface the positively and negatively saturated shielding currents do not cancel but both positively and negatively saturated regions appear next to each other, as was already mentioned. This increases the loss even more.

Notice that the first deviation between analytical and numerical results depends on the grid size Δr. When Δr decreases, saturation occurs at the outer grid cells for smaller values of B_0 so in the limit $\Delta r \to 0$ saturation occurs for $B_0 \to 0$.

For increasing amplitude B_0, the saturated regions penetrate more towards the centre of the wire. For $B_0 = B_2$ also a saturated region grows outward from the centre of the wire and the loss increases even more. The value of B_2 is $5T$ which follows from equation (3.3.5). For very large values of B_0 the term $E_{\|} \eta j_p$ can be neglected compared to the Ohmic loss terms. The linear behaviour is dominant and the deviation between numerical and analytical results decreases.

(B) 3-dimensional space-time grid
An infinitely long circular wire in an AC perpendicular magnetic field. The solution of the Maxwell equations for this z-invariant problem has been determined with boundary conditions $B_z^A(t) = 0$ and B_φ and B_r at $r = R$ using the previously mentioned series expansion, Section (4.2.2.1). All three components of **E** and **B** and the superconducting

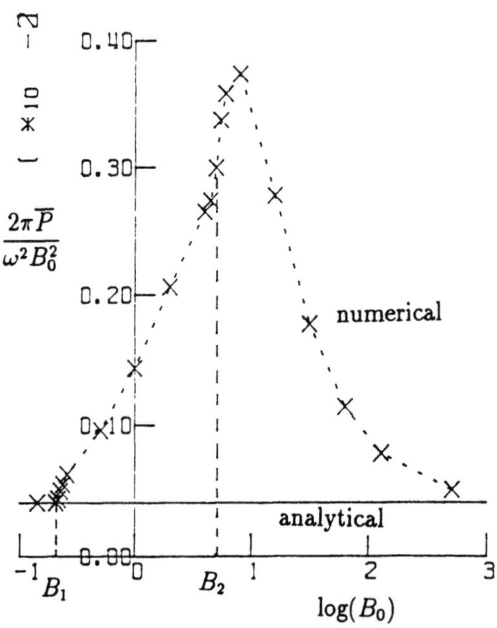

Fig. 4.12. $2\pi\bar{P}/(\omega^2 B_0^2)$ as function of $\log(B_0)$. The numerical and analytical results are indicated. For further explanation see text.

current density j_s must be calculated on an $r - \varphi - t$ grid. In Fig. 4.13 the degree of saturation is shown for 4 moments in time for an AC transport current in phase with a sinusoidal applied field [5].

In this class of problems we consider a second one. An infinitely long circular wire in a periodic applied magnetic field with periodicity length L_z:

$$(B_r^A, B_\varphi^A, B_z^A) = B^A(t)(I_1(pr)\sin pz, 0, I_0(pr)\cos pz)\,.$$

This problem is φ-invariant and periodic in z, resulting in an $r - z - t$ grid. We use the Maxwell equations with the boundary conditions:

(1) $B_\varphi(z, r = 0, t) = 0$,
(2) $B_\varphi(z, r = R, t) = 0$ (no applied current),

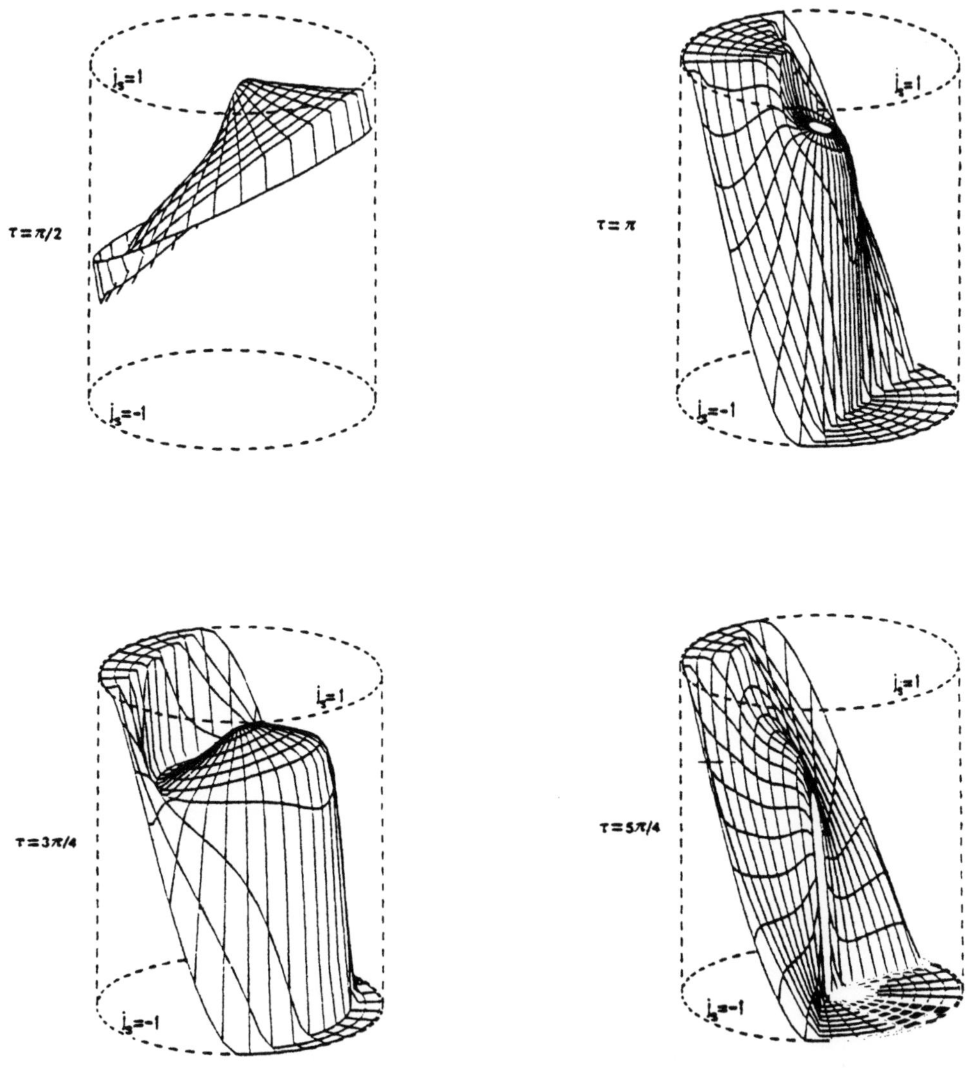

Fig. 4.13. The degree of saturation for 4 moments in time for an AC transport current in phase with a sinusoidal applied field [5].

(3) $E_\varphi(z, r=0, t) = 0$,
(4) $B_\varphi(z, R)$ and $B_z(z, R)$ are calculated using series expansions, see Section 4.2.2.2. (Notice that in the z-inv. problem (case 4.3.2A) we had $B_z(R) = B_z^A$.)
(5) at $z = \frac{1}{4}L_z$: $E_\varphi = E_z = B_\varphi = B_z = j_s = 0$,
(6) at $z = 0$: $\partial_z E_\varphi = \partial_z E_z = \partial_z B_\varphi = \partial_z B_z = \partial_z j_s = 0$.

So we have a grid for $0 \leq z \leq \frac{1}{4}L_z$. Not all 7 unknowns have to be calculated because we can very easily eliminate E_r and B_r using:

$$E_r = -\frac{\partial_z B_\varphi}{\mu_0 \sigma_\perp}$$

and $\dot{B}_r = \partial_z E_\varphi$ resulting in the set:

$$\frac{\partial_{zz} B_\varphi}{\mu_0 \sigma_\perp} + \partial_r E_z = \dot{B}_\varphi ,$$

$$\frac{1}{r}\, \partial_r r E_\varphi = -\dot{B}_z ,$$

$$\partial_{zz} E_\varphi - \partial_r \dot{B}_z = \mu_0 \dot{j}_\varphi ,$$

$$\frac{1}{r}\, \partial_r r B_\varphi = \mu_0 j_z ,$$

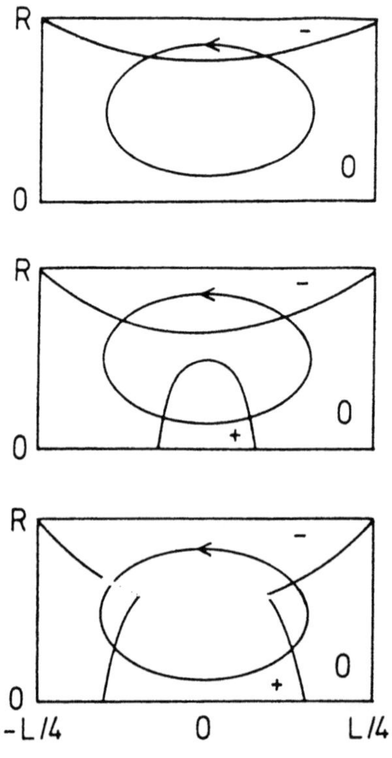

Fig. 4.14. The development of the saturated regions for periodic B_z^A and linear time dependence. For further explanation see text. The current pattern is indicated.

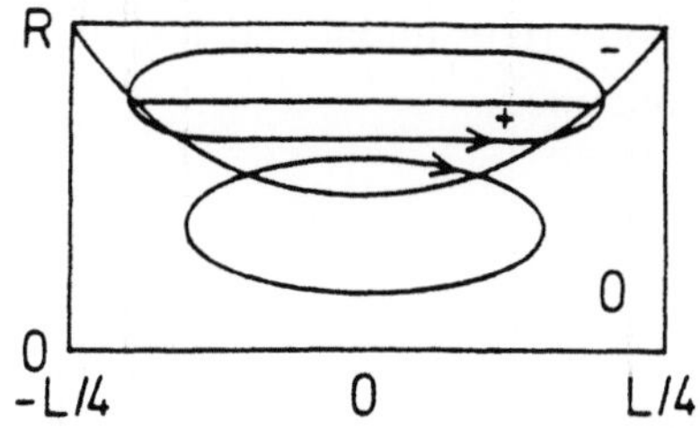

Fig. 4.15. Momentary solution of saturated regions for periodic B_z^A with sinusoidal time dependence.

plus constitutive equation (2.2.7) for the 5 unknowns B_φ, B_z, E_φ, E_z and j_s. This reduction from 7 to 5 unknowns is not considered in the first problem of this class because the elimination is not easy due to the φ dependence. Figure 4.14 shows the development of the saturated regions for $B^A(t) = \alpha t$. The upper picture shows the boundary of the outside region for small t. The middle picture shows the start of the development of the inner saturated region at $B_z^A \sim \mu_0 \eta j_c L_p / 4\pi$ [16] whereas the lower picture shows the final state of the saturated regions for $t \rightarrow \infty$. In Fig. 4.15 a schematic picture is given for $B^A(t) = \sin t$ at some time. As in the uniform case in the outer region saturation in both directions coexists. The minimal mesh size is about 20×10 grid points on the $r - z$ surface given by $0 \leqslant r \leqslant R$ and $0 \leqslant z \leqslant L_z/4$. Furthermore in one periodic cycle in time the solution is calculated in 20 time steps.

The figures 4.10 and 4.15 as well as the analytical analysis of Section 3.3 and Fig. 4.14 show that the z-invariant problem is a well defined limiting case of the problem periodic in z. The calculations show that superconducting twisted wires subjected to parallel magnetic fields can carry much less transport current than earlier reported in the literature [17].

5. Concluding remarks

In this article a framework has been presented concerning analytical and numerical calculations of the electromagnetic properties of composite superconducting wires. The basic analytical and numerical tools, applied to characteristic types of problems are presented. Furthermore a synopsis has been given of specific numerical calculations which enable the calculation of more complicated problems. As reported, numerical calculations sometimes show peculiar and unexpected effects.

The numerical and analytical problems considered are idealized cases. Numerical calculations of general cases are not performed due to large cpu times and high storage requirements. However, the response of a system subjected to general applied magnetic fields is not necessarily a linear combination of the responses of idealized cases. To find bounds for the use of idealized solutions in general situations, some complicated problems should be considered in the future.

A careful comparison of the analytical and numerical results for in essence the same setting can give a very good insight of what the different methods are capable of. Some comparisons are already given in this article but it will be worthwhile to make more comparisons in the future.

All calculations are performed under isothermal conditions. Taking into account also the heat balance equation with appropriate initial and boundary conditions, a more detailed description of the problem is obtained.

6. List of symbols

B, B_r, B_φ, B_z	= magnetic induction and its components
C, C_n	= constant
$\mathcal{D}$	= electric displacement
E, E_r, E_φ, E_z	= electric field and its components
I	= transport current
I_0, I_1	= Bessel functions of the second kind and order 0 or 1
K_0, K_1	= Bessel functions of the second kind and order 0 and 1
j, j_r, j_φ, j_z	= current density and its components
J_0, J_2	= Bessel functions of the first kind of order 0 and 2
j_s	= superconducting part of j_z
$J_\parallel, J_z, J_\varphi$	= components of the surface current
L	= half the length of the sample
L_z	= period length of the applied field
L_p	= twist length of the wire
M	= magnetization
n	= integer
N_f	= number of filaments
P	$= 2\pi/L_z$
R	= radius of the wire
R_f	= filament radius
R_0	= mean radius of the torus
u	$= npr = 2n\pi r/L_z$
U	$= 2\pi nR/L_z$
T	= temperature
α	= constant of functions of t only
β	$= 2\pi/L_p$
η	= fraction of the superconductor
$\sigma_r, \sigma_\perp, \sigma_\parallel, \sigma_{\varphi\varphi}, \sigma_{\varphi z}, \sigma_{zz}$	= components of the conductivity tensor in the $r, \perp, \parallel$ or r, φ, z system
τ	= time constant
ψ	= twist angle, tg $\psi = \beta r$

Subscripts:

n, m	= mode number
r, φ, z, t	= components in cylindrical coordinate system and time
x, y, z	= components in cartesian coordinates

Superscripts:

A	= applied
I	= induced

Operators:

$\nabla\cdot$	= divergence
$\nabla\times$	= curl
$\cdot$	= time derivative
$'$	= derivate w.r.t. the argument
Im	= the imaginary part of a complex number.

References

1. J.D. Jackson, *Classical Electrodynamics*. Wiley and Sons, New York (1962).
2. W.J. Carr Jr., Electromagnetic theory for filamentary superconductors. *Phys. Rev.* B11 (1975) 1547–1554.
3. W.J. Carr, *AC Loss and Macroscopic Theory of Superconductors*. Gordon and Breach (1983).
4. K. Kanbara, Y. Kubota and T. Ogasawara, *Proc. ICEC* (1981) 715.
5. P.C. Rem, Numerical models for AC superconductors. Enschede: PhD Thesis, University of Twente (1986).
6. E.M.J. Niessen and L.J.M. van de Klundert, The effective transverse resistivity in multifilamentary conductors, to be published.
7. G.L. Dorofejev, A.B. Imenitov and E.Yu. Klimenko, Voltage-current characteristics of type III superconductors. *Cryogenics* 20 (1980) 307–312.
8. E.Yu. Klimenko and N.N. Martovetsky, Analyses of conductor behaviour in the superconducting toroidal magnet system of the tokamak T-15 installation. *Cryogenics* 27 (1987) 238–242.
9. A.P. Martinelli and B. Turck, Some effects of field orientation on the magnetization of superconducting wires. *Cryogenics* 18 (1978) 155–161.
10. R.A. Hartmann, A contribution to the understanding of AC losses in composite superconductors. Enschede: PhD Thesis, University of Twente (1989).
11. M.N. Wilson, *Superconducting Magnets*. Oxford University Press, New York (1983).

12. J.A. Eikelboom and L.J.M. van de Klundert, Hysteresis losses in hollow superconducting filaments and in a multifilament system. *Cryogenics* 31 (1991) 354–362.
13. J.A. Eikelboom, AC losses in prototype conductors for the NET toroidal field coils. Enschede: PhD Thesis, University of Twente (1991).
14. E.M.J. Niessen and L.J.M. van de Klundert, Superconducting multifilamentary wires in inhomogeneous perpendicular a.c. fields. In: *Proc. of the International Symposium on AC Superconductors*. Smolenice, CSFR (1991) ISBN 80-900506-0-3.
15. L.J.M. van de Klundert, New developments on calculating and measuring AC losses in composite superconductors. In *Proc. MT12*, Leningrad (1991) Paper DB01.
16. E.M.J. Niessen, A. Nijhuis, E.J. Oetsen and L.J.M. van de Klundert, Current degradation and coupling losses in ring shaped wires due to applied magnetic field changes. In: *Proc. MT12*, Leningrad (1991) Paper DB03.
17. B. Turck, Courants de circulation et pertes dans les composites supraconducteurs soumis a une induction longitudinal variable. *Revue de physique appliquée* (1976) 369.
18. G. Ries and K.P. Jüngst, Filament coupling in multifilamentary superconductors in pulsed longitudinal fields. *Cryogenics* 26 (1976) 143–146.
19. *Fusion Technology* 14 (1988) no. 1.
20. E.M.J. Niessen, L.J.M. van de Klundert, R.M.J. van Damme, F.P.H. van Beckum and P.J. Zandbergen. Coupling losses in superconducting, torus shaped wires due to applied magnetic field changes. Submitted to *Journal of Eng. Mathematics* (1992).
21. M. Abramowitz and I.A. Stegun, *Handbook of Mathematical Functions with Formulas, Graphs and Mathematical tables*. New York: John Wiley & Sons (1972).
22. D. ter Avest, Properties of the superconductor in accelerator dipole magnets. Enschede: PhD Thesis, University of Twente (1991).
23. Y.B. Kim, C.F. Hempstead and A.R. Strnad, Critical persistent currents in hard superconductors. *Phys. Rev. Lett.* 9 (1962) 306–309.
24. H. Boschman, On the resistive transition of composite superconductors. Enschede: PhD Thesis, University of Twente (1991).

MIX
Papier aus verantwortungsvollen Quellen
Paper from responsible sources
FSC® C105338

If you have any concerns about our products,
you can contact us on
ProductSafety@springernature.com

In case Publisher is established outside the EU,
the EU authorized representative is:
Springer Nature Customer Service Center GmbH
Europaplatz 3, 69115 Heidelberg, Germany

Printed by Libri Plureos GmbH
in Hamburg, Germany